NEW FOR 2014

Cambridge
IGCSE®

Physics

Third Edition

Maddie
Chivers

ENDORSED BY

CAMBRIDGE
International Examinations

NEW FOR 2014

Cambridge IGCSE®

Physics

Third Edition

Tom Duncan
and Heather Kennett

HODDER
EDUCATION
AN HACHETTE UK COMPANY

® IGCSE is the registered trademark of Cambridge International Examinations. The questions, example answers, marks awarded and/or comments that appear in this book/CD were written by the authors. In examination the way marks would be awarded to answers like these may be different.

Past examination questions reproduced by permission of Cambridge International Examinations.

Cambridge International Examinations bears no responsibility for the example answers to questions taken from its past question papers which are contained in this publication.

Although every effort has been made to ensure that website addresses are correct at time of going to press, Hodder Education cannot be held responsible for the content of any website mentioned in this book. It is sometimes possible to find a relocated web page by typing in the address of the home page for a website in the URL window of your browser.

Hachette UK's policy is to use papers that are natural, renewable and recyclable products and made from wood grown in sustainable forests. The logging and manufacturing processes are expected to conform to the environmental regulations of the country of origin.

Orders: please contact Bookpoint Ltd, 130 Milton Park, Abingdon, Oxon OX14 4SB. Telephone: (44) 01235 827720. Fax: (44) 01235 400454. Lines are open 9.00–5.00, Monday to Saturday, with a 24-hour message answering service. Visit our website at www.hoddereducation.com

© Tom Duncan and Heather Kennett 2002

First published in 2002 by

Hodder Education, an Hachette UK Company,

338 Euston Road

London NW1 3BH

This third edition published 2014

Impression number 5 4 3 2 1

Year   2018 2017 2016 2015 2014

Cover photo © robertkoczera – Fotolia

Illustrations by Fakenham Prepress Solutions, Wearset and Integra Software Services Pvt. Ltd.

Typeset in 11/13pt ITC Galliard Std by Integra Software Services Pvt. Ltd., Pondicherry, India

Printed and bound in Italy.

A catalogue record for this title is available from the British Library

ISBN 978 1 4441 76421

Contents

Preface   vii

Physics and technology   viii

Scientific enquiry   x

Section 1   General physics

Measurements and motion

1 Measurements   2

2 Speed, velocity and acceleration   9

3 Graphs of equations   13

4 Falling bodies   17

5 Density   21

Forces and momentum

6 Weight and stretching   24

7 Adding forces   27

8 Force and acceleration   30

9 Circular motion   35

10 Moments and levers   39

11 Centres of mass   43

12 Momentum   47

Energy, work, power and pressure

13 Energy transfer   50

14 Kinetic and potential energy   56

15 Energy sources   60

16 Pressure and liquid pressure   66

Section 2   Thermal physics

Simple kinetic molecular model of matter

17 Molecules   72

18 The gas laws   76

Thermal properties and temperature

19 Expansion of solids, liquids and gases   81

20 Thermometers   85

21 Specific heat capacity   88

22 Specific latent heat   91

Thermal processes

23 Conduction and convection   97

24 Radiation   102

Section 3 Properties of waves

General wave properties
25 Mechanical waves   106

Light
26 Light rays   113
27 Reflection of light   116
28 Plane mirrors   119
29 Refraction of light   122
30 Total internal reflection   126
31 Lenses   129
32 Electromagnetic radiation   135

Sound
33 Sound waves   140

Section 4 Electricity and magnetism

Simple phenomena of magnetism
34 Magnetic fields   146

Electrical quantities and circuits
35 Static electricity   150
36 Electric current   157
37 Potential difference   162
38 Resistance   167
39 Capacitors   174
40 Electric power   177
41 Electronic systems   185
42 Digital electronics   193

Electromagnetic effects
43 Generators   199
44 Transformers   204
45 Electromagnets   209
46 Electric motors   215
47 Electric meters   219
48 Electrons   222

Section 5 Atomic physics
49 Radioactivity   230
50 Atomic structure   238

Revision questions   245
Cambridge IGCSE exam questions   251
Mathematics for physics   279
Further experimental investigations   283
Practical test questions   285
Alternative to practical test questions   291

Answers   299
Index   308
Photo acknowledgements   315

Preface

IGCSE Physics Third Edition aims to provide an up-to-date and comprehensive coverage of the Core and Extended curriculum in Physics specified in the current Cambridge International Examinations IGCSE syllabus.

As you read through the book, you will notice four sorts of shaded area in the text.

Material highlighted in green is for the Cambridge IGCSE Extended curriculum.

Areas highlighted in yellow contain material that is not part of the Cambridge IGCSE syllabus. It is extension work and will not be examined.

Areas highlighted in blue contain important facts.

Questions are highlighted by a box like this.

The book has been completely restructured to align chapters and sections with the order of the IGCSE syllabus. A new chapter on momentum has been included and the checklists at the end of each chapter are all aligned more closely with the syllabus requirements. New questions from recent exam papers are included at the end of the book in the sections entitled *Cambridge IGCSE exam questions, Practical test questions* and *Alternative to practical test questions*. These can be used for quick comprehensive revision before exams.

The accompanying **Revision CD-ROM** provides invaluable exam preparation and practice. Interactive tests, organised by syllabus topic, cover both the Core and Extended curriculum.

T.D. and H.K.

Physics and technology

Physicists explore the Universe. Their investigations range from particles that are smaller than atoms to stars that are millions and millions of kilometres away, as shown in Figures 1a and 1b.

As well as having to find the **facts** by observation and experiment, physicists also must try to discover the **laws** that summarise these facts (often as mathematical equations). They then have to make sense of the laws by thinking up and testing theories (thought-models) to explain the laws. The reward, apart from satisfied curiosity, is a better understanding of the physical world. Engineers and technologists use physics to solve practical problems for the benefit of people, though, in solving them, social, environmental and other problems may arise.

In this book we will study the behaviour of matter (the stuff things are made of) and the different kinds of energy (such as light, sound, heat, electricity). We will also consider the applications of physics in the home, in transport, medicine, research, industry,

energy production and electronics. Figure 2 shows some examples.

Mathematics is an essential tool of physics and a 'reference section' for some of the basic mathematics is given at the end of the book along with suggested methods for solving physics problems.

Figure 1a This image, produced by a scanning tunnelling microscope, shows an aggregate of gold just three atoms thick on a graphite substrate. Individual graphite (carbon) atoms are shown as green.

Figure 1b The many millions of stars in the Universe, of which the Sun is just one, are grouped in huge galaxies. This photograph of two interacting spiral galaxies was taken with the Hubble Space Telescope. This orbiting telescope is enabling astronomers to tackle one of the most fundamental questions in science, i.e. the age and scale of the Universe, by giving much more detailed information about individual stars than is possible with ground-based telescopes.

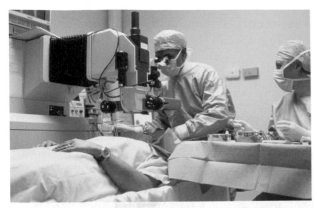

Figure 2a The modern technology of laser surgery enables very delicate operations to be performed. Here the surgeon is removing thin sheets of tissue from the surface of the patient's cornea, in order to alter its shape and correct severe short-sightedness.

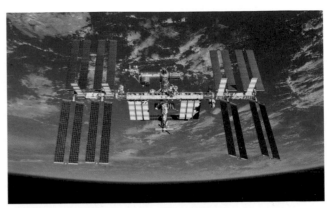

Figure 2c The manned exploration of space is such an expensive operation that international co-operation is seen as the way forward. This is the International Space Station, built module by module in orbit around the Earth. It is operated as a joint venture by the USA and Russia.

Figure 2b Mobile phones provide us with the convenience of instant communication wherever we are – but does the electromagnetic radiation they use pose a hidden risk to our health?

Figure 2d In the search for alternative energy sources, 'wind farms' of 20 to 100 wind turbines have been set up in suitable locations, such as this one in North Wales, to generate at least enough electricity for the local community.

Scientific enquiry

During your course you will have to carry out a few experiments and investigations aimed at encouraging you to develop some of the **skills** and **abilities** that scientists use to solve real-life problems.

Simple experiments may be designed to measure, for example, the temperature of a liquid or the electric current in a circuit. Longer investigations may be designed to establish or verify a relationship between two or more physical quantities.

Investigations may arise from the topic you are currently studying in class, or your teacher may provide you with suggestions to choose from, or you may have your own ideas. However an investigation arises, it will probably require at least one hour of laboratory time, but often longer, and will involve the following four aspects.

1 **Planning** how you are going to set about finding answers to the questions the problem poses. Making predictions and hypotheses (informed guesses) may help you to focus on what is required at this stage.
2 **Obtaining** the necessary experimental data safely and accurately. You will have to decide what equipment is needed, what observations and measurements have to be made and what variable quantities need to be manipulated. Do not dismantle the equipment until you have completed your analysis and you are sure you do not need to repeat any of the measurements!
3 **Presenting** and **interpreting** the evidence in a way that enables any relationships between quantities to be established.
4 **Considering** and **evaluating** the evidence by drawing conclusions, assessing the reliability of data and making comparisons with what was expected.

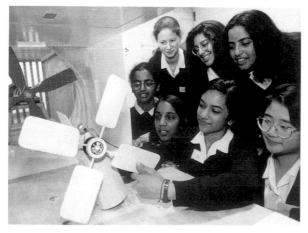

Figure 3 Girls from Copthall School, London, with their winning entry for a contest to investigate, design and build the most efficient, elegant and cost-effective windmill.

A **written report** of the investigation would normally be made. This should include:

- The **aim** of the work.
- A list of all items of **apparatus** used and a record of the smallest division of the scale of each measuring device. For example, the smallest division on a metre rule is 1 mm. The scale of the rule can be read to the nearest mm. So when used to measure a length of 100 mm (0.1 m), the length is measured to the nearest 1 mm, the degree of accuracy of the measurement being 1 part in 100. When used to measure 10 mm (0.01 m), the degree of accuracy of the measurement is 1 part in 10. A thermometer is calibrated in degrees Celsius and may be read to the nearest 1 °C. A temperature may be measured to the nearest 1 °C. So when used to measure a temperature of 20 °C, the degree of accuracy is 1 part in 20 (this is 5 parts in 100).
- Details of **procedures**, observations and measurements made. A clearly labelled diagram will be helpful here; any difficulties encountered or precautions taken to achieve accuracy should be mentioned.
- Presentation of **results** and **calculations**. If several measurements of a quantity are made, draw up a table in which to record your results. Use the column headings, or start of rows, to name the measurement and state its unit; for example 'Mass of load/kg'. Repeat the measurement of each observation; record each value in your table, then calculate an average value. Numerical values should be given to the number of significant figures appropriate to the measuring device (see Chapter 1).
  If you decide to make a graph of your results you will need at least eight data points taken over as large a range as possible; be sure to label each axis of a graph with the name and unit of the quantity being plotted (see Chapter 3).
- **Conclusions** which can be drawn from the evidence. These can take the form of a numerical value (and unit), the statement of a known law, a relationship between two quantities or a statement related to the aim of the experiment (sometimes experiments do not achieve the intended objective).
- An **evaluation** and discussion of the findings which should include:
  (i) a comparison with expected outcomes,
  (ii) a comment on the reliability of the readings, especially in relation to the scale of the measuring apparatus,

(iii) a reference to any apparatus that was unsuitable for the experiment,

(iv) a comment on any graph drawn, its shape and whether the graph points lie on the line,

(v) a comment on any trend in the readings, usually shown by the graph,

(vi) how the experiment might be modified to give more reliable results, for example in an electrical experiment by using an ammeter with a more appropriate scale.

● Suggestions for investigations

Investigations which extend the practical work or theory covered in some chapters are listed below. The section *Further experimental investigations* on p. 283 details how you can carry out some of these investigations.

1 Pitch of a note from a vibrating wire (Chapter 33).

2 Stretching of a rubber band (Chapter 6 and *Further experimental investigations*, p. 283).

3 Stretching of a copper wire – **wear safety glasses** (Chapter 6).

4 Toppling (*Further experimental investigations*, p. 283).

5 Friction – factors affecting (Chapter 7).

6 Energy values from burning fuel, e.g. a firelighter (Chapter 13).

7 Model wind turbine design (Chapter 15).

8 Speed of a bicycle and its stopping distance (Chapter 14).

9 Circular motion using a bung on a string (Chapter 9).

10 Heat loss using different insulating materials (Chapter 23).

11 Cooling and evaporation (*Further experimental investigations*, pp. 283–84).

12 Variation of the resistance of a thermistor with temperature (Chapter 38).

13 Variation of the resistance of a wire with length (*Further experimental investigations*, p. 284).

14 Heating effect of an electric current (Chapter 36).

15 Strength of an electromagnet (Chapter 45).

16 Efficiency of an electric motor (Chapter 46).

● Ideas and evidence in science

In some of the investigations you perform in the school laboratory, you may find that you do not interpret your data in the same way as your friends do; perhaps you will argue with them as to the best way to explain your results and try to convince them that your interpretation is right. Scientific controversy frequently arises through people interpreting evidence differently.

Observations of the heavens led the ancient Greek philosophers to believe that the Earth was at the centre of the planetary system, but a complex system of rotation was needed to match observations of the apparent movement of the planets across the sky. In 1543 Nicolaus Copernicus made the radical suggestion that all the planets revolved not around the Earth but around the Sun. (His book *On the Revolutions of the Celestial Spheres* gave us the modern usage of the word 'revolution'.) It took time for his ideas to gain acceptance. The careful astronomical observations of planetary motion documented by Tycho Brahe were studied by Johannes Kepler, who realised that the data could be explained if the planets moved in elliptical paths (not circular) with the Sun at one focus. Galileo's observations of the moons of Jupiter with the newly invented telescope led him to support this 'Copernican view' and to be imprisoned by the Catholic Church in 1633 for disseminating heretical views. About 50 years later, Isaac Newton introduced the idea of gravity and was able to explain the motion of all bodies, whether on Earth or in the heavens, which led to full acceptance of the Copernican model. Newton's mechanics were refined further at the beginning of the 20th century when Einstein developed his theories of relativity. Even today, data from the Hubble Space Telescope is providing new evidence which confirms Einstein's ideas.

Many other scientific theories have had to wait for new data, technological inventions, or time and the right social and intellectual climate for them to become accepted. In the field of health and medicine, for example, because cancer takes a long time to develop it was several years before people recognised that X-rays and radioactive materials could be dangerous (Chapter 49).

At the beginning of the 20th century scientists were trying to reconcile the wave theory and the particle theory of light by means of the new ideas of quantum mechanics.

Today we are collecting evidence on possible health risks from microwaves used in mobile phone networks. The cheapness and popularity of mobile phones may make the public and manufacturers reluctant to accept adverse findings, even if risks are made widely known in the press and on television. Although scientists can provide evidence and evaluation of that evidence, there may still be room for controversy and a reluctance to accept scientific findings, particularly if there are vested social or economic interests to contend with. This is most clearly shown today in the issue of global warming.

Section 1 General physics

Chapters

Measurements and motion
1 Measurements
2 Speed, velocity and acceleration
3 Graphs of equations
4 Falling bodies
5 Density

Forces and momentum
6 Weight and stretching
7 Adding forces

8 Force and acceleration
9 Circular motion
10 Moments and levers
11 Centres of mass
12 Momentum

Energy, work, power and pressure
13 Energy transfer
14 Kinetic and potential energy
15 Energy sources
16 Pressure and liquid pressure

① Measurements

- Units and basic quantities
- Powers of ten shorthand
- Length
- Significant figures
- Area
- Volume
- Mass
- Time
- Systematic errors
- Vernier scales and micrometers
- Practical work: Period of a simple pendulum

● Units and basic quantities

Before a measurement can be made, a standard or **unit** must be chosen. The size of the quantity to be measured is then found with an instrument having a scale marked in the unit.

Three basic quantities we measure in physics are **length**, **mass** and **time**. Units for other quantities are based on them. The SI (Système International d'Unités) system is a set of metric units now used in many countries. It is a decimal system in which units are divided or multiplied by 10 to give smaller or larger units.

Figure 1.1 Measuring instruments on the flight deck of a passenger jet provide the crew with information about the performance of the aircraft.

● Powers of ten shorthand

This is a neat way of writing numbers, especially if they are large or small. The example below shows how it works.

$$4000 = 4 \times 10 \times 10 \times 10 \quad = 4 \times 10^3$$
$$400 = 4 \times 10 \times 10 \quad\quad = 4 \times 10^2$$
$$40 = 4 \times 10 \quad\quad\quad = 4 \times 10^1$$
$$4 = 4 \times 1 \quad\quad\quad = 4 \times 10^0$$
$$0.4 = 4/10 \quad = 4/10^1 = 4 \times 10^{-1}$$
$$0.04 = 4/100 \quad = 4/10^2 = 4 \times 10^{-2}$$
$$0.004 = 4/1000 \quad = 4/10^3 = 4 \times 10^{-3}$$

The small figures 1, 2, 3, etc., are called **powers of ten**. The power shows how many times the number has to be multiplied by 10 if the power is greater than 0 or divided by 10 if the power is less than 0. Note that 1 is written as 10^0.

This way of writing numbers is called **standard notation**.

● Length

The unit of **length** is the **metre** (m) and is the distance travelled by light in a vacuum during a specific time interval. At one time it was the distance between two marks on a certain metal bar. Submultiples are:

$$1 \text{ decimetre (dm)} = 10^{-1} \text{ m}$$
$$1 \text{ centimetre (cm)} = 10^{-2} \text{ m}$$
$$1 \text{ millimetre (mm)} = 10^{-3} \text{ m}$$
$$1 \text{ micrometre (μm)} = 10^{-6} \text{ m}$$
$$1 \text{ nanometre (nm)} = 10^{-9} \text{ m}$$

A multiple for large distances is

$$1 \text{ kilometre (km)} = 10^3 \text{ m } (\tfrac{5}{8} \text{ mile approx.})$$

Many length measurements are made with rulers; the correct way to read one is shown in Figure 1.2. The reading is 76 mm or 7.6 cm. Your eye must be directly over the mark on the scale or the thickness of the ruler causes a parallax error.

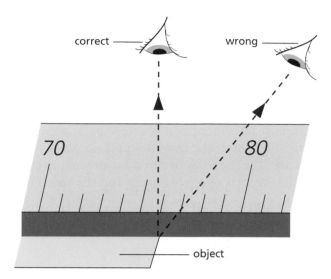

Figure 1.2 The correct way to measure with a ruler

To obtain an average value for a small distance, multiples can be measured. For example, in ripple tank experiments (Chapter 25) measure the distance occupied by five waves, then divide by 5 to obtain the average wavelength.

● Significant figures

Every measurement of a quantity is an attempt to find its true value and is subject to errors arising from limitations of the apparatus and the experimenter. The number of figures, called **significant figures**, given for a measurement indicates how accurate we think it is and more figures should not be given than is justified.

For example, a value of 4.5 for a measurement has two significant figures; 0.0385 has three significant figures, 3 being the most significant and 5 the least, i.e. it is the one we are least sure about since it might be 4 or it might be 6. Perhaps it had to be estimated by the experimenter because the reading was between two marks on a scale.

When doing a calculation your answer should have the same number of significant figures as the measurements used in the calculation. For example, if your calculator gave an answer of 3.4185062, this would be written as 3.4 if the measurements had two significant figures. It would be written as 3.42 for three significant figures. Note that in deciding the least significant figure you look at the next figure to the right. If it is less than 5 you leave the least significant figure as it is (hence 3.41 becomes 3.4) but if it equals or is greater than 5 you increase the least significant figure by 1 (hence 3.418 becomes 3.42).

If a number is expressed in standard notation, the number of significant figures is the number of digits before the power of ten. For example, 2.73×10^3 has three significant figures.

● Area

The **area** of the square in Figure 1.3a with sides 1 cm long is 1 square centimetre (1 cm²). In Figure 1.3b the rectangle measures 4 cm by 3 cm and has an area of $4 \times 3 = 12$ cm² since it has the same area as twelve squares each of area 1 cm². The area of a square or rectangle is given by

$$\text{area} = \text{length} \times \text{breadth}$$

The SI unit of area is the square metre (m²) which is the area of a square with sides 1 m long. Note that

$$1\,\text{cm}^2 = \frac{1}{100}\,\text{m} \times \frac{1}{100}\,\text{m} = \frac{1}{10\,000}\,\text{m}^2 = 10^{-4}\,\text{m}^2$$

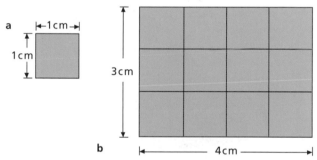

Figure 1.3

Sometimes we need to know the area of a triangle (Chapter 3). It is given by

$$\text{area of triangle} = \tfrac{1}{2} \times \text{base} \times \text{height}$$

For example in Figure 1.4

$$\text{area } \Delta\text{ABC} = \tfrac{1}{2} \times \text{AB} \times \text{AC}$$
$$= \tfrac{1}{2} \times 4\,\text{cm} \times 6\,\text{cm} = 12\,\text{cm}^2$$

and

$$\text{area } \Delta\text{PQR} = \tfrac{1}{2} \times \text{PQ} \times \text{SR}$$
$$= \tfrac{1}{2} \times 5\,\text{cm} \times 4\,\text{cm} = 10\,\text{cm}^2$$

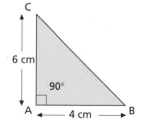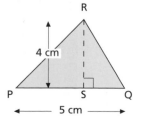

Figure 1.4

The area of a circle of radius r is πr^2 where $\pi = 22/7$ or 3.14; its circumference is $2\pi r$.

● Volume

Volume is the amount of space occupied. The unit of volume is the **cubic metre** (m^3) but as this is rather large, for most purposes the **cubic centimetre** (cm^3) is used. The volume of a cube with $1\,cm$ edges is $1\,cm^3$. Note that

$$1\,cm^3 = \frac{1}{100}\,m \times \frac{1}{100}\,m \times \frac{1}{100}\,m$$

$$= \frac{1}{1\,000\,000}\,m^3 = 10^{-6}\,m^3$$

For a regularly shaped object such as a rectangular block, Figure 1.5 shows that

$$\text{volume} = \text{length} \times \text{breadth} \times \text{height}$$

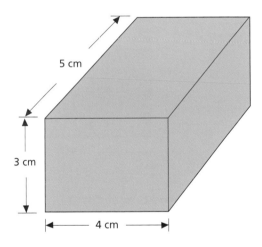

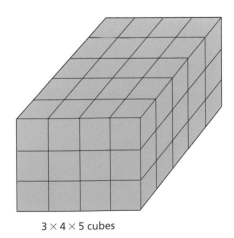

3 × 4 × 5 cubes

Figure 1.5

The volume of a sphere of radius r is $\frac{4}{3}\pi r^3$ and that of a cylinder of radius r and height h is $\pi r^2 h$.

The volume of a liquid may be obtained by pouring it into a measuring cylinder, Figure 1.6a. A known volume can be run off accurately from a burette, Figure 1.6b. When making a reading both vessels must be upright and your eye must be level with the bottom of the curved liquid surface, i.e. the **meniscus**. The meniscus formed by mercury is curved oppositely to that of other liquids and the top is read.

Liquid volumes are also expressed in litres (l); $1\,\text{litre} = 1000\,cm^3 = 1\,dm^3$. One millilitre ($1\,ml$) = $1\,cm^3$.

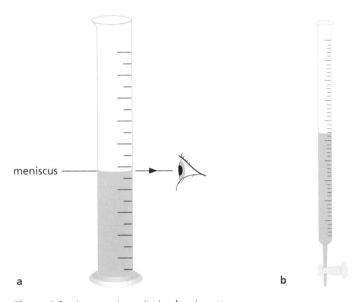

Figure 1.6a A measuring cylinder; **b** a burette

● Mass

The **mass** of an object is the measure of the amount of matter in it. The unit of mass is the kilogram (kg) and is the mass of a piece of platinum–iridium alloy at the Office of Weights and Measures in Paris. The gram (g) is one-thousandth of a kilogram.

$$1\,g = \frac{1}{1000}\,kg = 10^{-3}\,kg = 0.001\,kg$$

The term **weight** is often used when mass is really meant. In science the two ideas are distinct and have different units, as we shall see later. The confusion is not helped by the fact that mass is found on a balance by a process we unfortunately call 'weighing'!

There are several kinds of balance. In the **beam balance** the unknown mass in one pan is balanced against known masses in the other pan. In the **lever balance** a system of levers acts against the mass when it is placed in the pan. A direct reading is obtained from the position on a scale of a pointer joined to the lever system. A digital **top-pan balance** is shown in Figure 1.7.

Figure 1.7 A digital top-pan balance

● Time

The unit of **time** is the **second** (s) which used to be based on the length of a day, this being the time for the Earth to revolve once on its axis. However, days are not all of exactly the same duration and the second is now defined as the time interval for a certain number of energy changes to occur in the caesium atom.

Time-measuring devices rely on some kind of constantly repeating oscillation. In traditional clocks and watches a small wheel (the balance wheel) oscillates to and fro; in digital clocks and watches the oscillations are produced by a tiny quartz crystal. A swinging pendulum controls a pendulum clock.

To measure an interval of time in an experiment, first choose a timer that is accurate enough for the task. A stopwatch is adequate for finding the period in seconds of a pendulum, see Figure 1.8, but to measure the speed of sound (Chapter 33), a clock that can time in milliseconds is needed. To measure very short time intervals, a digital clock that can be triggered to start and stop by an electronic signal from a microphone, photogate or mechanical switch is useful. Tickertape timers or dataloggers are often used to record short time intervals in motion experiments (Chapter 2).

Accuracy can be improved by measuring longer time intervals. Several oscillations (rather than just one) are timed to find the period of a pendulum. 'Tenticks' (rather than 'ticks') are used in tickertape timers.

Practical work

Period of a simple pendulum

In this investigation you have to make time measurements using a stopwatch or clock.

Attach a small metal ball (called a bob) to a piece of string, and suspend it as shown in Figure 1.8. Pull the bob a small distance to one side, and then release it so that it oscillates to and fro through a small angle.

Find the time for the bob to make several complete oscillations; one oscillation is from A to O to B to O to A (Figure 1.8). Repeat the timing a few times for the same number of oscillations and work out the average. The time for one oscillation is the **period** T. What is it for your system? The **frequency** f of the oscillations is the number of complete oscillations per second and equals 1/T. Calculate f.

How does the amplitude of the oscillations change with time?

Investigate the effect on T of **(i)** a longer string, **(ii)** a heavier bob. A motion sensor connected to a datalogger and computer (Chapter 2) could be used instead of a stopwatch for these investigations.

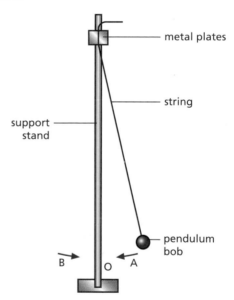

Figure 1.8

● Systematic errors

Figure 1.9 shows a part of a rule used to measure the height of a point P above the bench. The rule chosen has a space before the zero of the scale. This is shown as the length x. The height of the point P is given by the scale reading added to the value of x. The equation for the height is

$$\text{height} = \text{scale reading} + x$$
$$\text{height} = 5.9 + x$$

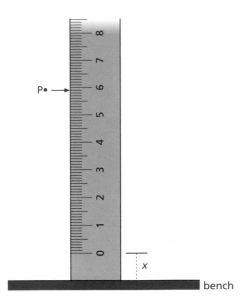

Figure 1.9

By itself the scale reading is not equal to the height. It is too small by the value of x.

This type of error is known as a **systematic error**. The error is introduced by the system. A half-metre rule has the zero at the *end of the rule* and so can be used without introducing a systematic error.

When using a rule to determine a height, the rule must be held so that it is vertical. If the rule is at an angle to the vertical, a systematic error is introduced.

● Vernier scales and micrometers

Lengths can be measured with a ruler to an accuracy of about 1 mm. Some investigations may need a more accurate measurement of length, which can be achieved by using **vernier calipers** (Figure 1.10) or **a micrometer screw gauge**.

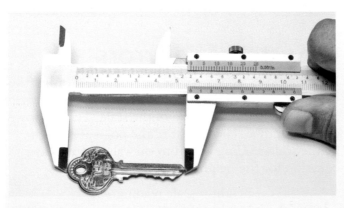

Figure 1.10 Vernier calipers in use

a) Vernier scale

The calipers shown in Figure 1.10 use a vernier scale. The simplest type enables a length to be measured to 0.01 cm. It is a small sliding scale which is 9 mm long but divided into 10 equal divisions (Figure 1.11a) so

$$1 \text{ vernier division} = \frac{9}{10} \text{ mm}$$
$$= 0.9 \text{ mm}$$
$$= 0.09 \text{ cm}$$

One end of the length to be measured is made to coincide with the zero of the millimetre scale and the other end with the zero of the vernier scale. The length of the object in Figure 1.11b is between 1.3 cm and 1.4 cm. The reading to the second place of decimals is obtained by finding the vernier mark which is exactly opposite (or nearest to) a mark on the millimetre scale. In this case it is the 6th mark and the length is 1.36 cm, since

$$OA = OB - AB$$
$$\therefore \quad OA = (1.90 \text{ cm}) - (6 \text{ vernier divisions})$$
$$= 1.90 \text{ cm} - 6(0.09) \text{ cm}$$
$$= (1.90 - 0.54) \text{ cm}$$
$$= 1.36 \text{ cm}$$

Vernier scales are also used on barometers, travelling microscopes and spectrometers.

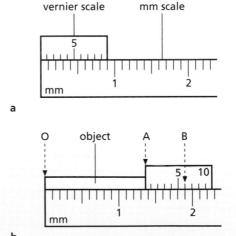

Figure 1.11 Vernier scale

b) Micrometer screw gauge

This measures very small objects to 0.001 cm. One revolution of the drum opens the accurately flat, parallel jaws by one division on the scale on the shaft of the gauge; this is usually $\frac{1}{2}$ mm, i.e. 0.05 cm. If the drum has a scale of 50 divisions round it, then rotation of the drum by one division opens the jaws by 0.05/50 = 0.001 cm (Figure 1.12). A friction clutch ensures that the jaws exert the same force when the object is gripped.

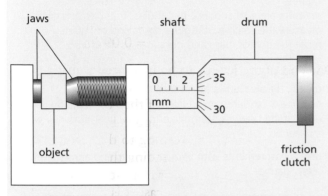

Figure 1.12 Micrometer screw gauge

The object shown in Figure 1.12 has a length of

2.5 mm on the shaft scale +
33 divisions on the drum scale
$$= 0.25\,\text{cm} + 33(0.001)\,\text{cm}$$
$$= 0.283\,\text{cm}$$

Before making a measurement, check to ensure that the reading is zero when the jaws are closed. Otherwise the zero error must be allowed for when the reading is taken.

Questions

1 How many millimetres are there in
   a 1 cm, b 4 cm, c 0.5 cm, d 6.7 cm, e 1 m?

2 What are these lengths in metres:
   a 300 cm,   b 550 cm,   c 870 cm,
   d 43 cm,   e 100 mm?

3 a Write the following as powers of ten with one figure before the decimal point:
      100 000   3500   428 000 000   504   27 056
   b Write out the following in full:
      10^3 2×10^6 6.92×10^4 1.34×10^2 10^9

4 a Write these fractions as powers of ten:
      1/1000   7/100 000   1/10 000 000   3/60 000
   b Express the following decimals as powers of ten with one figure before the decimal point:
      0.5   0.084   0.000 36   0.001 04

5 The pages of a book are numbered 1 to 200 and each leaf is 0.10 mm thick. If each cover is 0.20 mm thick, what is the thickness of the book?

6 How many significant figures are there in a length measurement of:
   a 2.5 cm,   b 5.32 cm,   c 7.180 cm,   d 0.042 cm?

7 A rectangular block measures 4.1 cm by 2.8 cm by 2.1 cm. Calculate its volume giving your answer to an appropriate number of significant figures.

8 A metal block measures 10 cm × 2 cm × 2 cm. What is its volume? How many blocks each 2 cm × 2 cm × 2 cm have the same total volume?

9 How many blocks of ice cream each 10 cm × 10 cm × 4 cm can be stored in the compartment of a freezer measuring 40 cm × 40 cm × 20 cm?

10 A Perspex container has a 6 cm square base and contains water to a height of 7 cm (Figure 1.13).
   a What is the volume of the water?
   b A stone is lowered into the water so as to be completely covered and the water rises to a height of 9 cm. What is the volume of the stone?

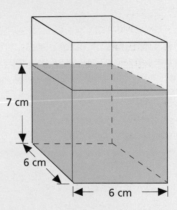

Figure 1.13

11 What are the readings on the vernier scales in Figures 1.14a and b?

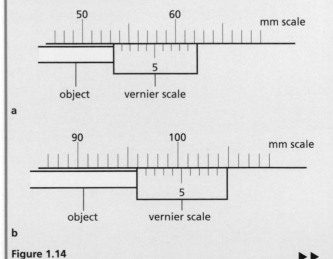

Figure 1.14 ▶▶

12 What are the readings on the micrometer screw gauges in Figures 1.15a and b?

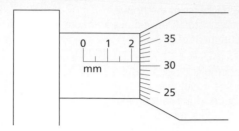

a

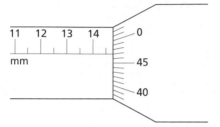

b

Figure 1.15

13 a Name the basic units of: length, mass, time.
   b What is the difference between two measurements of the same object with values of 3.4 and 3.42?
   c Write expressions for
      (i)   the area of a circle,
      (ii) the volume of a sphere,
      (iii) the volume of a cylinder.

Checklist

After studying this chapter you should be able to

- recall three basic quantities in physics,
- write a number in powers of ten (standard notation),
- recall the unit of length and the meaning of the prefixes kilo, centi, milli, micro, nano,
- use a ruler to measure length so as to minimise errors,
- give a result to an appropriate number of significant figures,
- measure areas of squares, rectangles, triangles and circles,
- measure the volume of regular solids and of liquids,
- recall the unit of mass and how mass is measured,
- recall the unit of time and how time is measured,
- describe the use of clocks and devices, both analogue and digital, for measuring an interval of time,
- describe an experiment to find the period of a pendulum,
- understand how a systematic error may be introduced when measuring,

- take measurements with vernier calipers and a micrometer screw gauge.

(2) Speed, velocity and acceleration

- Speed
- Velocity
- Acceleration

- Timers
- Practical work: Analysing motion

● Speed

If a car travels 300 km from Liverpool to London in five hours, its **average speed** is 300 km/5 h = 60 km/h. The speedometer would certainly not read 60 km/h for the whole journey but might vary considerably from this value. That is why we state the average speed. If a car could travel at a constant speed of 60 km/h for five hours, the distance covered would still be 300 km. It is *always* true that

$$\text{average speed} = \frac{\text{distance moved}}{\text{time taken}}$$

To find the actual speed at any instant we would need to know the distance moved in a very short interval of time. This can be done by multiflash photography. In Figure 2.1 the golfer is photographed while a flashing lamp illuminates him 100 times a second. The speed of the club-head as it hits the ball is about 200 km/h.

Figure 2.1 Multiflash photograph of a golf swing

● Velocity

Speed is the distance travelled in unit time; **velocity is the distance travelled in unit time in a stated direction**. If two trains travel due north at 20 m/s, they have the same speed of 20 m/s and the same velocity of 20 m/s *due north*. If one travels north and the other south, their speeds are the same but not their velocities since their directions of motion are different. Speed is a **scalar** quantity and velocity a **vector** quantity (see Chapter 7).

$$\text{velocity} = \frac{\text{distance moved in a stated direction}}{\text{time taken}}$$

The velocity of a body is uniform or constant if it moves with a steady speed in a straight line. It is not uniform if it moves in a curved path. Why?

The units of speed and velocity are the same, km/h, m/s.

$$60 \text{ km/h} = \frac{6000 \text{ m}}{3600 \text{ s}} = 17 \text{ m/s}$$

Distance moved in a stated direction is called the **displacement**. It is a vector, unlike distance which is a scalar. Velocity may also be defined as

$$\text{velocity} = \frac{\text{displacement}}{\text{time taken}}$$

● Acceleration

When the velocity of a body changes we say the body **accelerates**. If a car starts from rest and moving due north has velocity 2 m/s after 1 second, its velocity has increased by 2 m/s in 1 s and its acceleration is 2 m/s per second due north. We write this as 2 m/s².

Acceleration is the change of velocity in unit time, or

$$\text{acceleration} = \frac{\text{change of velocity}}{\text{time taken for change}}$$

For a steady increase of velocity from 20 m/s to 50 m/s in 5 s

$$\text{acceleration} = \frac{(50-20)\,\text{m/s}}{5\,\text{s}} = 6\,\text{m/s}^2$$

Acceleration is also a vector and both its magnitude and direction should be stated. However, at present we will consider only motion in a straight line and so the magnitude of the velocity will equal the speed, and the magnitude of the acceleration will equal the change of speed in unit time.

The speeds of a car accelerating on a straight road are shown below.

Time/s	0	1	2	3	4	5	6
Speed/m/s	0	5	10	15	20	25	30

The speed increases by 5 m/s every second and the acceleration of 5 m/s² is said to be **uniform**.

An acceleration is positive if the velocity increases and negative if it decreases. A negative acceleration is also called a **deceleration** or **retardation**.

● Timers

A number of different devices are useful for analysing motion in the laboratory.

a) Motion sensors

Motion sensors use the ultrasonic echo technique (see p. 143) to determine the distance of an object from the sensor. Connection of a datalogger and computer to the motion sensor then enables a distance–time graph to be plotted directly (see Figure 2.6). Further data analysis by the computer allows a velocity–time graph to be obtained, as in Figures 3.1 and 3.2, p. 13.

b) Tickertape timer: tape charts

A **tickertape timer** also enables us to measure speeds and hence accelerations. One type, Figure 2.2, has

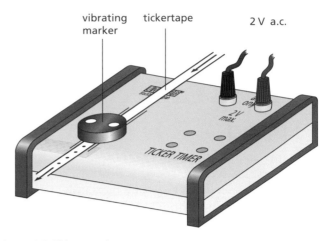

Figure 2.2 Tickertape timer

a marker that vibrates 50 times a second and makes dots at $\frac{1}{50}$ s intervals on the paper tape being pulled through it; $\frac{1}{50}$ s is called a '**tick**'.

The distance between successive dots equals the average speed of whatever is pulling the tape in, say, cm per $\frac{1}{50}$ s, i.e. cm per tick. The '**tentick**' ($\frac{1}{5}$ s) is also used as a unit of time. Since ticks and tenticks are small we drop the 'average' and just refer to the 'speed'.

Tape charts are made by sticking successive strips of tape, usually tentick lengths, side by side. That in Figure 2.3a represents a body moving with **uniform speed** since equal distances have been moved in each tentick interval.

The chart in Figure 2.3b is for **uniform acceleration**: the 'steps' are of equal size showing that the speed increased by the same amount in every tentick ($\frac{1}{5}$ s). The acceleration (average) can be found from the chart as follows.

The speed during the *first* tentick is 2 cm for every $\frac{1}{5}$ s, or 10 cm/s. During the *sixth* tentick it is 12 cm per $\frac{1}{5}$ s or 60 cm/s. And so during this interval of 5 tenticks, i.e. 1 second, the change of speed is $(60-10)\,\text{cm/s} = 50\,\text{cm/s}$.

$$\text{acceleration} = \frac{\text{change of speed}}{\text{time taken}}$$
$$= \frac{50\,\text{cm/s}}{1\,\text{s}}$$
$$= 50\,\text{cm/s}^2$$

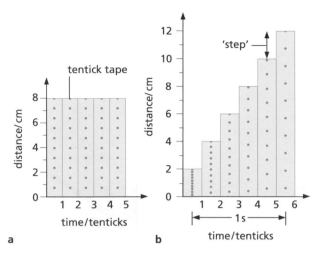

a

b

Figure 2.3 Tape charts: **a** uniform speed; **b** uniform acceleration

c) Photogate timer

Photogate timers may be used to record the time taken for a trolley to pass through the gate, Figure 2.4. If the length of the 'interrupt card' on the trolley is measured, the velocity of the trolley can then be calculated. Photogates are most useful in experiments where the velocity at only one or two positions is needed.

Figure 2.4 Use of a photogate timer

Practical work

Analysing motion

a) Your own motion

Pull a 2 m length of tape through a tickertape timer as you walk away from it quickly, then slowly, then speeding up again and finally stopping.

Cut the tape into tentick lengths and make a tape chart. Write labels on it to show where you speeded up, slowed down, etc.

b) Trolley on a sloping runway

Attach a length of tape to a trolley and release it at the top of a runway (Figure 2.5). The dots will be very crowded at the start – ignore those; but beyond them cut the tape into tentick lengths.

Make a tape chart. Is the acceleration uniform? What is its average value?

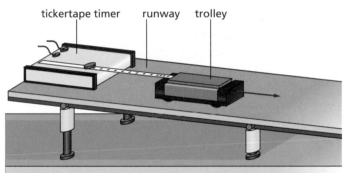

Figure 2.5

c) Datalogging

Replace the tickertape timer with a motion sensor connected to a datalogger and computer (Figure 2.6). Repeat the experiments in a) and b) and obtain distance–time and velocity–time graphs for each case; identify regions where you think the acceleration changes or remains uniform.

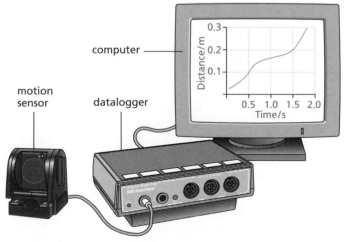

Figure 2.6 Use of a motion sensor

Questions

1 What is the average speed of
  a a car that travels 400 m in 20 s,
  b an athlete who runs 1500 m in 4 minutes?

2 A train increases its speed steadily from 10 m/s to 20 m/s in 1 minute.
  a What is its average speed during this time, in m/s?
  b How far does it travel while increasing its speed?

3 A motorcyclist starts from rest and reaches a speed of 6 m/s after travelling with uniform acceleration for 3 s. What is his acceleration?

4 An aircraft travelling at 600 km/h accelerates steadily at 10 km/h per second. Taking the speed of sound as 1100 km/h at the aircraft's altitude, how long will it take to reach the 'sound barrier'?

5 A vehicle moving with a uniform acceleration of 2 m/s² has a velocity of 4 m/s at a certain time. What will its velocity be
  a 1 s later,
  b 5 s later?

6 If a bus travelling at 20 m/s is subject to a steady deceleration of 5 m/s², how long will it take to come to rest?

7 The tape in Figure 2.7 was pulled through a timer by a trolley travelling down a runway. It was marked off in tentick lengths.
  a What can you say about the trolley's motion?
  b Find its acceleration in cm/s².

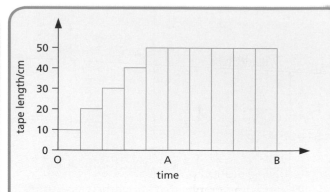

Figure 2.8

9 The speeds of a car travelling on a straight road are given below at successive intervals of 1 second.

Time/s	0	1	2	3	4
Speed/m/s	0	2	4	6	8

The car travels
1 with an average velocity of 4 m/s
2 16 m in 4 s
3 with a uniform acceleration of 2 m/s².
Which statement(s) is (are) correct?
A 1, 2, 3   B 1, 2   C 2, 3   D 1   E 3

10 If a train travelling at 10 m/s starts to accelerate at 1 m/s² for 15 s on a straight track, its final velocity in m/s is
  A 5   B 10   C 15   D 20   E 25

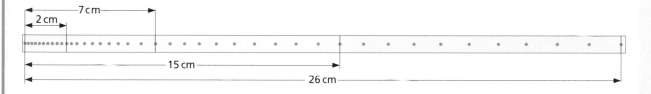

Figure 2.7

8 Each strip in the tape chart of Figure 2.8 is for a time interval of 1 tentick.
  a If the timer makes 50 dots per second, what time intervals are represented by OA and AB?
  b What is the acceleration between O and A in
    (i) cm/tentick²,
    (ii) cm/s per tentick,
    (iii) cm/s²?
  c What is the acceleration between A and B?

Checklist

After studying this chapter you should be able to

• explain the meaning of the terms speed and acceleration,

• distinguish between speed and velocity,

• describe how speed and acceleration may be found using tape charts and motion sensors.

③ Graphs of equations

- Velocity–time graphs
- Distance–time graphs

- Equations for uniform acceleration

● Velocity–time graphs

If the velocity of a body is plotted against the time, the graph obtained is a **velocity–time graph**. It provides a way of solving motion problems. Tape charts are crude velocity–time graphs that show the velocity changing in jumps rather than smoothly, as occurs in practice. A motion sensor gives a smoother plot.

> The area under a velocity–time graph measures the distance travelled.

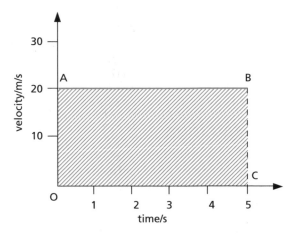

Figure 3.1 Uniform velocity

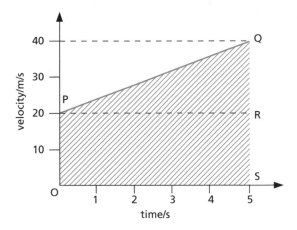

Figure 3.2a Uniform acceleration

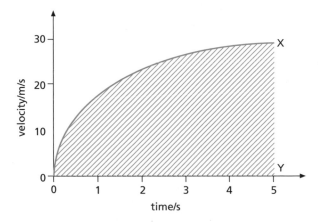

Figure 3.2b Non-uniform acceleration

In Figure 3.1, AB is the velocity–time graph for a body moving with a **uniform velocity** of 20 m/s. Since distance = average velocity × time, after 5 s it will have moved 20 m/s × 5 s = 100 m. This is the shaded area under the graph, i.e. rectangle OABC.

In Figure 3.2a, PQ is the velocity–time graph for a body moving with **uniform acceleration**. At the start of the timing the velocity is 20 m/s but it increases steadily to 40 m/s after 5 s. If the distance covered equals the area under PQ, i.e. the shaded area OPQS, then

$$\text{distance} = \text{area of rectangle OPRS} + \text{area of triangle PQR}$$

$$= \text{OP} \times \text{OS} + \tfrac{1}{2} \times \text{PR} \times \text{QR}$$
(area of a triangle $= \tfrac{1}{2}$ base × height)

$$= 20 \, \text{m/s} \times 5 \, \text{s} + \tfrac{1}{2} \times 5 \, \text{s} \times 20 \, \text{m/s}$$

$$= 100 \, \text{m} + 50 \, \text{m} = 150 \, \text{m}$$

Notes

1 When calculating the area from the graph, the unit of time must be the same on both axes.
2 This rule for finding distances travelled is true even if the acceleration is not uniform. In Figure 3.2b, the distance travelled equals the shaded area OXY.

> The slope or gradient of a velocity–time graph represents the acceleration of the body.

In Figure 3.1, the slope of AB is zero, as is the acceleration. In Figure 3.2a, the slope of PQ is QR/PR = 20/5 = 4: the acceleration is 4 m/s². In Figure 3.2b, when the slope along OX changes, so does the acceleration.

Distance–time graphs

A body travelling with uniform velocity covers equal distances in equal times. Its **distance–time graph** is a straight line, like OL in Figure 3.3 for a velocity of $10\,\text{m/s}$. The slope of the graph is $\text{LM/OM} = 40\,\text{m}/4\,\text{s} = 10\,\text{m/s}$, which is the value of the velocity. The following statement is true in general:

> The slope or gradient of a distance–time graph represents the velocity of the body.

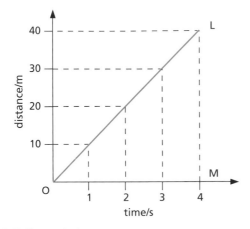

Figure 3.3 Uniform velocity

When the velocity of the body is changing, the slope of the distance–time graph varies, as in Figure 3.4, and at any point equals the slope of the tangent. For example, the slope of the tangent at T is $\text{AB/BC} = 40\,\text{m}/2\,\text{s} = 20\,\text{m/s}$. The velocity at the instant corresponding to T is therefore $20\,\text{m/s}$.

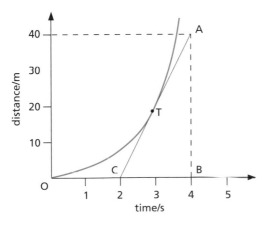

Figure 3.4 Non-uniform velocity

Equations for uniform acceleration

Problems involving bodies moving with **uniform acceleration** can often be solved quickly using the **equations of motion**.

First equation

If a body is moving with uniform acceleration a and its velocity increases from u to v in time t, then

$$a = \frac{\text{change of velocity}}{\text{time taken}} = \frac{v - u}{t}$$

$\therefore$ $\qquad at = v - u$

or

$$v = u + at \qquad (1)$$

Note that the initial velocity u and the final velocity v refer to the start and the finish of the *timing* and do not necessarily mean the start and finish of the motion.

Second equation

The velocity of a body moving with uniform acceleration increases steadily. Its average velocity therefore equals half the sum of its initial and final velocities, that is,

$$\text{average velocity} = \frac{u + v}{2}$$

If s is the distance moved in time t, then since average velocity = distance/time = s/t,

$$\frac{s}{t} = \frac{u + v}{2}$$

or

$$s = \frac{(u + v)}{2}t \qquad (2)$$

Third equation

From equation (1), $v = u + at$
From equation (2),

$$\frac{s}{t} = \frac{u + v}{2}$$

$$\frac{s}{t} = \frac{u + u + at}{2} = \frac{2u + at}{2}$$

$$= u + \frac{1}{2}at$$

and so

$$s = ut + \frac{1}{2}at^2 \qquad (3)$$

Fourth equation

This is obtained by eliminating t from equations (1) and (3). Squaring equation (1) we have

$$v^2 = (u + at)^2$$

$$\therefore \qquad v^2 = u^2 + 2uat + a^2t^2$$

$$= u^2 + 2a\left(ut + \frac{1}{2}at^2\right)$$

But $\qquad s = ut + \frac{1}{2}at^2$

$$\therefore \qquad v^2 = u^2 + 2as$$

If we know any *three* of u, v, a, s and t, the others can be found from the equations.

● Worked example

A sprint cyclist starts from rest and accelerates at $1\,\text{m/s}^2$ for 20 seconds. He then travels at a constant speed for 1 minute and finally decelerates at $2\,\text{m/s}^2$ until he stops. Find his maximum speed in km/h and the total distance covered in metres.

First stage

$$u = 0 \qquad a = 1\,\text{m/s}^2 \qquad t = 20\,\text{s}$$

We have $\qquad v = u + at = 0 + 1\,\text{m/s}^2 \times 20\,\text{s}$

$$= 20\,\text{m/s}$$

$$= \frac{20}{1000} \times 60 \times 60 = 72\,\text{km/h}$$

The distance s moved in the first stage is given by

$$s = ut + \frac{1}{2}at^2 = 0 \times 20\,\text{s} + \frac{1}{2} \times 1\,\text{m/s}^2 \times 20^2\,\text{s}^2$$

$$= \frac{1}{2} \times 1\,\text{m/s}^2 \times 400\,\text{s}^2 = 200\,\text{m}$$

Second stage

$$u = 20\,\text{m/s (constant)} \qquad t = 60\,\text{s}$$

$$\text{distance moved} = \text{speed} \times \text{time} = 20\,\text{m/s} \times 60\,\text{s}$$

$$= 1200\,\text{m}$$

Third stage

$$u = 20\,\text{m/s} \quad v = 0 \quad a = -2\,\text{m/s}^2 \text{ (a deceleration)}$$

We have
$$v^2 = u^2 + 2as$$

$$\therefore \quad s = \frac{v^2 - u^2}{2a} = \frac{0 - 20^2\,\text{m}^2/\text{s}^2}{2 \times (-2)\,\text{m/s}^2} = \frac{-400\,\text{m}^2/\text{s}^2}{-4\,\text{m/s}^2}$$

$$= 100\,\text{m}$$

Answers

Maximum speed = $72\,\text{km/h}$
Total distance covered = $200\,\text{m} + 1200\,\text{m} + 100\,\text{m}$
$$= 1500\,\text{m}$$

Questions

1 The distance–time graph for a girl on a cycle ride is shown in Figure 3.5.
  a How far did she travel?
  b How long did she take?
  c What was her average speed in km/h?
  d How many stops did she make?
  e How long did she stop for altogether?
  f What was her average speed *excluding* stops?
  g How can you tell from the shape of the graph when she travelled fastest? Over which stage did this happen?

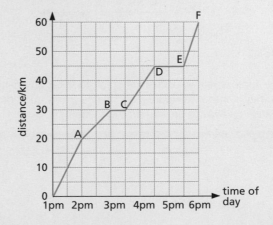

Figure 3.5

▶▶

15

2 The graph in Figure 3.6 represents the distance travelled by a car plotted against time.
  a How far has the car travelled at the end of 5 seconds?
  b What is the speed of the car during the first 5 seconds?
  c What has happened to the car after A?
  d Draw a graph showing the speed of the car plotted against time during the first 5 seconds.

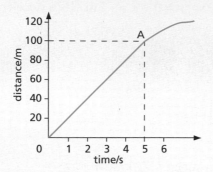

Figure 3.6

3 Figure 3.7 shows an incomplete velocity–time graph for a boy running a distance of 100 m.
  a What is his acceleration during the first 4 seconds?
  b How far does the boy travel during (i) the first 4 seconds, (ii) the next 9 seconds?
  c Copy and complete the graph showing clearly at what time he has covered the distance of 100 m. Assume his speed remains constant at the value shown by the horizontal portion of the graph.

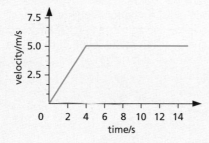

Figure 3.7

4 The approximate velocity–time graph for a car on a 5-hour journey is shown in Figure 3.8. (There is a very quick driver change midway to prevent driving fatigue!)

  a State in which of the regions OA, AB, BC, CD, DE the car is (i) accelerating, (ii) decelerating, (iii) travelling with uniform velocity.
  b Calculate the value of the acceleration, deceleration or constant velocity in each region.
  c What is the distance travelled over each region?
  d What is the total distance travelled?
  e Calculate the average velocity for the whole journey.

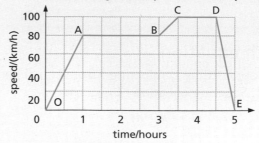

Figure 3.8

5 The distance–time graph for a motorcyclist riding off from rest is shown in Figure 3.9.
  a Describe the motion.
  b How far does the motorbike move in 30 seconds?
  c Calculate the speed.

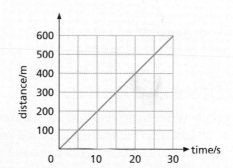

Figure 3.9

Checklist

After studying this chapter you should be able to

• draw, interpret and use velocity–time and distance–time graphs to solve problems.

4 Falling bodies

- Acceleration of free fall
- Measuring *g*
- Distance–time graphs

- Projectiles
- Practical work: Motion of a falling body

In air, a coin falls faster than a small piece of paper. In a vacuum they fall at the same rate, as may be shown with the apparatus of Figure 4.1. The difference in air is due to **air resistance** having a greater effect on light bodies than on heavy bodies. The air resistance to a light body is large when compared with the body's weight. With a dense piece of metal the resistance is negligible at low speeds.

There is a story, untrue we now think, that in the 16th century the Italian scientist Galileo dropped a small iron ball and a large cannonball ten times heavier from the top of the Leaning Tower of Pisa (Figure 4.2). And we are told that, to the surprise of onlookers who expected the cannonball to arrive first, they reached the ground almost simultaneously. You will learn more about air resistance in Chapter 8.

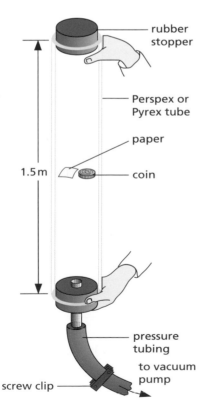

Figure 4.1 A coin and a piece of paper fall at the same rate in a vacuum.

Figure 4.2 The Leaning Tower of Pisa, where Galileo is said to have experimented with falling objects

Practical work

Motion of a falling body

Arrange things as shown in Figure 4.3 and investigate the motion of a 100 g mass falling from a height of about 2 m.

Construct a tape chart using *one-tick* lengths. Choose as dot '0' the first one you can distinguish clearly. What does the tape chart tell you about the motion of the falling mass? Repeat the experiment with a 200 g mass; what do you notice?

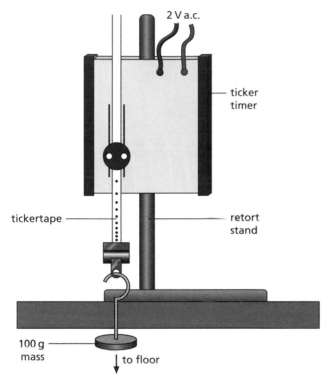

2 V a.c.

ticker timer

tickertape

retort stand

100 g mass

to floor

Figure 4.3

● Acceleration of free fall

All bodies falling freely under the force of gravity do so with uniform acceleration if air resistance is negligible (i.e. the 'steps' in the tape chart from the practical work should all be equal).

This acceleration, called the **acceleration of free fall,** is denoted by the italic letter g. Its value varies slightly over the Earth but is constant in each place; in India for example, it is about $9.8\,\text{m/s}^2$ or near enough $10\,\text{m/s}^2$. The velocity of a free-falling body therefore increases by $10\,\text{m/s}$ every second. A ball shot straight upwards with a velocity of $30\,\text{m/s}$ decelerates by $10\,\text{m/s}$ every second and reaches its highest point after 3 s.

In calculations using the equations of motion, g replaces a. It is given a positive sign for falling bodies

(i.e. $a = g = +10\,\text{m/s}^2$) and a negative sign for rising bodies since they are decelerating (i.e. $a = -g = -10\,\text{m/s}^2$).

● Measuring *g*

Using the arrangement in Figure 4.4 the time for a steel ball-bearing to fall a known distance is measured by an electronic timer.

When the two-way switch is changed to the 'down' position, the electromagnet releases the ball and simultaneously the clock starts. At the end of its fall the ball opens the 'trap-door' on the impact switch and the clock stops.

The result is found from the third equation of motion $s = ut + \frac{1}{2}at^2$, where s is the distance fallen (in m), t is the time taken (in s), $u = 0$ (the ball starts from rest) and $a = g$ (in m/s^2). Hence

$$s = \frac{1}{2}gt^2$$

or

$$g = 2s/t^2$$

Air resistance is negligible for a dense object such as a steel ball-bearing falling a short distance.

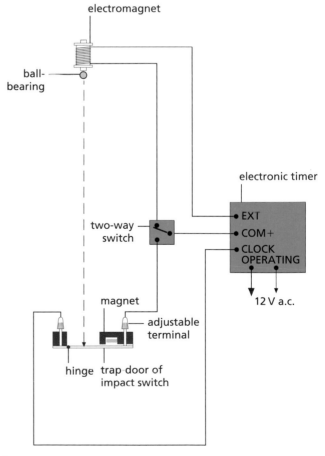

electromagnet

ball-bearing

two-way switch

electronic timer

EXT
COM+
CLOCK OPERATING

magnet

12 V a.c.

adjustable terminal

hinge trap-door of impact switch

Figure 4.4

Worked example

A ball is projected vertically upwards with an initial velocity of 30 m/s. Find **a** its maximum height and **b** the time taken to return to its starting point. Neglect air resistance and take $g = 10\,\text{m/s}^2$.

a We have $u = 30\,\text{m/s}$, $a = -10\,\text{m/s}^2$ (a deceleration) and $v = 0$ since the ball is momentarily at rest at its highest point. Substituting in $v^2 = u^2 + 2as$,

$$0 = 30^2\,\text{m}^2/\text{s}^2 + 2(-10\,\text{m/s}^2) \times s$$

or

$$-900\,\text{m}^2/\text{s}^2 = -s \times 20\,\text{m/s}^2$$

$$\therefore \quad s = \frac{-900\,\text{m}^2/\text{s}^2}{-20\,\text{m/s}^2} = 45\,\text{m}$$

b If t is the time to reach the highest point, we have, from $v = u + at$,

$$0 = 30\,\text{m/s} + (-10\,\text{m/s}^2) \times t$$

or

$$-30\,\text{m/s} = -t \times 10\,\text{m/s}^2$$

$$\therefore \quad t = \frac{-30\,\text{m/s}}{-10\,\text{m/s}^2} = 3\,\text{s}$$

The downward trip takes exactly the same time as the upward one and so the answer is 6 s.

Distance–time graphs

For a body falling freely from rest we have

$$s = \tfrac{1}{2}gt^2$$

A graph of distance s against time t is shown in Figure 4.5a and for s against t^2 in Figure 4.5b. The second graph is a straight line through the origin since $s \propto t^2$ (g being constant at one place).

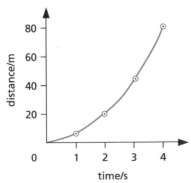

Figure 4.5a A graph of distance against time for a body falling freely from rest

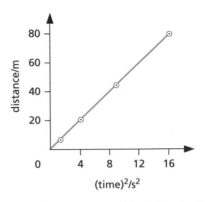

Figure 4.5b A graph of distance against (time)² for a body falling freely from rest

Projectiles

The photograph in Figure 4.6 was taken while a lamp emitted regular flashes of light. One ball was *dropped from rest* and the other, a '**projectile**', was *thrown sideways* at the same time. Their vertical accelerations (due to gravity) are equal, showing that a projectile falls like a body which is dropped from rest. Its horizontal velocity does not affect its vertical motion.

> The horizontal and vertical motions of a body are independent and can be treated separately.

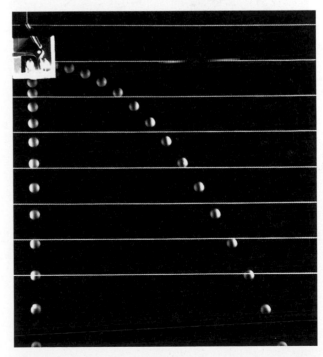

Figure 4.6 Comparing free fall and projectile motion using multiflash photography

For example if a ball is thrown horizontally from the top of a cliff and takes 3 s to reach the beach below, we can calculate the height of the cliff by considering the vertical motion only. We have $u = 0$ (since the ball has no vertical velocity initially), $a = g = +10 \, \text{m/s}^2$ and $t = 3 \, \text{s}$. The height s of the cliff is given by

$$s = ut + \frac{1}{2} at^2$$
$$= 0 \times 3 \, \text{s} + \frac{1}{2}(+10 \, \text{m/s}^2)3^2 \, \text{s}^2$$
$$= 45 \, \text{m}$$

Projectiles such as cricket balls and explosive shells are projected from near ground level and at an angle. The horizontal distance they travel, i.e. their range, depends on

(i) the speed of projection – the greater this is, the greater the range, and

(ii) the angle of projection – it can be shown that, neglecting air resistance, the range is a maximum when the angle is 45° (Figure 4.7).

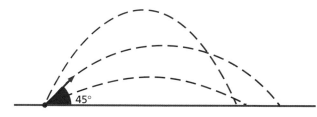

Figure 4.7 The range is greatest for an angle of projection of 45°

Questions

1 A stone falls from rest from the top of a high tower. Ignore air resistance and take $g = 10 \, \text{m/s}^2$.
  a What is its velocity after
    (i)   1 s,
    (ii) 2 s,
    (iii) 3 s,
    (iv) 5 s?
  b How far has it fallen after
    (i)   1 s,
    (ii) 2 s,
    (iii) 3 s,
    (iv) 5 s?
2 An object falls from a hovering helicopter and hits the ground at a speed of 30 m/s. How long does it take the object to reach the ground and how far does it fall? Sketch a velocity–time graph for the object (ignore air resistance).

Checklist

After studying this chapter you should be able to

• describe the behaviour of falling objects,

• state that the acceleration of free fall for a body near the Earth is constant.

5 Density

- Calculations
- Simple density measurements
- Floating and sinking

In everyday language, lead is said to be 'heavier' than wood. By this it is meant that a certain volume of lead is heavier than the same volume of wood. In science such comparisons are made by using the term **density**. This is the **mass per unit volume** of a substance and is calculated from

$$\text{density} = \frac{\text{mass}}{\text{volume}}$$

The density of lead is 11 grams per cubic centimetre (11 g/cm³) and this means that a piece of lead of volume 1 cm³ has mass 11 g. A volume of 5 cm³ of lead would have mass 55 g. If the density of a substance is known, the mass of *any* volume of it can be calculated. This enables engineers to work out the weight of a structure if they know from the plans the volumes of the materials to be used and their densities. Strong enough foundations can then be made.

The SI unit of density is the **kilogram per cubic metre**. To convert a density from g/cm³, normally the most suitable unit for the size of sample we use, to kg/m³, we multiply by 10³. For example the density of water is 1.0 g/cm³ or 1.0 × 10³ kg/m³.

The approximate densities of some common substances are given in Table 5.1.

Table 5.1 Densities of some common substances

Solids	Density/g/cm³	Liquids	Density/g/cm³
aluminium	2.7	paraffin	0.80
copper	8.9	petrol	0.80
iron	7.9	pure water	1.0
gold	19.3	mercury	13.6
glass	2.5	**Gases**	**Density/kg/m³**
wood (teak)	0.80	air	1.3
ice	0.92	hydrogen	0.09
polythene	0.90	carbon dioxide	2.0

● Calculations

Using the symbols ρ (rho) for density, m for mass and V for volume, the expression for density is

$$\rho = \frac{m}{V}$$

Rearranging the expression gives

$$m = V \times \rho \qquad \text{and} \qquad V = \frac{m}{\rho}$$

These are useful if ρ is known and m or V have to be calculated. If you do not see how they are obtained refer to the *Mathematics for physics* section on p. 279. The triangle in Figure 5.1 is an aid to remembering them. If you cover the quantity you want to know with a finger, such as m, it equals what you can still see, i.e. $\rho \times V$. To find V, cover V and you get $V = m/\rho$.

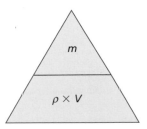

Figure 5.1

● Worked example

Taking the density of copper as 9 g/cm³, find **a** the mass of 5 cm³ and **b** the volume of 63 g.

a $\rho = 9$ g/cm³, $V = 5$ cm³ and m is to be found.
$m = V \times \rho = 5\,\text{cm}^3 \times 9\,\text{g/cm}^3 = 45\,\text{g}$

b $\rho = 9$ g/cm³, $m = 63$ g and V is to be found.

$$\therefore \qquad V = \frac{m}{\rho} = \frac{63\,\text{g}}{9\,\text{g/cm}^3} = 7\,\text{cm}^3$$

● Simple density measurements

If the mass m and volume V of a substance are known, its density can be found from $\rho = m/V$.

a) Regularly shaped solid

The mass is found on a balance and the volume by measuring its dimensions with a ruler.

b) Irregularly shaped solid, such as a pebble or glass stopper

The mass of the solid is found on a balance. Its volume is measured by one of the methods shown in Figures 5.2a and b. In Figure 5.2a the volume is the difference between the first and second readings. In Figure 5.2b it is the volume of water collected in the measuring cylinder.

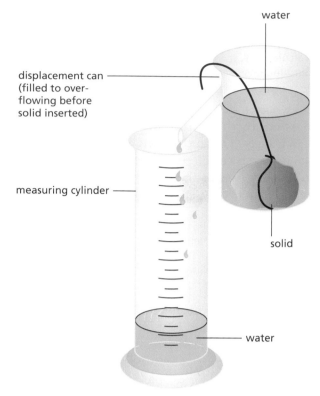

Figure 5.2b Measuring the volume of an irregular solid: method 2

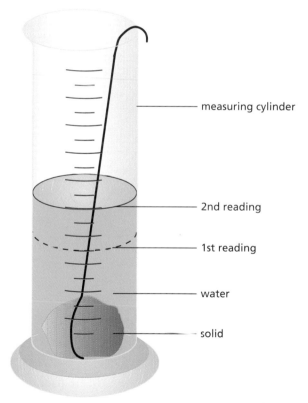

Figure 5.2a Measuring the volume of an irregular solid: method 1

c) Liquid

The mass of an empty beaker is found on a balance. A known volume of the liquid is transferred from a burette or a measuring cylinder into the beaker. The mass of the beaker plus liquid is found and the mass of liquid is obtained by subtraction.

d) Air

Using a balance, the mass of a $500\,cm^3$ round-bottomed flask full of air is found and again after removing the air with a vacuum pump; the difference gives the mass of air in the flask. The volume of air is found by filling the flask with water and pouring it into a measuring cylinder.

● Floating and sinking

An object sinks in a liquid of lower density than its own; otherwise it floats, partly or wholly submerged. For example, a piece of glass of density $2.5\,g/cm^3$ sinks in water (density $1.0\,g/cm^3$) but floats in mercury (density $13.6\,g/cm^3$). An iron nail sinks in water but an iron ship floats because its average density is less than that of water.

Figure 5.3 Why is it easy to float in the Dead Sea?

Checklist

After studying this chapter you should be able to

• define density and perform calculations using $\rho = m/V$,
• describe experiments to measure the density of solids, liquids and air,
• predict whether an object will float based on density data.

Questions

1 a If the density of wood is 0.5 g/cm³ what is the mass of
      (i)   1 cm³,
      (ii) 2 cm³,
      (iii) 10 cm³?
   b What is the density of a substance of
      (i)   mass 100 g and volume 10 cm³,
      (ii) volume 3 m³ and mass 9 kg?
   c The density of gold is 19 g/cm³. Find the volume of
      (i)   38 g,
      (ii) 95 g of gold.
2 A piece of steel has a volume of 12 cm³ and a mass of 96 g.
   What is its density in
   a g/cm³,
   b kg/m³?
3 What is the mass of 5 m³ of cement of density 3000 kg/m³?
4 What is the mass of air in a room measuring 10 m × 5.0 m × 2.0 m if the density of air is 1.3 kg/m³?
5 When a golf ball is lowered into a measuring cylinder of water, the water level rises by 30 cm³ when the ball is completely submerged. If the ball weighs 33 g in air, find its density.
6 Why does ice float on water?

6 Weight and stretching

- Force
- Weight
- The newton

- Hooke's law
- Practical work: Stretching a spring

Force

A **force** is a push or a pull. It can cause a body at rest to move, or if the body is already moving it can change its speed or direction of motion. A force can also change a body's shape or size.

Figure 6.1 A weightlifter in action exerts first a pull and then a push.

Weight

We all constantly experience the force of **gravity**, in other words. the pull of the Earth. It causes an unsupported body to fall from rest to the ground.

The weight of a body is the force of gravity on it.

For a body above or on the Earth's surface, the nearer it is to the centre of the Earth, the more the Earth attracts it. Since the Earth is not a perfect sphere but is flatter at the poles, the weight of a body varies over the Earth's surface. It is greater at the poles than at the equator.

Gravity is a force that can act through space, i.e. there does not need to be contact between the Earth and the object on which it acts as there does when we push or pull something. Other action-at-a-distance forces which, like gravity, decrease with distance are:

(i) **magnetic** forces between magnets, and
(ii) **electric** forces between electric charges.

The newton

The unit of force is the **newton** (N). It will be defined later (Chapter 8); the definition is based on the change of speed a force can produce in a body. Weight is a force and therefore should be measured in newtons.

The weight of a body can be measured by hanging it on a spring balance marked in newtons (Figure 6.2) and letting the pull of gravity stretch the spring in the balance. The greater the pull, the more the spring stretches.

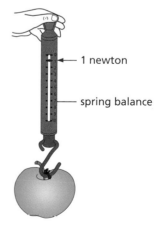

Figure 6.2 The weight of an average-sized apple is about 1 newton.

On most of the Earth's surface:

The weight of a body of mass 1 kg is 9.8 N.

24

Often this is taken as 10 N. A mass of 2 kg has a weight of 20 N, and so on. The mass of a body is the same wherever it is and, unlike weight, does not depend on the presence of the Earth.

Practical work

Stretching a spring

Arrange a steel spring as in Figure 6.3. Read the scale opposite the bottom of the hanger. Add 100 g loads one at a time (thereby increasing the stretching force by steps of 1 N) and take the readings after each one. Enter the readings in a table for loads up to 500 g.

Note that at the head of columns (or rows) in data tables it is usual to give the name of the quantity or its symbol followed by / and the unit.

Stretching force/N	Scale reading/mm	Total extension/mm

Do the results suggest any rule about how the spring behaves when it is stretched?

Sometimes it is easier to discover laws by displaying the results on a graph. Do this on graph paper by plotting stretching force readings along the x-axis (horizontal axis) and total extension readings along the y-axis (vertical axis). Every pair of readings will give a point; mark them by small crosses and draw a smooth line through them. What is its shape?

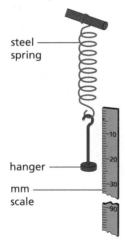

steel spring

hanger

mm scale

Figure 6.3

● Hooke's law

Springs were investigated by Robert Hooke nearly 350 years ago. He found that the extension was proportional to the stretching force provided the spring was not permanently stretched. This means that doubling the force doubles the extension, trebling the force trebles the extension, and so on.

Using the sign for proportionality, ∝, we can write **Hooke's law** as

> extension ∝ stretching force

It is true only if the **elastic limit** or 'limit of proportionality' of the spring is not exceeded. In other words, the spring returns to its original length when the force is removed.

The graph of Figure 6.4 is for a spring stretched beyond its elastic limit, E. OE is a straight line passing through the origin O and is graphical proof that Hooke's law holds over this range. If the force for point A on the graph is applied to the spring, the proportionality limit is passed and on removing the force some of the extension (OS) remains. Over which part of the graph does a spring balance work?

The **force constant**, k, of a spring is the force needed to cause unit extension, i.e. 1 m. If a force F produces extension x then

$$k = \frac{F}{x}$$

Rearranging the equation gives

$$F = kx$$

This is the usual way of writing Hooke's law in symbols.

Hooke's law also holds when a force is applied to a straight metal wire or an elastic band, provided they are not permanently stretched. Force–extension graphs similar to Figure 6.4 are obtained. You should label each axis of your graph with the name of the quantity or its symbol followed by / and the unit, as shown in Figure 6.4.

For a rubber band, a small force causes a large extension.

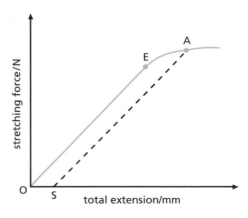

Figure 6.4

● Worked example

A spring is stretched 10 mm (0.01 m) by a weight of 2.0 N. Calculate: **a** the force constant k, and **b** the weight W of an object that causes an extension of 80 mm (0.08 m).

a $k = \dfrac{F}{x} = \dfrac{2.0\text{ N}}{0.01\text{ m}} = 200\text{ N/m}$

b W = stretching force F
  $= k \times x$
  $= 200\text{ N/m} \times 0.08\text{ m}$
  $= 16\text{ N}$

Checklist

After studying this chapter you should be able to

- recall that a force can cause a change in the motion, size or shape of a body,
- recall that the weight of a body is the force of gravity on it,
- recall the unit of force and how force is measured,
- describe an experiment to study the relation between force and extension for springs,
- draw conclusions from force–extension graphs,
- recall Hooke's law and solve problems using it,
- recognise the significance of the term limit of proportionality.

Questions

1 A body of mass 1 kg has weight 10 N at a certain place. What is the weight of
  a 100 g, 1 N
  b 5 kg, 50 N
  c 50 g? 0.5 N
2 The force of gravity on the Moon is said to be one-sixth of that on the Earth. What would a mass of 12 kg weigh
  a on the Earth, and 120 N
  b on the Moon? 20 N
3 What is the force constant of a spring which is stretched
  a 2 mm by a force of 4 N, 2.
  b 4 cm by a mass of 200 g? 500
4 The spring in Figure 6.5 stretches from 10 cm to 22 cm when a force of 4 N is applied. If it obeys Hooke's law, its total length in cm when a force of 6 N is applied is
  A 28   B 42   C 50   D 56   E 100

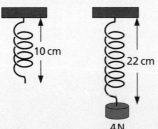

4 N

Figure 6.5

7 Adding forces

- ● Forces and resultants
- ● Examples of addition of forces
- ● Vectors and scalars
- ● Friction
- ● Practical work: Parallelogram law

● Forces and resultants

Force has both magnitude (size) and direction. It is represented in diagrams by a straight line with an arrow to show its direction of action.

Usually more than one force acts on an object. As a simple example, an object resting on a table is pulled downwards by its weight W and pushed upwards by a force R due to the table supporting it (Figure 7.1). Since the object is at rest, the forces must balance, i.e. $R = W$.

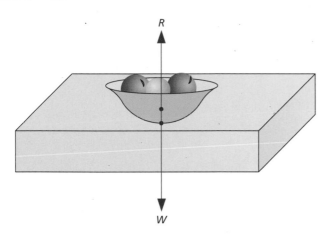

Figure 7.1

In structures such as a giant oil platform (Figure 7.2), two or more forces may act at the same point. It is then often useful for the design engineer to know the value of the single force, i.e. the **resultant**, which has exactly the same effect as these forces. If the forces act in the same straight line, the resultant is found by simple addition or subtraction as shown in Figure 7.3; if they do not they are added by using the **parallelogram law**.

Practical work

Parallelogram law

Arrange the apparatus as in Figure 7.4a with a sheet of paper behind it on a vertical board. We have to find the resultant of forces P and Q.

Read the values of P and Q from the spring balances. Mark on the paper the directions of P, Q and W as shown by the strings.

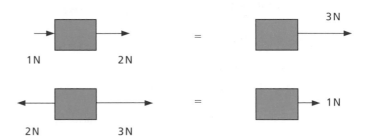

Figure 7.2 The design of an offshore oil platform requires an understanding of the combination of many forces.

Figure 7.3 The resultant of forces acting in the same straight line is found by addition or subtraction.

Remove the paper and, using a scale of 1 cm to represent 1 N, draw OA, OB and OD to represent the three forces P, Q and W which act at O, as in Figure 7.4b. (W = weight of the 1 kg mass = 9.8 N; therefore OD = 9.8 cm.)

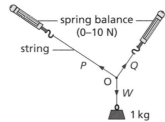

Figure 7.4a

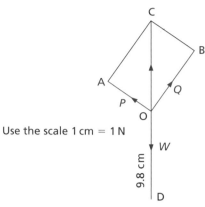

Use the scale 1 cm = 1 N

Figure 7.4b Finding a resultant by the parallelogram law

P and *Q* together are balanced by *W* and so their resultant must be a force equal and opposite to *W*.

Complete the parallelogram OACB. Measure the diagonal OC; if it is equal in size (i.e. 9.8 cm) and opposite in direction to *W* then it represents the resultant of *P* and *Q*.

The parallelogram law for adding two forces is:

If two forces acting at a point are represented in size and direction by the sides of a parallelogram drawn from the point, their resultant is represented in size and direction by the diagonal of the parallelogram drawn from the point.

Worked example

Find the resultant of two forces of 4.0 N and 5.0 N acting at an angle of 45° to each other.

Using a scale of 1.0 cm = 1.0 N, draw parallelogram ABDC with AB = 5.0 cm, AC = 4.0 N and angle CAB = 45° (Figure 7.5). By the parallelogram law, the diagonal AD represents the resultant in magnitude and direction; it measures 8.3 cm, and angle BAD = 20°.

∴ Resultant is a force of 8.3 N acting at an angle of 20° to the force of 5.0 N.

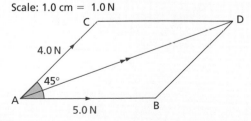

Scale: 1.0 cm = 1.0 N

Figure 7.5

Examples of addition of forces

1 **Two people carrying a heavy bucket.** The weight of the bucket is balanced by the force *F*, the resultant of F_1 and F_2 (Figure 7.6a).

2 **Two tugs pulling a ship.** The resultant of T_1 and T_2 is forwards in direction (Figure 7.6b), and so the ship moves forwards (as long as the resultant is greater than the resistance to motion of the sea and the wind).

a

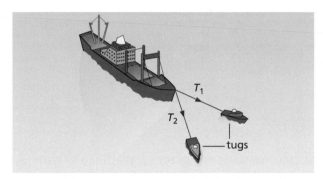

b

Figure 7.6

Vectors and scalars

A **vector** quantity is one such as force which is described completely only if both its size (magnitude) and direction are stated. It is not enough to say, for example, a force of 10 N, but rather a force of 10 N acting vertically downwards.

A vector can be represented by a straight line whose length represents the magnitude of the quantity and whose direction gives its line of action. An arrow on the line shows which way along the line it acts.

A **scalar** quantity has magnitude only. Mass is a scalar and is completely described when its value is known. Scalars are added by ordinary arithmetic; vectors are added geometrically, taking account of their directions as well as their magnitudes.

● Friction

Friction is the force that opposes one surface moving, or trying to move, over another. It can be a help or a hindrance. We could not walk if there was no friction between the soles of our shoes and the ground. Our feet would slip backwards, as they tend to if we walk on ice. On the other hand, engineers try to reduce friction to a minimum in the moving parts of machinery by using lubricating oils and ball-bearings.

When a gradually increasing force P is applied through a spring balance to a block on a table (Figure 7.7), the block does not move at first. This is because an equally increasing but opposing frictional force F acts where the block and table touch. At any instant P and F are equal and opposite.

If P is increased further, the block eventually moves; as it does so F has its maximum value, called **starting** or **static friction**. When the block is moving at a steady speed, the balance reading is slightly less than that for starting friction. **Sliding** or **dynamic friction** is therefore less than starting or static friction.

Placing a mass on the block increases the force pressing the surfaces together and increases friction.

When work is done against friction, the temperatures of the bodies in contact rise (as you can test by rubbing your hands together); mechanical energy is being changed into heat energy (see Chapter 13).

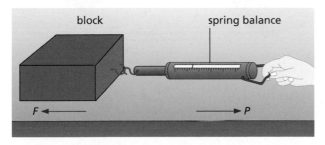

Figure 7.7 Friction opposes motion between surfaces in contact.

1 Jo, Daniel and Helen are pulling a metal ring. Jo pulls with a force of 100 N in one direction and Daniel with a force of 140 N in the opposite direction. If the ring does not move, what force does Helen exert if she pulls in the same direction as Jo?
2 A boy drags a suitcase along the ground with a force of 100 N. If the frictional force opposing the motion of the suitcase is 50 N, what is the resultant forward force on the suitcase?
3 A picture is supported by two vertical strings; if the weight of the picture is 50 N what is the force exerted by each string?
4 Using a scale of 1 cm to represent 10 N, find the size and direction of the resultant of forces of 30 N and 40 N acting at right angles to each other.
5 Find the size of the resultant of two forces of 5 N and 12 N acting
  a in opposite directions to each other,
  b at 90° to each other.

Checklist

After studying this chapter you should be able to

• combine forces acting along the same straight line to find their resultant,

• add vectors graphically to determine a resultant,

• distinguish between vectors and scalars and give examples of each,

• understand friction as the force between two surfaces that impedes motion and results in heating.

8 Force and acceleration

- Newton's first law
- Mass and inertia
- Newton's second law
- Weight and gravity
- Gravitational field
- Newton's third law
- Air resistance: terminal velocity
- Practical work: Effect of force and mass on acceleration

● Newton's first law

Friction and air resistance cause a car to come to rest when the engine is switched off. If these forces were absent we believe that a body, once set in motion, would go on moving forever with a constant speed in a straight line. That is, force is not needed to keep a body moving with uniform velocity provided that no opposing forces act on it.

This idea was proposed by Galileo and is summed up in **Newton's first law of motion**:

> A body stays at rest, or if moving it continues to move with uniform velocity, unless an external force makes it behave differently.

It seems that the question we should ask about a moving body is not 'what keeps it moving' but 'what changes or stops its motion'.

The smaller the external forces opposing a moving body, the smaller is the force needed to keep it moving with uniform velocity. An 'airboard', which is supported by a cushion of air (Figure 8.1), can skim across the ground with little frictional opposition, so that relatively little power is needed to maintain motion.

Figure 8.1 Friction is much reduced for an airboard.

● Mass and inertia

Newton's first law is another way of saying that all matter has a built-in opposition to being moved if it is at rest or, if it is moving, to having its motion changed. This property of matter is called **inertia** (from the Latin word for laziness).

Its effect is evident on the occupants of a car that stops suddenly; they lurch forwards in an attempt to continue moving, and this is why seat belts are needed. The reluctance of a stationary object to move can be shown by placing a large coin on a piece of card on your finger (Figure 8.2). If the card is flicked *sharply* the coin stays where it is while the card flies off.

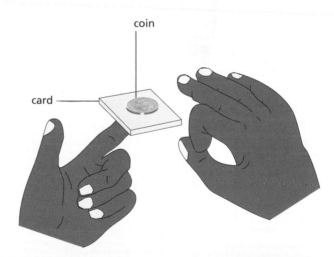

Figure 8.2 Flick the card sharply

The larger the mass of a body, the greater is its inertia, i.e. the more difficult it is to move it when at rest and to stop it when in motion. Because of this we consider that **the mass of a body measures its inertia**. This is a better definition of mass than the one given earlier (Chapter 1) in which it was stated to be the 'amount of matter' in a body.

Practical work

Effect of force and mass on acceleration

The apparatus consists of a trolley to which a force is applied by a stretched length of elastic (Figure 8.3). The velocity of the trolley is found from a tickertape timer or a motion sensor, datalogger and computer (see Figure 2.6, p.11).

First compensate the runway for friction: raise one end until the trolley runs down with uniform velocity when given a push. The dots on the tickertape should be equally spaced, or a horizontal trace obtained on a velocity–time graph. There is now no resultant force on the trolley and any acceleration produced later will be due only to the force caused by the stretched elastic.

a) Force and acceleration (mass constant)

Fix one end of a short length of elastic to the rod at the back of the trolley and stretch it until the other end is level with the front of the trolley. Practise pulling the trolley down the runway, keeping the same stretch on the elastic. After a few trials you should be able to produce a steady accelerating force.

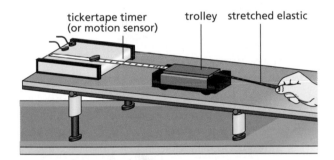
tickertape timer (or motion sensor)   trolley   stretched elastic

Figure 8.3

Repeat using first two and then three *identical* pieces of elastic, stretched side by side by the same amount, to give two and three units of force.

If you are using tickertape, make a tape chart for each force and use it to find the acceleration produced in cm/tentick² (see Chapter 2). Ignore the start of the tape (where the dots are too close) and the end (where the force may not be steady). If you use a motion sensor and computer to plot a velocity–time graph, the acceleration can be obtained in m/s² from the slope of the graph (Chapter 3).

Does a steady force cause a steady acceleration? Put the results in a table. Do they suggest any relationship between acceleration, *a*, and force *F*?

Force (*F*)/(no. of pieces of elastic)	1	2	3
Acceleration (*a*)/cm/tentick² or m/s²			

b) Mass and acceleration (force constant)

Do the experiment as in a) using two pieces of elastic (i.e. constant *F*) to accelerate first one trolley, then two (stacked one above the other) and finally three. Check the friction compensation of the runway each time.

Find the accelerations from the tape charts or computer plots and tabulate the results. Do they suggest any relationship between *a* and *m*?

Mass (*m*)/(no. of trolleys)	1	2	3
Acceleration (*a*)/cm/tentick² or m/s²			

● Newton's second law

The previous experiment should show roughly that the acceleration *a* is

(i) directly proportional to the applied force *F* for a fixed mass, i.e., $a \propto F$, and

(ii) inversely proportional to the mass *m* for a fixed force, i.e., $a \propto 1/m$.

Combining the results into one equation, we get

$$a \propto \frac{F}{m} \quad \text{or} \quad F \propto ma$$

Therefore

$$F = kma$$

where *k* is the constant of proportionality.

> One newton is defined as the force which gives a mass of 1 kg an acceleration of 1 m/s², i.e., $1\,N = 1\,kg\,m/s^2$.

So if $m = 1\,kg$ and $a = 1\,m/s^2$, then $F = 1\,N$. Substituting in $F = kma$, we get $k = 1$ and so we can write

> $$F = ma$$

This is **Newton's second law of motion**. When using it two points should be noted. First, *F* is the resultant (or unbalanced) force causing the acceleration *a*. Second, *F* must be in newtons, *m* in kilograms and *a* in metres per second squared, otherwise *k* is not 1. The law shows that *a* will be largest when *F* is large and *m* small.

You should now appreciate that when the forces acting on a body do not balance there is a net (resultant) force which causes a change of motion, i.e. the body accelerates or decelerates. If the forces balance, there is no change in the motion of the body. However, there may be a change of shape, in which case internal forces in the body (i.e. forces between neighbouring atoms) balance the external forces.

● Worked example

A block of mass 2 kg has a constant velocity when it is pushed along a table by a force of 5 N. When the push is increased to 9 N what is

a the resultant force,
b the acceleration?

When the block moves with constant velocity the forces acting on it are balanced. The force of friction opposing its motion must therefore be 5 N.

a When the push is increased to 9 N the resultant (unbalanced) force F on the block is $(9-5)\,N = 4\,N$ (since the frictional force is still 5 N).

b The acceleration a is obtained from $F = ma$ where $F = 4\,N$ and $m = 2\,kg$.

$$\therefore\ a = \frac{F}{m} = \frac{4\,N}{2\,kg} = \frac{4\,kg\,m/s^2}{2\,kg} = 2\,m/s^2$$

● Weight and gravity

The weight W of a body is the force of gravity acting on it which gives it an acceleration g when it is falling freely near the Earth's surface. If the body has mass m, then W can be calculated from $F = ma$. We put $F = W$ and $a = g$ to give

$$W = mg$$

Taking $g = 9.8\,m/s^2$ and $m = 1\,kg$, this gives $W = 9.8\,N$, i.e. a body of mass 1 kg has weight 9.8 N, or near enough 10 N. Similarly a body of mass 2 kg has weight of about 20 N, and so on. While the mass of a body is always the same, its weight varies depending on the value of g. On the Moon the acceleration of free fall is only about $1.6\,m/s^2$, and so a mass of 1 kg has a weight of just 1.6 N there.

The weight of a body is directly proportional to its mass, which explains why g is the same for all bodies. The greater the mass of a body, the greater is the force of gravity on it but it does not accelerate faster when falling because of its greater inertia (i.e. its greater resistance to acceleration).

● Gravitational field

The force of gravity acts through space and can cause a body, not in contact with the Earth, to fall to the ground. It is an invisible, action-at-a-distance force. We try to 'explain' its existence by saying that the Earth is surrounded by a **gravitational field** which exerts a force on any body in the field. Later, magnetic and electric fields will be considered.

The strength of a gravitational field is defined as the force acting on unit mass in the field.

Measurement shows that on the Earth's surface a mass of 1 kg experiences a force of 9.8 N, i.e. its weight is 9.8 N. The strength of the Earth's field is therefore 9.8 N/kg (near enough 10 N/kg). It is denoted by g, the letter also used to denote the acceleration of free fall. Hence

$$g = 9.8\,N/kg = 9.8\,m/s^2$$

We now have two ways of regarding g. When considering bodies *falling freely*, we can think of it as an acceleration of $9.8\,m/s^2$. When a body of known mass is *at rest* and we wish to know the force of gravity (in N) acting on it we think of g as the Earth's gravitational field strength of 9.8 N/kg.

● Newton's third law

If a body A exerts a force on body B, then body B exerts an equal but opposite force on body A.

This is **Newton's third law of motion** and states that forces never occur singly but always in pairs as a result of the action between two bodies. For example, when you step forwards from rest your foot pushes backwards on the Earth, and the Earth exerts an equal and opposite force forward on you. Two bodies and two forces are involved. The small force you exert on the large mass of the Earth gives no noticeable acceleration to the Earth but the equal force it exerts on your very much smaller mass causes you to accelerate.

Note that the pair of equal and opposite forces **do not act on the same body**; if they did, there could never be any resultant forces and acceleration would be impossible. For a book resting on a table, the book exerts a downward force on the table and the table exerts an equal and opposite upward force on the book; this pair of forces act on different objects and are represented by the red arrows in Figure 8.4. The weight of the book (blue arrow) does not form a pair with the upward force on the book (although they are equal numerically) as these two forces act on the same body.

An appreciation of the third law and the effect of friction is desirable when stepping from a rowing boat (Figure 8.5). You push backwards on the boat and, although the boat pushes you forwards with an equal force, it is itself now moving backwards

(because friction with the water is slight). This reduces your forwards motion by the same amount – so you may fall in!

Figure 8.5 The boat moves backwards when you step forwards!

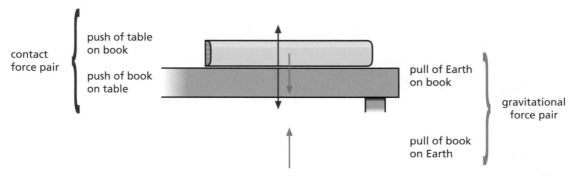

Figure 8.4 Forces between book and table

● Air resistance: terminal velocity

When an object falls in air, the air resistance (fluid friction) opposing its motion **increases as its speed rises**, so reducing its acceleration. Eventually, air resistance acting upwards equals the weight of the object acting downwards. The resultant force on the object is then zero since the gravitational force balances the frictional force. The object falls at a constant velocity, called its **terminal velocity**, whose value depends on the size, shape and weight of the object.

A small dense object, such as a steel ball-bearing, has a high terminal velocity and falls a considerable distance with a constant acceleration of $9.8\,\mathrm{m/s^2}$ before air resistance equals its weight. A light object, like a raindrop, or an object with a large surface area, such as a parachute, has a low terminal velocity and only accelerates over a comparatively short distance before air resistance equals its weight. A skydiver (Figure 8.6) has a terminal velocity of more than $50\,\mathrm{m/s}$ ($180\,\mathrm{km/h}$) before the parachute is opened.

Objects falling in liquids behave similarly to those falling in air.

Figure 8.6 Synchronised skydivers

Questions

1 Which one of the diagrams in Figure 8.7 shows the arrangement of forces that gives the block of mass *M* the greatest acceleration?

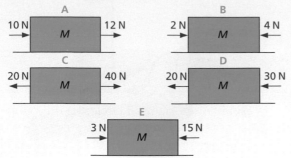

Figure 8.7

2 In Figure 8.8 if *P* is a force of 20 N and the object moves with constant velocity, what is the value of the opposing force *F*?

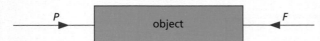

Figure 8.8

3 a What resultant force produces an acceleration of 5 m/s² in a car of mass 1000 kg?
  b What acceleration is produced in a mass of 2 kg by a resultant force of 30 N?
4 A block of mass 500 g is pulled from rest on a horizontal frictionless bench by a steady force *F* and travels 8 m in 2 s. Find
  a the acceleration,
  b the value of *F*.
5 Starting from rest on a level road a girl can reach a speed of 5 m/s in 10 s on her bicycle. Find
  a the acceleration,
  b the average speed during the 10 s,
  c the distance she travels in 10 s.
  Eventually, even though she is still pedalling as fast as she can, she stops accelerating and her speed reaches a maximum value. Explain in terms of the forces acting why this happens.
6 What does an astronaut of mass 100 kg weigh
  a on Earth where the gravitational field strength is 10 N/kg,
  b on the Moon where the gravitational field strength is 1.6 N/kg?

7 A rocket has a mass of 500 kg.
  a What is its weight on Earth where *g* = 10 N/kg?
  b At lift-off the rocket engine exerts an upward force of 25 000 N. What is the resultant force on the rocket? What is its initial acceleration?
8 Figure 8.9 shows the forces acting on a raindrop which is falling to the ground.
  a (i) *A* is the force which causes the raindrop to fall. What is this force called?
    (ii) *B* is the total force opposing the motion of the drop. State *one* possible cause of this force.
  b What happens to the drop when force *A* = force *B*?

Figure 8.9

9 Explain the following using *F* = *ma*.
  a A racing car has a powerful engine and is made of strong but lightweight material.
  b A car with a small engine can still accelerate rapidly.

Checklist

After studying this chapter you should be able to

• describe an experiment to investigate the relationship between force, mass and acceleration,
• state the unit of force,

• state Newton's second law of motion and use it to solve problems,
• recall and use the equation *W* = *mg*

• define the strength of the Earth's gravitational field,
• describe the motion of an object falling in air.

Circular motion

- Centripetal force
- Rounding a bend
- Looping the loop

- Satellites
- Practical work: Investigating circular motion

There are many examples of bodies moving in circular paths – rides at a funfair, clothes being spun dry in a washing machine, the planets going round the Sun and the Moon circling the Earth. When a car turns a corner it may follow an arc of a circle. 'Throwing the hammer' is a sport practised at highland games in Scotland (Figure 9.1), in which the hammer is whirled round and round before it is released.

Figure 9.1 'Throwing the hammer'

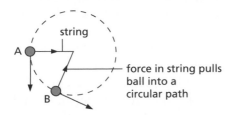

Figure 9.2

● Centripetal force

In Figure 9.2 a ball attached to a string is being whirled round in a horizontal circle. Its direction of motion is constantly changing. At A it is along the tangent at A; shortly afterwards, at B, it is along the tangent at B; and so on.

Velocity has both size and direction; speed has only size. Velocity is speed in a stated direction and if the direction of a moving body changes, even if its speed does not, then its velocity has changed. A change of velocity is an acceleration, and so during its whirling motion the ball is accelerating.

It follows from Newton's first law of motion that if we consider a body moving in a circle to be accelerating then there must be a force acting on it to cause the acceleration. In the case of the whirling ball it is reasonable to say the force is provided by the string pulling inwards on the ball. Like the acceleration, the force acts towards the centre of the circle and keeps the body at a fixed distance from the centre.

A larger force is needed if

(i) the speed v of the ball is increased,

(ii) the radius r of the circle is decreased,

(iii) the mass m of the ball is increased.

The rate of change of direction, and so the acceleration a, is increased by **(i)** and **(ii)**. It can be shown that $a = v^2/r$ and so, from $F = ma$, we can write

$$F = \frac{mv^2}{r}$$

This force, which acts **towards the centre** and keeps a body moving in a circular path, is called the **centripetal force** (centre-seeking force).

Should the force be greater than the string can bear, the string breaks and the ball flies off with

▶▶

steady speed in a straight line **along the tangent**, i.e. in the direction of travel when the string broke (as Newton's first law of motion predicts). It is not thrown outwards.

Whenever a body moves in a circle (or circular arc) there must be a centripetal force acting on it. In throwing the hammer it is the pull of the athlete's arms acting on the hammer towards the centre of the whirling path. When a car rounds a bend a frictional force is exerted inwards by the road on the car's tyres.

Practical work

Investigating circular motion

Use the apparatus in Figure 9.3 to investigate the various factors that affect circular motion. Make sure the rubber bung is **tied securely** to the string and that **the area around you is clear of other students**. The paper clip acts as an indicator to aid keeping the radius of the circular motion constant.

Spin the rubber bung at a constant speed while adding more weights to the holder; it will be found that the radius of the orbit decreases. Show that if the rubber bung is spun faster, more weights must be added to the holder to keep the radius constant. Are these findings in agreement with the formula given on p. 35 for the centripetal force, $F = mv^2/r$?

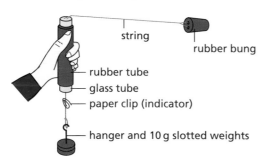

string
rubber bung
— rubber tube
— glass tube
— paper clip (indicator)
— hanger and 10 g slotted weights

Figure 9.3

● Rounding a bend

When a car rounds a bend, a frictional force is exerted *inwards* by the road on the car's tyres, so providing the centripetal force needed to keep it in its curved path (Figure 9.4a). Here friction acts as an accelerating force (towards the centre of the circle) rather than a retarding force (p. 29). The successful negotiation of a bend on a flat road,

therefore, depends on the tyres and the road surface being in a condition that enables them to provide a sufficiently large frictional force – otherwise skidding occurs.

Safe cornering that does not rely entirely on friction is achieved by 'banking' the road as in Figure 9.4b. Some of the centripetal force is then supplied by the part of the contact force N, from the road surface on the car, that acts horizontally. A bend in a railway track is banked, so that the outer rail is not strained by having to supply the centripetal force by pushing inwards on the wheel flanges.

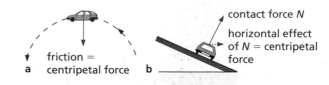

contact force N

horizontal effect of N = centripetal force

friction = centripetal force

a   b

Figure 9.4

● Looping the loop

A pilot who is not strapped into his aircraft can still loop the loop without falling downwards at the top of the loop. A bucket of water can be swung round in a vertical circle without spilling. Some amusement park rides (Figure 9.5) give similar effects. Can you suggest what provides the centripetal force for each of these three cases (i) at the top of the loop and (ii) at the bottom of the loop?

● Satellites

For a **satellite of** mass m orbiting the Earth at radius r with orbital speed v, the centripetal force, $F = mv^2/r$, is provided by gravity.

To put an artificial satellite in orbit at a certain height above the Earth it must enter the orbit at the correct speed. If it does not, the force of gravity, which decreases as height above the Earth increases, will not be equal to the centripetal force needed for the orbit.

This can be seen by imagining a shell fired horizontally from the top of a very high mountain (Figure 9.6). If gravity did not pull it towards the centre of the Earth it would continue to travel horizontally, taking path A. In practice it might take path B. A second shell fired faster might take path C

The orbital period T (the time for one orbit) of a satellite = distance/velocity. So for a circular orbit

$$T = \frac{2\pi r}{v}$$

Satellites in high orbits have longer periods than those in low orbits.

The Moon is kept in a circular orbit round the Earth by the force of gravity between it and the Earth. It has an orbital period of 27 days.

a) Communication satellites

These circle the Earth in orbits above the equator. **Geostationary satellites** have an orbit high above the equator (36 000 km); they travel with the same speed as the Earth rotates, so appear to be stationary at a particular point above the Earth's surface – their orbital period is 24 hours. They are used for transmitting television, intercontinental telephone and data signals. Geostationary satellites need to be well separated so that they do not interfere with each other; there is room for about 400.

Mobile phone networks use many satellites in much lower equatorial orbits; they are slowed by the Earth's atmosphere and their orbit has to be regularly adjusted by firing a rocket engine. Eventually they run out of fuel and burn up in the atmosphere as they fall to Earth.

b) Monitoring satellites

These circle the Earth rapidly in low polar orbits, i.e. passing over both poles; at a height of 850 km the orbital period is only 100 minutes. The Earth rotates below them so they scan the whole surface at short range in a 24-hour period and can be used to map or monitor regions of the Earth's surface which may be inaccessible by other means. They are widely used in weather forecasting to continuously transmit infrared pictures of cloud patterns down to Earth (Figure 9.7), which are picked up in turn by receiving stations around the world.

▶▶

Figure 9.5 Looping the loop at an amusement park

and travel further. If a third shell is fired even faster, it might never catch up with the rate at which the Earth's surface is falling away. It would remain at the same height above the Earth (path D) and return to the mountain top, behaving like a satellite.

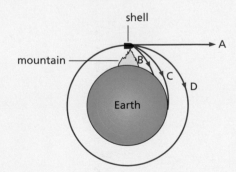

Figure 9.6

Figure 9.7 Satellite image of cloud over Europe

Questions

1 An apple is whirled round in a horizontal circle on the end of a string which is tied to the stalk. It is whirled faster and faster and at a certain speed the apple is torn from the stalk. Why?

2 A car rounding a bend travels in an arc of a circle.
   a   What provides the centripetal force?
   b   Is a larger or a smaller centripetal force required if
      (i)   the car travels faster,
      (ii) the bend is less curved,
      (iii) the car has more passengers?

3 Racing cars are fitted with tyres called 'slicks', which have no tread pattern, for dry tracks, and with 'tread' tyres for wet tracks. Why?

4 A satellite close to the Earth (at a height of about 200 km) has an orbital speed of 8 km/s. Taking the radius of the orbit as approximately equal to the Earth's radius of 6400 km, calculate the time it takes to make one orbit.

10 Moments and levers

- Moment of a force
- Balancing a beam
- Levers
- Conditions for equilibrium
- Practical work: Law of moments

● Moment of a force

The handle on a door is at the outside edge so that it opens and closes easily. A much larger force would be needed if the handle were near the hinge. Similarly it is easier to loosen a nut with a long spanner than with a short one.

The **turning effect of a force** is called the **moment of the force**. It depends on both the size of the force and how far it is applied from the pivot or **fulcrum**. It is measured by multiplying the force by the perpendicular distance of the line of action of the force from the fulcrum. The unit is the **newton metre** (N m).

> moment of a force = force × perpendicular distance of the line of action of the force from fulcrum

In Figure 10.1a, a force F acts on a gate at its edge, and in Figure 10.1b it acts at the centre.
In Figure 10.1a:

moment of F about O = $5\,N \times 3\,m = 15\,N\,m$

In Figure 10.1b:

moment of F about O = $5\,N \times 1.5\,m = 7.5\,N\,m$

The turning effect of F is greater in the first case; this agrees with the fact that a gate opens most easily when pushed or pulled at the edge furthest from the hinge.

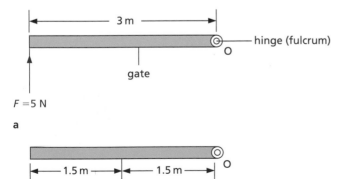

Figure 10.1

● Balancing a beam

To balance a beam about a pivot, like the ruler in Figure 10.2, the weights must be moved so that the clockwise turning effect equals the anticlockwise turning effect and the net moment on the beam becomes zero. If the beam tends to swing clockwise, m_1 can be moved further from the pivot to increase its turning effect; alternatively m_2 can be moved nearer to the pivot to reduce its turning effect. What adjustment would you make to the position of m_2 to balance the beam if it is tending to swing anticlockwise?

Practical work

Law of moments

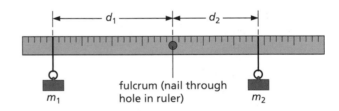

Figure 10.2

Balance a half-metre ruler at its centre, adding Plasticine to one side or the other until it is horizontal.

Hang unequal loads m_1 and m_2 from either side of the fulcrum and alter their distances d_1 and d_2 from the centre until the ruler is again balanced (Figure 10.2). Forces F_1 and F_2 are exerted by gravity on m_1 and m_2 and so on the ruler; the force on 100 g is 1 N. Record the results in a table and repeat for other loads and distances.

m_1/g	F_1/N	d_1/cm	$F_1 \times d_1$ /N cm	m_2/g	F_2/N	d_2/cm	$F_2 \times d_2$ / N cm

F_1 is trying to turn the ruler anticlockwise and $F_1 \times d_1$ is its moment. F_2 is trying to cause clockwise turning and its moment is $F_2 \times d_2$. When the ruler is balanced or, as we say, **in equilibrium**, the results should show that the anticlockwise moment $F_1 \times d_1$ equals the clockwise moment $F_2 \times d_2$.

The **law of moments** (also called the **law of the lever**) is stated as follows.

> When a body is in equilibrium the sum of the clockwise moments about any point equals the sum of the anticlockwise moments about the same point. There is no *net* moment on a body which is in equilibrium.

● Worked example

The see-saw in Figure 10.3 balances when Shani of weight 320 N is at A, Tom of weight 540 N is at B and Harry of weight W is at C. Find W.

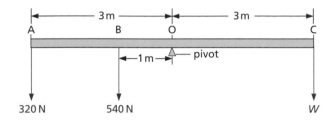

Figure 10.3

Taking moments about the fulcrum, O:

anticlockwise moment $= (320\,\text{N} \times 3\,\text{m}) + (540\,\text{N} \times 1\,\text{m})$
$= 960\,\text{N m} + 540\,\text{N m}$
$= 1500\,\text{N m}$

clockwise moment $= W \times 3\,\text{m}$

By the law of moments,

clockwise moments = anticlockwise moments

∴ $\qquad W \times 3\,\text{m} = 1500\,\text{N m}$

∴ $\qquad W = \dfrac{1500\,\text{N m}}{3\,\text{m}} = 500\,\text{N}$

● Levers

A **lever** is any device which can turn about a pivot. In a working lever a force called the **effort** is used to overcome a resisting force called the **load**. The pivotal point is called the fulcrum.

If we use a crowbar to move a heavy boulder (Figure 10.4), our hands apply the effort at one end of the bar and the load is the force exerted by the boulder on the other end. If distances from the fulcrum O are as shown and the load is 1000 N (i.e. the part of the weight of the boulder supported by the crowbar), the effort

can be calculated from the law of moments. As the boulder just begins to move we can say, taking moments about O, that

clockwise moment = anticlockwise moment

$$\text{effort} \times 200\,\text{cm} = 1000\,\text{N} \times 10\,\text{cm}$$

$$\text{effort} = \frac{10\,000\,\text{N cm}}{200\,\text{cm}} = 50\,\text{N}$$

Examples of other levers are shown in Figure 10.5. How does the effort compare with the load for scissors and a spanner in Figures 10.5c and d?

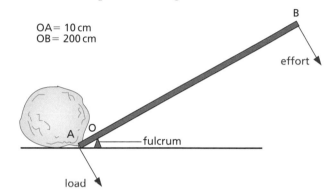

OA = 10 cm
OB = 200 cm

Figure 10.4 Crowbar

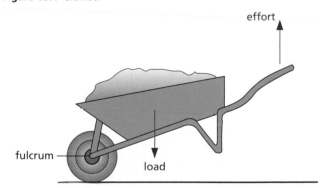

Figure 10.5a Wheelbarrow

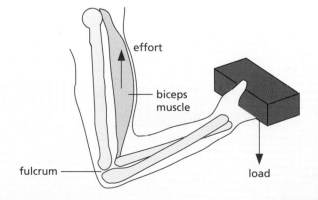

Figure 10.5b Forearm

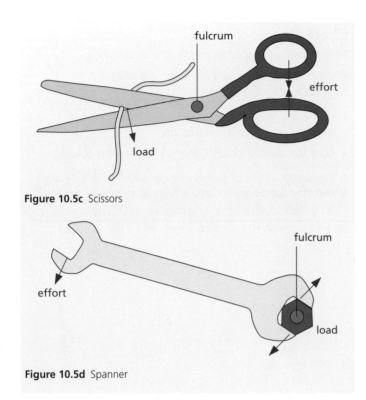

Figure 10.5c Scissors

Figure 10.5d Spanner

● Conditions for equilibrium

Sometimes a number of parallel forces act on a body so that it is in equilibrium. We can then say:

(i) The sum of the forces in one direction equals the sum of the forces in the opposite direction.
(ii) The law of moments must apply.

A body is in equilibrium when there is no resultant force and no resultant turning effect acting on it.

As an example consider a heavy plank resting on two trestles, as in Figure 10.6. In the next chapter we will see that the whole weight of the plank (400 N) may be taken to act vertically downwards at its centre, O. If P and Q are the upward forces exerted by the trestles on the plank (called reactions) then we have from (i) above:

$$P + Q = 400\,\text{N} \qquad (1)$$

Moments can be taken about any point but if we take them about C, the moment due to force Q is zero.

clockwise moment = $P \times 5\,\text{m}$

anticlockwise moment = $400\,\text{N} \times 2\,\text{m}$
$$= 800\,\text{N}\,\text{m}$$

Since the plank is in equilibrium we have from (ii) above:

$$P \times 5\,\text{m} = 800\,\text{N}\,\text{m}$$

∴
$$P = \frac{800\,\text{N}\,\text{m}}{5\,\text{m}} = 160\,\text{N}$$

From equation (1)

$$Q = 240\,\text{N}$$

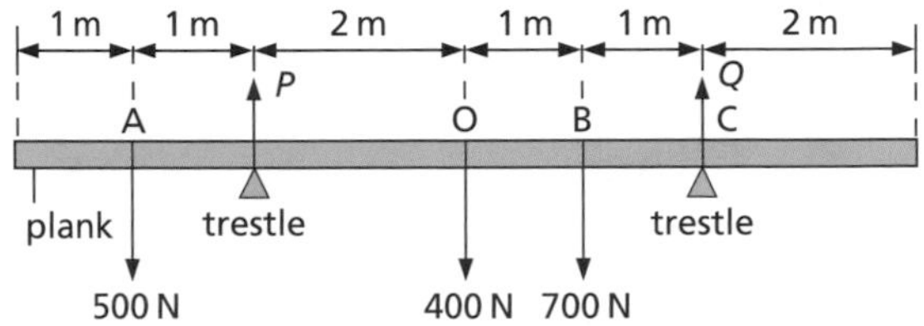

Figure 10.6

Questions

1 The metre rule in Figure 10.7 is pivoted at its centre. If it balances, which of the following equations gives the mass of M?

A $M + 50 = 40 + 100$
B $M \times 40 = 100 \times 50$
C $M/50 = 100/40$
D $M/50 = 40/100$
E $M \times 50 = 100 \times 40$

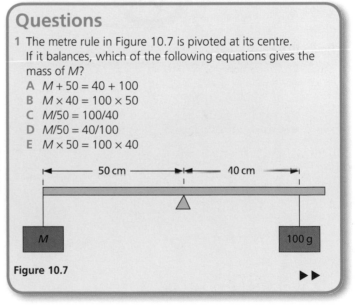

Figure 10.7

▶▶

2 Figure 10.8 shows three positions of the pedal on a bicycle which has a crank 0.20 m long. If the cyclist exerts the same vertically downward push of 25 N with his foot, in which case, A, B or C, is the turning effect
(i) $25 \times 0.2 = 5 \, N \, m$,
(ii) 0,
(iii) between 0 and 5 N m?
Explain your answers.

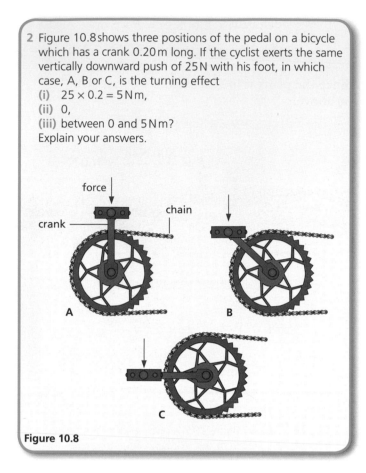

Figure 10.8

Checklist

After studying this chapter you should be able to

- define the moment of a force about a point,
- describe qualitatively the balancing of a beam about a pivot,

- describe an experiment to verify that there is no net moment on a body in equilibrium,

- state the law of moments and use it to solve problems,

- explain the action of common tools and devices as levers,
- state the conditions for equilibrium when parallel forces act on a body.

11 Centres of mass

- Toppling
- Stability
- Balancing tricks and toys
- Practical work: Centre of mass using a plumb line

A body behaves as if its whole mass were concentrated at one point, called its **centre of mass** or **centre of gravity**, even though the Earth attracts every part of it. The body's weight can be considered to act at this point. The centre of mass of a uniform ruler is at its centre and when supported there it can be balanced, as in Figure 11.1a. If it is supported at any other point it topples because the moment of its weight W about the point of support is not zero, as in Figure 11.1b.

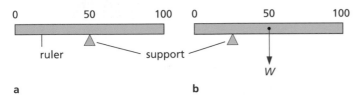

a
b

Figure 11.1

Your centre of mass is near the centre of your body and the vertical line from it to the floor must be within the area enclosed by your feet or you will fall over. You can test this by standing with one arm and the side of one foot pressed against a wall (Figure 11.2). Now try to raise the other leg sideways.

raise this leg

Figure 11.2 Can you do this without falling over?

A tightrope walker has to keep his centre of mass exactly above the rope. Some carry a long pole to help them to balance (Figure 11.3). The combined weight of the walker and pole is then spread out more and if the walker begins to topple to one side, he moves the pole to the other side.

Figure 11.3 A tightrope walker using a long pole

The centre of mass of a regularly shaped body that has the same density throughout is at its centre. In other cases it can be found by experiment.

Practical work

Centre of mass using a plumb line

Suppose we have to find the centre of mass of an irregularly shaped lamina (a thin sheet) of cardboard.

Make a hole A in the lamina and hang it so that it can *swing freely* on a nail clamped in a stand. It will come to rest with its centre of mass vertically below A. To locate the vertical line through A, tie a plumb line (a thread and a weight) to the nail (Figure 11.4), and mark its position AB on the lamina. The centre of mass lies somewhere on AB.

Hang the lamina from another position, C, and mark the plumb line position CD. The centre of mass lies on CD and must be at the point of intersection of AB and CD. Check this by hanging the lamina from a third hole. Also try balancing it at its centre of mass on the tip of your forefinger.

Devise a method using a plumb line for finding the centre of mass of a tripod.

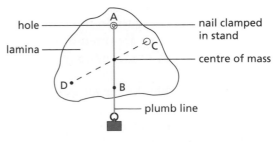

hole — nail clamped in stand
lamina — centre of mass
plumb line

Figure 11.4

● Toppling

The position of the centre of mass of a body affects whether or not it **topples** over easily. This is important in the design of such things as tall vehicles (which tend to overturn when rounding a corner), racing cars, reading lamps and even drinking glasses.

A body topples when the vertical line through its centre of mass falls outside its base, as in Figure 11.5a. Otherwise it remains stable, as in Figure 11.5b, where the body will not topple.

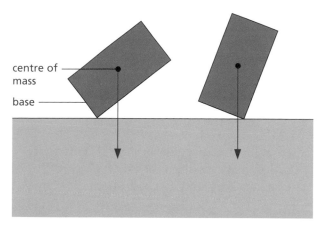

centre of mass

base

a **Topples**

b **Will not topple (stable)**

Figure 11.5

Toppling can be investigated by placing an empty can on a plank (with a rough surface to prevent slipping) which is slowly tilted. The angle of tilt is noted when the can falls over. This is repeated with a mass of 1 kg in the can. How does this affect the position of the centre of mass? The same procedure is followed with a second can of the same height as the first but of greater width. It will be found that the second can with the mass in it can be tilted through the greater angle.

The stability of a body is therefore increased by

(i) lowering its centre of mass, and
(ii) increasing the area of its base.

In Figure 11.6a the centre of mass of a tractor is being found. It is necessary to do this when testing a new design since tractors are often driven over sloping surfaces and any tendency to overturn must be discovered.

The stability of double-decker buses is being tested in Figure 11.6b. When the top deck only is fully laden with passengers (represented by sand bags in the test), it must not topple if tilted through an angle of 28°.

Racing cars have a low centre of mass and a wide wheelbase for maximum stability.

Figure 11.6a A tractor under test to find its centre of mass

Figure 11.6b A double-decker bus being tilted to test its stability

● Stability

Three terms are used in connection with **stability**.

a) Stable equilibrium

A body is in **stable equilibrium** if when slightly displaced and then released it returns to its previous position. The ball at the bottom of the dish in Figure 11.7a is an example. Its centre of mass rises when it is displaced. It rolls back because its weight has a moment about the point of contact that acts to reduce the displacement.

b) Unstable equilibrium

A body is in **unstable equilibrium** if it moves further away from its previous position when slightly displaced and released. The ball in Figure 11.7b behaves in this way. Its centre of mass falls when it is displaced slightly because there is a moment which increases the displacement. Similarly in Figure 11.1a (p. 43) the balanced ruler is in unstable equilibrium.

c) Neutral equilibrium

A body is in **neutral equilibrium** if it stays in its new position when displaced (Figure 11.7c). Its centre of mass does not rise or fall because there is no moment to increase or decrease the displacement.

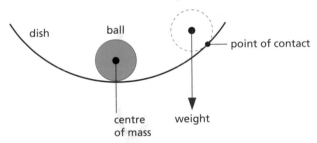

a **Stable**

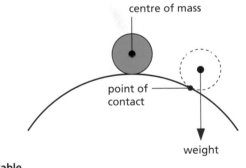

b **Unstable**

c **Neutral**

Figure 11.7 States of equilibrium

● Balancing tricks and toys

Some tricks that you can try or toys you can make are shown in Figure 11.8. In each case the centre of mass is vertically below the point of support and equilibrium is stable.

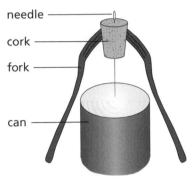

a **Balancing a needle on its point**

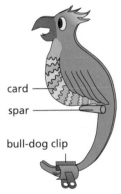

b **The perched parrot**

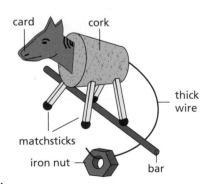

c **A rocking horse**

Figure 11.8 Balancing tricks

A self-righting toy (Figure 11.9) has a heavy base and, when tilted, the weight acting through the centre of mass has a moment about the point of contact. This restores it to the upright position.

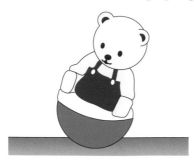

Figure 11.9 A self-righting toy

Questions

1 Figure 11.10 shows a Bunsen burner in three different positions. State in which position it is in
- **a** stable equilibrium,
- **b** unstable equilibrium,
- **c** neutral equilibrium.

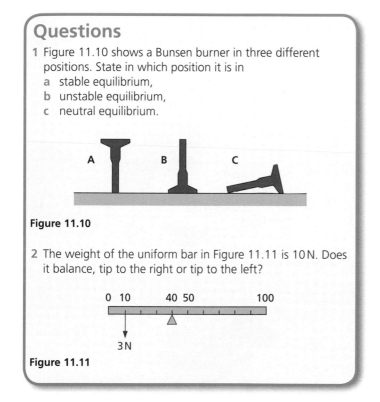

Figure 11.10

2 The weight of the uniform bar in Figure 11.11 is 10 N. Does it balance, tip to the right or tip to the left?

Figure 11.11

Checklist

After studying this chapter you should be able to

- recall that an object behaves as if its whole mass were concentrated at its centre of mass,
- recall that an object's weight acts through the centre of mass (or centre of gravity),
- describe an experiment to find the centre of mass of an object,
- connect the stability of an object to the position of its centre of mass.

12 Momentum

- Conservation of momentum
- Explosions
- Rockets and jets
- Force and momentum
- Sport: impulse and collision time
- Practical work: Collisions and momentum

Momentum is a useful quantity to consider when bodies are involved in collisions and explosions. It is defined as the **mass of the body multiplied by its velocity** and is measured in kilogram metre per second (kg m/s) or newton second (N s).

> momentum = mass × velocity

A 2 kg mass moving at 10 m/s has momentum 20 kg m/s, the same as the momentum of a 5 kg mass moving at 4 m/s.

Practical work

Collisions and momentum

Figure 12.1 shows an arrangement which can be used to find the velocity of a trolley before and after a collision. If a trolley of length l takes time t to pass through a photogate, its velocity = distance/time = l/t. Two photogates are needed, placed each side of the collision point, to find the velocities before and after the collision. Set them up so that they will record the time taken for the passage of a trolley.

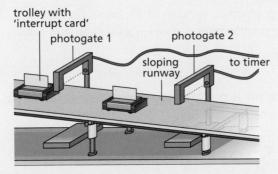

Figure 12.1

A tickertape timer or motion sensor, placed at the top end of the runway, could be used instead of the photogates if preferred.

Attach a strip of Velcro to each trolley so that they 'stick' to each other on collision and compensate the runway for friction (see Chapter 8). Place one trolley at rest halfway down the runway and another at the top; give the top trolley a push. It will move forwards with uniform velocity and should hit the second trolley so that they travel on as one. Using the times recorded by the photogate timer, calculate the velocity of the moving trolley before the collision and the common velocity of both trolleys after the collision.

Repeat the experiment with another trolley stacked on top of the one to be pushed so that two are moving before the collision and three after.

Copy and complete the tables of results.

Before collision (m_2 at rest)

Mass m_1 (no. of trolleys)	Velocity v/m/s	Momentum $m_1 v$
1		
2		

After collision (m_1 and m_2 together)

Mass $m_1 + m_2$ (no. of trolleys)	Velocity v_1/m/s	Momentum $(m_1 + m_2)v_1$
2		
3		

Do the results suggest any connection between the momentum before the collision and after it in each case?

● Conservation of momentum

> When two or more bodies act on one another, as in a collision, the total momentum of the bodies remains constant, provided no external forces act (e.g. friction).

This statement is called the **principle of conservation of momentum**. Experiments like those in the *Practical work* section show that it is true for all types of collisions.

As an example, suppose a truck of mass 60 kg moving with velocity 3 m/s collides and couples with a stationary truck of mass 30 kg (Figure 12.2a). The two move off together with the same velocity v which we can find as follows (Figure 12.2b).

Total momentum before is

$$(60\,\text{kg} \times 3\,\text{m/s}) + (30\,\text{kg} \times 0\,\text{m/s}) = 180\,\text{kg m/s}$$

▶▶

Total momentum after is

$$(60\,\text{kg} + 30\,\text{kg}) \times v = 90\,\text{kg} \times v$$

Since momentum is not lost

$$90\,\text{kg} \times v = 180\,\text{kg}\,\text{m/s} \quad \text{or} \quad v = 2\,\text{m/s}$$

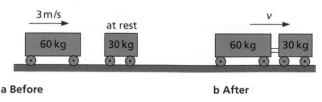

a Before **b After**

Figure 12.2

● Explosions

Momentum, like velocity, is a vector since it has both magnitude and direction. Vectors cannot be added by ordinary addition unless they act in the same direction. If they act in exactly opposite directions, such as east and west, the smaller subtracts from the greater, or if the same they cancel out.

Momentum is conserved in an **explosion** such as occurs when a rifle is fired. Before firing, the total momentum is zero since both rifle and bullet are at rest. During the firing the rifle and bullet receive **equal** but **opposite** amounts of momentum so that the **total momentum** after firing is zero. For example, if a rifle fires a bullet of mass 0.01 kg with a velocity of 300 m/s,

forward momentum of bullet = 0.01 kg × 300 m/s
= 3 kg m/s

∴   backward momentum of rifle = 3 kg m/s

If the rifle has mass m, it recoils (kicks back) with a velocity v such that

$$mv = 3\,\text{kg}\,\text{m/s}$$

Taking $m = 6\,\text{kg}$ gives $v = 3/6\,\text{m/s} = 0.5\,\text{m/s}$.

● Rockets and jets

If you release an inflated balloon with its neck open, it flies off in the opposite direction to that of the escaping air. In Figure 12.3 the air has momentum to the left and the balloon moves to the right with equal momentum.

This is the principle of rockets and jet engines. In both, a high-velocity stream of hot gas is produced

by burning fuel and leaves the exhaust with large momentum. The rocket or jet engine itself acquires an equal forward momentum. Space rockets carry their own oxygen supply; jet engines use the surrounding air.

Figure 12.3 A deflating balloon demonstrates the principle of a rocket or a jet engine.

● Force and momentum

If a steady force F acting on a body of mass m increases its velocity from u to v in time t, the acceleration a is given by

$$a = (v - u)/t \quad \text{(from } v = u + at\text{)}$$

Substituting for a in $F = ma$,

$$F = \frac{m(v - u)}{t} = \frac{mv - mu}{t}$$

Therefore

$$\text{force} = \frac{\text{change of momentum}}{\text{time}} = \frac{\text{rate of change of}}{\text{momentum}}$$

This is another version of Newton's second law. For some problems it is more useful than $F = ma$.

We also have

$$Ft = mv - mu$$

where mv is the final momentum, mu the initial momentum and Ft is called the **impulse**.

● Sport: impulse and collision time

The good cricketer or tennis player 'follows through' with the bat or racket when striking the ball (Figure 12.4a). The force applied then acts for a longer time, the impulse is greater and so also is the gain of momentum (and velocity) of the ball.

When we want to stop a moving ball such as a cricket ball, however, its momentum has to be reduced to zero. An impulse is then required in the form of an opposing force acting for a certain time. While any number of combinations of force and time will give a particular impulse, the 'sting' can be removed from the catch by drawing back the hands as the ball is caught (Figure 12.4b). A smaller average force is then applied for a longer time.

Figure 12.5 Sand reduces the athlete's momentum more gently.

Figure 12.4a Batsman 'following through' after hitting the ball

Figure 12.4b Cricketer drawing back the hands to catch the ball

The use of sand gives a softer landing for long-jumpers (Figure 12.5), as a smaller stopping force is applied over a longer time. In a car crash the car's momentum is reduced to zero in a very short time. If the time of impact can be extended by using crumple zones (see Figure 14.6, p. 58) and extensible seat belts, the average force needed to stop the car is reduced so the injury to passengers should also be less.

Questions

1 What is the momentum in kg m/s of a 10 kg truck travelling at
   a 5 m/s,
   b 20 cm/s,
   c 36 km/h?
2 A ball X of mass 1 kg travelling at 2 m/s has a head-on collision with an identical ball Y at rest. X stops and Y moves off. What is Y's velocity?
3 A boy with mass 50 kg running at 5 m/s jumps on to a 20 kg trolley travelling in the same direction at 1.5 m/s. What is their common velocity?
4 A girl of mass 50 kg jumps out of a rowing boat of mass 300 kg on to the bank, with a horizontal velocity of 3 m/s. With what velocity does the boat begin to move backwards?
5 A truck of mass 500 kg moving at 4 m/s collides with another truck of mass 1500 kg moving in the same direction at 2 m/s. What is their common velocity just after the collision if they move off together?
6 The velocity of a body of mass 10 kg increases from 4 m/s to 8 m/s when a force acts on it for 2 s.
   a What is the momentum before the force acts?
   b What is the momentum after the force acts?
   c What is the momentum gain per second?
   d What is the value of the force?
7 A rocket of mass 10 000 kg uses 5.0 kg of fuel and oxygen to produce exhaust gases ejected at 5000 m/s. Calculate the increase in its velocity.

Checklist

After studying this chapter you should be able to

* define momentum,
* describe experiments to demonstrate the principle of conservation of momentum,
* state and use the principle of conservation of momentum to solve problems,
* understand the action of rocket and jet engines,
* state the relationship between force and rate of change of momentum and use it to solve problems,
* use the definition of impulse to explain how the time of impact affects the force acting in a collision.

13 Energy transfer

- Forms of energy
- Energy transfers
- Energy measurements
- Energy conservation

- Energy of food
- Combustion of fuels
- Practical work: Measuring power

Energy is a theme that pervades all branches of science. It links a wide range of phenomena and enables us to explain them. It exists in different forms and when something happens, it is likely to be due to energy being transferred from one form to another. Energy transfer is needed to enable people, computers, machines and other devices to work and to enable processes and changes to occur. For example, in Figure 13.1, the water skier can only be pulled along by the boat if there is energy transfer in its engine from the burning petrol to its rotating propeller.

Figure 13.1 Energy transfer in action

● Forms of energy

a) Chemical energy

Food and fuels, like oil, gas, coal and wood, are concentrated stores of **chemical energy** (see Chapter 15). The energy of food is released by chemical reactions in our bodies, and during the transfer to other forms we are able to do useful jobs. Fuels cause energy transfers when they are burnt in an engine or a boiler. Batteries are compact sources of chemical energy, which in use is transferred to electrical energy.

b) Potential energy (p.e.)

This is the energy a body has because of its position or condition. A body above the Earth's surface, like water in a mountain reservoir, has **potential energy** (p.e.) stored in the form of **gravitational potential energy**.

Work has to be done to compress or stretch a spring or elastic material and energy is transferred to potential energy; the p.e. is stored in the form of **strain energy** (or **elastic potential energy**). If the catapult in Figure 13.3c were released, the strain energy would be transferred to the projectile.

c) Kinetic energy (k.e.)

Any moving body has **kinetic energy** (k.e.) and the faster it moves, the more k.e. it has. As a hammer drives a nail into a piece of wood, there is a transfer of energy from the k.e. of the moving hammer to other forms of energy.

d) Electrical energy

Electrical energy is produced by energy transfers at power stations and in batteries. It is the commonest form of energy used in homes and industry because of the ease of transmission and transfer to other forms.

e) Heat energy

This is also called **thermal** or **internal energy** and is the final fate of other forms of energy. It is transferred by conduction, convection or radiation.

f) Other forms

These include **light energy** and other forms of **electromagnetic radiation**, **sound** and **nuclear energy**.

● Energy transfers

a) Demonstration

The apparatus in Figure 13.2 can be used to show a battery changing **chemical energy** to **electrical energy** which becomes **kinetic energy** in the electric motor. The motor raises a weight, giving it **potential energy**. If the changeover switch is joined to the lamp and the weight allowed to fall, the motor acts as a generator in which there is an energy transfer from **kinetic energy** to **electrical energy**. When this is supplied to the lamp, it produces a transfer to **heat** and **light energy**.

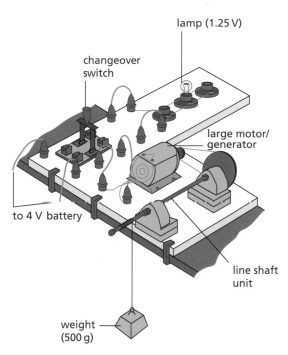

Figure 13.2 Demonstrating energy transfers

b) Other examples

Study the energy transfers shown in Figures 13.3a to d. Some devices have been invented to cause particular energy transfers. For example, a **microphone** changes sound energy into electrical energy; a **loudspeaker** does the reverse. Belts, chains or gears are used to transfer energy between moving parts, such as those in a bicycle.

a Potential energy to kinetic energy

b Electrical energy to heat and light energy

c Chemical energy (from muscles in the arm) to p.e. (strain energy of catapult)

d Potential energy of water to kinetic energy of turbine to electrical energy from generator

Figure 13.3 Some energy transfers

● Energy measurements

a) Work

In science the word **work** has a different meaning from its everyday use. **Work is done when a force moves.** No work is done in the scientific sense by someone standing still holding a heavy pile of books: an upward force is exerted, but no motion results.

If a building worker carries ten bricks up to the first floor of a building, he does more work than if he carries only one brick because he has to exert a larger force. Even more work is required if he carries the ten bricks to the second floor. The amount of work done depends on the size of the force applied and the distance it moves. We therefore measure work by

work = force × distance moved in direction of force   (1)

The unit of work is the **joule** (J); it is the work done when a force of 1 newton (N) moves through 1 metre (m). For example, if you have to pull with a force of 50 N to move a crate steadily 3 m in the direction of the force (Figure 13.4a), the work done is 50 N × 3 m = 150 N m = 150 J. That is

joules = newtons × metres

If you lift a mass of 3 kg vertically through 2 m (Figure 13.4b), you have to exert a vertically upward force equal to the weight of the body, i.e. 30 N (approximately) and the work done is 30 N × 2 m = 60 N m = 60 J.

Note that we must always take the distance in the direction in which the force acts.

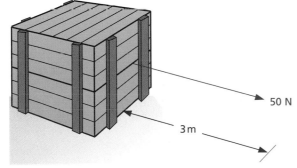

Figure 13.4a

Figure 13.4b

b) Measuring energy transfers

In an energy transfer, work is done. **The work done is a measure of the amount of energy transferred.** For example, if you have to exert an upward force of 10 N to raise a stone steadily through a vertical distance of 1.5 m, the work done is 15 J. This is also the amount of chemical energy transferred from your muscles to potential energy of the stone. All forms of energy, as well as work, are measured in joules.

c) Power

The more powerful a car is, the faster it can accelerate or climb a hill, i.e. the more rapidly it does work. The **power** of a device is the work it does per second, i.e. the rate at which it does work. This is **the same as the rate at which it transfers energy from one form to another.**

$$\text{power} = \frac{\text{work done}}{\text{time taken}} = \frac{\text{energy transfer}}{\text{time taken}} \qquad (2)$$

The unit of power is the **watt** (W) and is **a rate of working of 1 joule per second**, i.e. 1 W = 1 J/s. Larger units are the **kilowatt** (kW) and the **megawatt** (MW):

$$1\,\text{kW} = 1000\,\text{W} = 10^3\,\text{W}$$

$$1\,\text{mW} = 1\,000\,000\,\text{W} = 10^6\,\text{W}$$

If a machine does 500 J of work in 10 s, its power is 500 J/10 s = 50 J/s = 50 W. A small car develops a maximum power of about 25 kW.

Practical work

Measuring power

a) Your own power

Get someone with a stopwatch to time you running up a flight of stairs, the more steps the better. Find your weight (in newtons). Calculate the total vertical height (in metres) you have climbed by measuring the height of one step and counting the number of steps.

The work you do (in joules) in lifting your weight to the top of the stairs is (your weight) × (vertical height of stairs). Calculate your power (in watts) from equation (2). About 0.5 kW is good.

b) Electric motor

This experiment is described in Chapter 40.

● Energy conservation

a) Principle of conservation of energy

This is one of the basic laws of physics and is stated as follows.

Energy cannot be created or destroyed; it is always conserved.

However, energy is continually being transferred from one form to another. Some forms, such as electrical and chemical energy, are more easily transferred than others, such as heat, for which it is hard to arrange a useful transfer.

Ultimately all energy transfers result in the surroundings being heated (as a result of doing work against friction) and the energy is wasted, i.e. spread out and increasingly more difficult to use. For example, when a brick falls its potential energy becomes kinetic energy; as it hits the ground, its temperature rises and heat and sound are produced. If it seems in a transfer that some energy has disappeared, the 'lost' energy is often converted into non-useful heat. This appears to be the fate of all energy in the Universe and is one reason why new sources of useful energy have to be developed (Chapter 15).

b) Efficiency of energy transfers

The **efficiency** of a device is the percentage of the energy supplied to it that is usefully transferred. It is calculated from the expression:

$$\text{efficiency} = \frac{\text{useful energy output}}{\text{total energy input}} \times 100\%$$

For example, for the lever shown in Figure 10.4 (p. 40)

$$\text{efficiency} = \frac{\text{work done on load}}{\text{work done by effort}} \times 100\%$$

This will be less than 100% if there is friction in the fulcrum.

Table 13.1 lists the efficiencies of some devices and the energy transfers involved.

Table 13.1

Device	% Efficiency	Energy transfer
large electric motor	90	electrical to k.e.
large electric generator	90	k.e. to electrical
domestic gas boiler	75	chemical to heat
compact fluorescent lamp	50	electrical to light
steam turbine	45	heat to k.e.
car engine	25	chemical to k.e.
filament lamp	10	electrical to light

A device is efficient if it transfers energy mainly to useful forms and the 'lost' energy is small.

● Energy of food

When food is eaten it reacts with the oxygen we breathe into our lungs and is slowly 'burnt'. As a result chemical energy stored in food becomes thermal energy to warm the body and mechanical energy for muscular movement.

The **energy value** of a food substance is the amount of energy released when 1 kg is completely oxidised.

Energy value is measured in J/kg. The energy values of some foods are given in Figure 13.5 in megajoules per kilogram.

▶▶

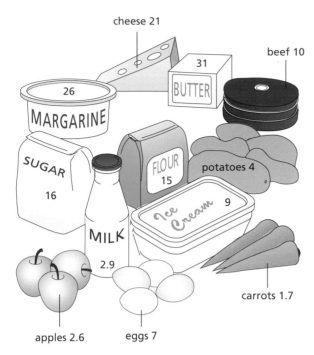

cheese 21

31 BUTTER

beef 10

26 MARGARINE

SUGAR 16

FLOUR 15

potatoes 4

Ice Cream 9

MILK 2.9

carrots 1.7

apples 2.6

eggs 7

Figure 13.5 Energy values of some foods in MJ/kg

Foods with high values are 'fattening' and if more food is eaten than the body really needs, the extra is stored as fat. The average adult requires about 10 MJ per day.

Our muscles change chemical energy into mechanical energy when we exert a force – to lift a weight, for example. Unfortunately, they are not very good at doing this; of every 100 J of chemical energy they use, they can convert only 25 J into mechanical energy – that is, they are only 25% efficient at changing chemical energy into mechanical energy. The other 75 J becomes thermal energy, much of which the body gets rid of by sweating.

● Combustion of fuels

Fuels can be solids such as wood and coal, liquids such as fuel oil and paraffin, or gases such as methane and butane.

Some fuels are better than others for certain jobs. For example, fuels for cooking or keeping us warm should, as well as being cheap, have a high **heating value**. This means that every gram of fuel should produce a large amount of heat energy when burnt.

A fuel for a space rocket (e.g. liquid hydrogen) must also burn very quickly so that the gases created expand rapidly and leave the rocket at high speed.

The heating values of some fuels are given in Table 13.2 in kilojoules per gram.

Table 13.2 Heating values of fuels in kJ/g

Solids	Value	Liquids	Value	Gases	Value
wood	17	fuel oil	45	methane	55
coal	25–33	paraffin	48	butane	50

The thick dark liquid called **petroleum** or **crude oil** is the source of most liquid and gaseous fuels. It is obtained from underground deposits at oil wells in many parts of the world. Natural gas (methane) is often found with it. In an oil refinery different fuels are obtained from petroleum, including fuel oil for industry, diesel oil for lorries, paraffin (kerosene) for jet engines, and petrol for cars, as well as butane (bottled gas).

Questions

1 Name the energy transfers which occur when
   a an electric bell rings,
   b someone speaks into a microphone,
   c a ball is thrown upwards,
   d there is a picture on a television screen,
   e a torch is on.
2 Name the forms of energy represented by the letters A, B, C and D in the following statement.
   In a coal-fired power station, the (A) energy of coal becomes (B) energy which changes water into steam. The steam drives a turbine which drives a generator. A generator transfers (C) energy into (D) energy.
3 How much work is done when a mass of 3 kg (weighing 30 N) is lifted vertically through 6 m?
4 A hiker climbs a hill 300 m high. If she has a mass of 50 kg calculate the work she does in lifting her body to the top of the hill.
5 In loading a lorry a man lifts boxes each of weight 100 N through a height of 1.5 m.
   a How much work does he do in lifting one box?
   b How much energy is transferred when one box is lifted?
   c If he lifts four boxes per minute at what power is he working?
6 A boy whose weight is 600 N runs up a flight of stairs 10 m high in 12 s. What is his average power?

7 a When the energy input to a gas-fired power station is 1000 MJ, the electrical energy output is 300 MJ. What is the efficiency of the power station in changing the energy in gas into electrical energy?
  b What form does the 700 MJ of 'lost' energy take?
  c What is the fate of the 'lost' energy?
8 State what energy transfers occur in
  a a hairdryer,
  b a refrigerator,
  c an audio system.
9 An escalator carries 60 people of average mass 70 kg to a height of 5 m in one minute. Find the power needed to do this.

Checklist

After studying this chapter you should be able to

- recall the different forms of energy,
- describe energy transfers in given examples,
- relate work done to the magnitude of a force and the distance moved,

- use the relation work done = force × distance moved to calculate energy transfer,

- define the unit of work,

- relate power to work done and time taken, and give examples,

- recall that power is the rate of energy transfer, give its unit and solve problems,

- describe an experiment to measure your own power,

- state the principle of conservation of energy,
- understand qualitatively the meaning of efficiency,

- recall and use the equation

$$\text{efficiency} = \frac{\text{useful energy output}}{\text{energy input}} \times 100\%$$

(14) Kinetic and potential energy

- Kinetic energy (k.e.)
- Potential energy (p.e.)
- Conservation of energy
- Elastic and inelastic collisions

- Driving and car safety
- Practical work: Change of p.e. to k.e.

Energy and its different forms were discussed earlier (Chapter 13). Here we will consider **kinetic energy** (k.e.) and **potential energy** (p.e.) in more detail.

● Kinetic energy (k.e.)

Kinetic energy is the energy a body has because of its motion.

For a body of mass m travelling with velocity v,

$$\text{kinetic energy} = E_k = \frac{1}{2}mv^2$$

If m is in kg and v in m/s, then kinetic energy is in J. For example, a football of mass $0.4\,\text{kg}$ ($400\,\text{g}$) moving with velocity $20\,\text{m/s}$ has

$$\text{k.e.} = \frac{1}{2}mv^2 = \frac{1}{2} \times 0.4\,\text{kg} \times (20)^2\,\text{m}^2/\text{s}^2$$

$$= 0.2 \times 400\,\text{kg m/s}^2 \times \text{m}$$

$$= 80\,\text{N m} = 80\,\text{J}$$

Since k.e. depends on v^2, a high-speed vehicle travelling at $1000\,\text{km/h}$ (Figure 14.1), has one hundred times the k.e. it has at $100\,\text{km/h}$.

● Potential energy (p.e.)

Potential energy is the energy a body has because of its position or condition.

A body above the Earth's surface is considered to have an amount of gravitational potential energy equal to the work that has been done against gravity by the force used to raise it. To lift a body of mass m through a vertical height h at a place where the Earth's gravitational field strength is g, needs a force equal and opposite to the weight mg of the body. Hence

$$\text{work done by force} = \text{force} \times \text{vertical height}$$
$$= mg \times h$$

$$\therefore \quad \text{potential energy} = E_p = mgh$$

When m is in kg, g in N/kg (or m/s²) and h in m, the potential energy is in J. For example, if $g = 10\,\text{N/kg}$, the potential energy gained by a $0.1\,\text{kg}$ ($100\,\text{g}$) mass raised vertically by $1\,\text{m}$ is

$$0.1\,\text{kg} \times 10\,\text{N/kg} \times 1\,\text{m} = 1\,\text{N m} = 1\,\text{J}$$

Note Strictly speaking we are concerned with *changes* in potential energy from that which a body has at the Earth's surface, rather than with actual values. The expression for potential energy is therefore more correctly written

$$\Delta E_p = mgh$$

where Δ (pronounced 'delta') stands for 'change in'.

Figure 14.1 Kinetic energy depends on the square of the velocity.

Practical work

Change of p.e. to k.e.

Friction-compensate a runway and arrange the apparatus as in Figure 14.2 with the bottom of the 0.1 kg (100 g) mass 0.5 m from the floor.

Start the timer and release the trolley. It will accelerate until the falling mass reaches the floor; after that it moves with *constant* velocity v.

From your results calculate v in m/s (on the tickertape 50 ticks = 1 s). Find the mass of the trolley in kg. Work out:

k.e. gained by trolley and 0.1 kg mass = ___ J
p.e. lost by 0.1 kg mass = ___ J

Compare and comment on the results.

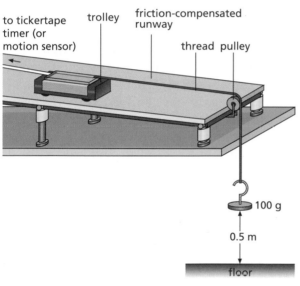

Figure 14.2

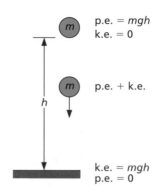

Figure 14.3 Loss of p.e. = gain of k.e.

This is an example of the **principle of conservation of energy** which was discussed in Chapter 13.

In the case of a pendulum (Figure 14.4), kinetic and potential energy are interchanged continually. The energy of the bob is all potential energy at the end of the swing and all kinetic energy as it passes through its central position. In other positions it has both potential and kinetic energy. Eventually all the energy is changed to heat as a result of overcoming air resistance.

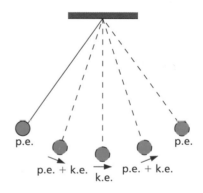

Figure 14.4 Interchange of p.e. and k.e. for a simple pendulum

● Conservation of energy

A mass m at height h above the ground has potential energy = mgh (Figure 14.3). When it falls, its velocity increases and it gains kinetic energy at the expense of its potential energy. If it starts from rest and air resistance is negligible, its kinetic energy on reaching the ground equals the potential energy lost by the mass

$$\frac{1}{2}mv^2 = mgh$$

or

loss of p.e. = gain of k.e.

● Elastic and inelastic collisions

In all collisions (where no external force acts) there is normally a loss of kinetic energy, usually to heat energy and to a small extent to sound energy. The greater the proportion of kinetic energy lost, the less **elastic** is the collision, i.e. the more **inelastic** it is. In a perfectly elastic collision, kinetic energy is conserved.

Figure 14.5 Newton's cradle is an instructive toy for studying collisions and conservation of energy.

Table 14.1

Speed/mph	20	40	60	80
Thinking distance/metres	6	12	18	24
Braking distance/metres	6	24	54	96
Total stopping distance/metres	12	36	72	120

● Driving and car safety

a) Braking distance and speed

For a car moving with speed v, the brakes must be applied over a braking distance s to bring the car to rest. The **braking distance is directly proportional to the square of the speed**, i.e. if v is doubled, s is quadrupled. The **thinking distance** (i.e. the distance travelled while the driver is reacting before applying the brakes) has to be added to the **braking distance** to obtain the **overall stopping distance**, in other words

> stopping distance = thinking distance + braking distance

Typical values taken from the Highway Code are given in Table 14.1 for different speeds. The greater the speed, the greater the stopping distance for a given braking force. (To stop the car in a given distance a greater braking force is needed for higher speeds.)

Thinking distance depends on the driver's reaction time – this will vary with factors such as the driver's degree of tiredness, use of alcohol or drugs, eyesight and the visibility of the hazard. Braking distance varies with both the road conditions and the state of the car; it is longer when the road is wet or icy, when friction between the tyres and the road is low, than when conditions are dry. Efficient brakes and deep tyre tread help to reduce the braking distance.

b) Car design and safety

When a car stops rapidly in a collision, large forces are produced on the car and its passengers, and their kinetic energy has to be dissipated.

Crumple zones at the front and rear collapse in such a way that the kinetic energy is absorbed gradually (Figure 14.6). As we saw in Chapter 12 this extends the collision time and reduces the decelerating force and hence the potential for injury to the passengers.

Extensible seat belts exert a backwards force (of 10 000 N or so) over about 0.5 m, which is roughly the distance between the front seat occupants and the windscreen. In a car travelling at 15 m/s (34 mph), the effect felt by anyone *not* using a seat belt is the same as that produced by jumping off a building 12 m high!

Air bags in some cars inflate and protect the driver from injury by the steering wheel.

Head restraints ensure that if the car is hit from behind, the head goes forwards with the body and not backwards over the top of the seat. This prevents damage to the top of the spine.

All these are **secondary** safety devices which aid **survival** in the event of an accident. **Primary** safety factors help to **prevent** accidents and depend on the car's roadholding, brakes, steering, handling and above all on the driver since most accidents are due to driver error.

The chance of being killed in an accident is about **five times less** if seat belts are worn and head restraints are installed.

Figure 14.6 Cars in an impact test showing the collapse of the front crumple zone

● Worked example

A boulder of mass 4 kg rolls over a cliff and reaches the beach below with a velocity of 20 m/s. Find: **a** the kinetic energy of the boulder as it lands; **b** the potential energy of the boulder when it was at the top of the cliff and **c** the height of the cliff.

a Mass of boulder = m = 4 kg

Velocity of boulder as it lands = v = 20 m/s

$\therefore$ k.e. of boulder as it lands = $E_k = \dfrac{1}{2}mv^2$

$$= \dfrac{1}{2} \times 4 \text{ kg} \times (20)^2 \text{ m}^2/\text{s}^2$$

$$= 800 \text{ kg m/s}^2 \times \text{m}$$

$$= 800 \text{ N m}$$

$$= 800 \text{ J}$$

b Applying the principle of conservation of energy (and neglecting energy lost in overcoming air resistance),

p.e. of boulder on cliff = k.e. as it lands

$\therefore$ $\Delta E_p = E_k = 800$ J

c If h is the height of the cliff,

$$\Delta E_p = mgh$$

$\therefore$ $h = \dfrac{\Delta E_p}{mg} = \dfrac{800 \text{ J}}{4 \text{ kg} \times 10 \text{ m/s}^2} = \dfrac{800 \text{ N m}}{40 \text{ kg m/s}^2}$

$$= \dfrac{800 \text{ kg m/s}^2 \times \text{m}}{40 \text{ kg m/s}^2} = 20 \text{ m}$$

Questions

1 Calculate the k.e. of
 a a 1 kg trolley travelling at 2 m/s,
 b a 2 g (0.002 kg) bullet travelling at 400 m/s,
 c a 500 kg car travelling at 72 km/h.
2 a What is the velocity of an object of mass 1 kg which has 200 J of k.e.?
 b Calculate the p.e. of a 5 kg mass when it is (i) 3 m, (ii) 6 m, above the ground. (g = 10 N/kg)
3 A 100 g steel ball falls from a height of 1.8 m on to a metal plate and rebounds to a height of 1.25 m. Find
 a the p.e. of the ball before the fall (g = 10 m/s²),
 b its k.e. as it hits the plate,
 c its velocity on hitting the plate,
 d its k.e. as it leaves the plate on the rebound,
 e its velocity of rebound.
4 It is estimated that 7×10^6 kg of water pours over the Niagara Falls every second. If the falls are 50 m high, and if all the energy of the falling water could be harnessed, what power would be available? (g = 10 N/kg)

Checklist

After studying this chapter you should be able to

* define kinetic energy (k.e.),
* perform calculations using $E_k = \frac{1}{2}mv^2$,
* define potential energy (p.e.),
* calculate changes in p.e. using $\Delta E_p = mgh$,
* apply the principle of conservation of energy to simple mechanical systems, such as a pendulum,
* recall the effect of speed on the braking distance of a vehicle,
* describe secondary safety devices in cars.

15 Energy sources

- Non-renewable energy sources
- Renewable energy sources
- Power stations
- Economic, environmental and social issues

Energy is needed to heat buildings, to make cars move, to provide artificial light, to make computers work, and so on. The list is endless. This 'useful' energy needs to be produced in controllable energy transfers (Chapter 13). For example, in power stations a supply of useful energy in the form of electricity is produced. The 'raw materials' for energy production are **energy sources**. These may be **non-renewable** or **renewable**. Apart from nuclear, geothermal, hydroelectric or tidal energy, the Sun is the source for all our energy resources.

● Non-renewable energy sources

Once used up these cannot be replaced.

a) Fossil fuels

These include coal, oil and natural gas, formed from the remains of plants and animals which lived millions of years ago and obtained energy originally from the Sun. At present they are our main energy source. Predictions vary as to how long they will last since this depends on what reserves are recoverable and on the future demands of a world population expected to increase from about 7000 million in 2011 to at least 7600 million by the year 2050. Some estimates say oil and gas will run low early in the present century but coal should last for 200 years or so.

Burning fossil fuels in power stations and in cars pollutes the atmosphere with harmful gases such as carbon dioxide and sulfur dioxide. Carbon dioxide emission aggravates the greenhouse effect (Chapter 24) and increases global warming. It is not immediately feasible to prevent large amounts of carbon dioxide entering the atmosphere, but less is produced by burning natural gas than by burning oil or coal; burning coal produces most carbon dioxide for each unit of energy produced. When coal and oil are burnt they also produce sulfur dioxide which causes acid rain. The sulfur dioxide can be extracted from the waste gases so it does not enter the atmosphere or the sulfur can be removed

from the fuel before combustion, but these are both costly processes which increase the price of electricity produced using these measures.

b) Nuclear fuels

The energy released in a nuclear reactor (Chapter 50) from **uranium**, found as an ore in the ground, can be used to produce electricity. Nuclear fuels do not pollute the atmosphere with carbon dioxide or sulfur dioxide but they do generate radioactive waste materials with very long half-lives (Chapter 49); safe ways of storing this waste for perhaps thousands of years must be found. As long as a reactor is operating normally it does not pose a radiation risk, but if an accident occurs, dangerous radioactive material can leak from the reactor and spread over a large area.

Two advantages of all non-renewable fuels are

(i) their high **energy density** (i.e. they are concentrated sources) and the relatively small size of the energy transfer device (e.g. a furnace) which releases their energy, and

(ii) their ready **availability** when energy demand increases suddenly or fluctuates seasonally.

● Renewable energy sources

These cannot be exhausted and are generally non-polluting.

a) Solar energy

The energy falling on the Earth from the Sun is mostly in the form of light and in an hour equals the total energy used by the world in a year. Unfortunately its low energy density requires large collecting devices and its availability varies. Its greatest potential use is as an energy source for low-temperature water heating. This uses **solar panels** as the energy transfer devices, which convert light into heat energy. They are used increasingly to produce domestic hot water at about 70 °C and to heat swimming pools.

Solar energy can also be used to produce high-temperature heating, up to 3000 °C or so, if a large curved mirror (a **solar furnace**) focuses the Sun's rays on to a small area. The energy can then be used to turn water to steam for driving the turbine of an electric generator in a power station.

Figure 15.1 Solar cells on a house provide electricity.

Solar cells, made from semiconducting materials, convert sunlight into electricity directly. A number of cells connected together can be used to supply electricity to homes (Figure 15.1) and to the electronic equipment in communication and other satellites. They are also used for small-scale power generation in remote areas of developing countries where there is no electricity supply. Recent developments have made large-scale generation more cost effective and there is now a large solar power plant in California. There are many designs for prototype light vehicles run on solar power (Figure 15.2).

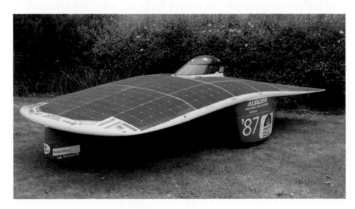

Figure 15.2 Solar-powered car

b) Wind energy

Giant windmills called **wind turbines** with two or three blades each up to 30 m long drive electrical generators. 'Wind farms' of 20 to 100 turbines spaced about 400 m apart (Figure 2d, p. ix), supply about 400 MW (enough electricity for 250 000 homes) in the UK and provide a useful 'top-up' to the National Grid.

Wind turbines can be noisy and may be considered unsightly so there is some environmental objection to wind farms, especially as the best sites are often in coastal or upland areas of great natural beauty.

c) Wave energy

The rise and fall of sea waves has to be transferred by some kind of wave-energy converter into the rotary motion required to drive a generator. It is a difficult problem and the large-scale production of electricity by this means is unlikely in the near future, but small systems are being developed to supply island communities with power.

d) Tidal and hydroelectric energy

The flow of water from a higher to a lower level from behind a **tidal barrage** (barrier) or the dam of a **hydroelectric scheme** is used to drive a water turbine (water wheel) connected to a generator.

One of the largest working tidal schemes is the La Grande I project in Canada (Figure 15.3). Feasibility studies have shown that a 10-mile-long barrage across the Severn Estuary could produce about 7% of today's electrical energy consumption in England and Wales. Such schemes have significant implications for the environment, as they may destroy wildlife habitats of wading birds for example, and also for shipping routes.

In the UK, hydroelectric power stations generate about 2% of the electricity supply. Most are located in Scotland and Wales where the average rainfall is higher than in other areas. With good management hydroelectric energy is a reliable energy source, but there are risks connected with the construction of dams, and a variety of problems may result from the impact of a dam on the environment. Land previously used for forestry or farming may have to be flooded.

Figure 15.3 Tidal barrage in Canada

Figure 15.4 Filling up with biofuel in Brazil

e) Geothermal energy

If cold water is pumped down a shaft into hot rocks below the Earth's surface, it may be forced up another shaft as steam. This can be used to drive a turbine and generate electricity or to heat buildings. The energy that heats the rocks is constantly being released by radioactive elements deep in the Earth as they decay (Chapter 49).

Geothermal power stations are in operation in the USA, New Zealand and Iceland.

f) Biomass (vegetable fuels)

These include cultivated crops (e.g. oilseed rape), crop residues (e.g. cereal straw), natural vegetation (e.g. gorse), trees (e.g. spruce) grown for their wood, animal dung and sewage. **Biofuels** such as alcohol (ethanol) and methane gas are obtained from them by fermentation using enzymes or by decomposition by bacterial action in the absence of air. Liquid biofuels can replace petrol (Figure 15.4); although they have up to 50% less energy per litre, they are lead- and sulfur-free and so cleaner. **Biogas** is a mix of methane and carbon dioxide with an energy content about two-thirds that of natural gas. In developing countries it is produced from animal and human waste in 'digesters' (Figure 15.5) and used for heating and cooking.

Figure 15.5 Feeding a biogas digester in rural India

● Power stations

The processes involved in the production of electricity at **power stations** depend on the energy source being used.

a) Non-renewable sources

These are used in **thermal** power stations to produce heat energy that turns water into steam. The steam drives turbines which in turn drive the generators that produce electrical energy as described in Chapter 43. If fossil fuels are the energy source (usually coal but natural gas is favoured in new stations), the steam is obtained from a boiler. If nuclear fuel is used, such as uranium or plutonium, the steam is produced in a heat exchanger as explained in Chapter 50.

The action of a **steam turbine** resembles that of a water wheel but moving steam not moving water causes the motion. Steam enters the turbine and is directed by the **stator** or **diaphragm** (sets of fixed blades) on to the **rotor** (sets of blades on a shaft that can rotate) (Figure 15.6). The rotor revolves and drives the electrical generator. The steam expands as it passes through the turbine and the size of the blades increases along the turbine to allow for this.

Figure 15.6 The rotor of a steam turbine

The overall efficiency of thermal power stations is only about 30%. They require cooling towers to condense steam from the turbine to water and this is a waste of energy. A block diagram and an energy-transfer diagram for a thermal power station are given in Figure 15.7.

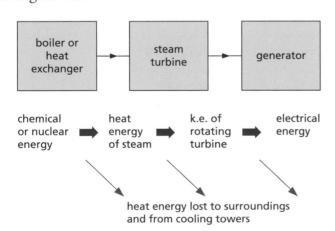

Figure 15.7 Energy transfers in a thermal power station

In gas-fired power stations, natural gas is burnt in a **gas turbine** linked directly to an electricity generator. The hot exhaust gases from the turbine are not released into the atmosphere but used to produce steam in a boiler. The steam is then used to generate more electricity from a steam turbine driving another generator. The efficiency is claimed to be over 50% without any extra fuel consumption. Furthermore, the gas turbines have a near 100% combustion efficiency so very little harmful exhaust gas (i.e. unburnt methane) is produced, and natural gas is almost sulfur-free so the environmental pollution caused is much less than for coal.

b) Renewable sources

In most cases the renewable energy source is used to drive turbines directly, as explained earlier in the cases of hydroelectric, wind, wave, tidal and geothermal schemes.

The block diagram and energy-transfer diagram for a hydroelectric scheme like that in Figure 13.3d (p. 51) are shown in Figure 15.8. The efficiency of a large installation can be as high as 85–90% since many of the causes of loss in thermal power stations (e.g. water cooling towers) are absent. In some cases the generating costs are half those of thermal stations.

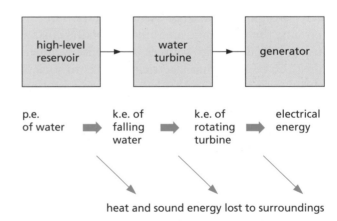

Figure 15.8 Energy transfers in a hydroelectric power station

A feature of some hydroelectric stations is **pumped storage**. Electrical energy cannot be stored on a large scale but must be used as it is generated. The demand varies with the time of day and the season (Figure 15.9), so in a pumped-storage system electricity generated at off-peak periods is used to pump water back up from a low-level reservoir to a higher-level one. It is easier to do this than to reduce the output of the generator. At peak times the potential energy of the water in the high-level reservoir is converted back into electrical energy; three-quarters of the electrical energy that was used to pump the water is generated.

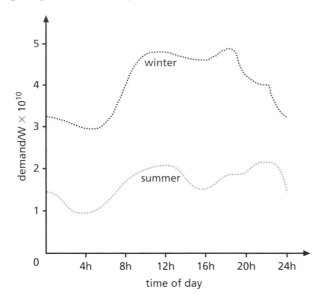

Figure 15.9 Variation in power demand

● Economic, environmental and social issues

When considering the large-scale generation of electricity, the economic and environmental costs of using various energy sources have to be weighed against the benefits that electricity brings to society as a 'clean', convenient and fairly 'cheap' energy supply.

Environmental problems such as polluting emissions that arise with different energy sources were outlined when each was discussed previously. Apart from people using less energy, how far pollution can be reduced by, for example, installing desulfurisation processes in coal-fired power stations, is often a matter of cost.

Although there are no fuel costs associated with electricity generation from renewable energy sources such as wind power, the energy is so 'dilute' that the capital costs of setting up the generating installation are high. Similarly, although fuel costs for nuclear power stations are relatively low, the costs of building the stations and of dismantling them at the end of their useful lives is higher than for gas- or coal-fired stations.

It has been estimated that currently it costs between 6p and 15p to produce a unit of electricity in a gas- or coal-fired power station in the UK.

The cost for a nuclear power station is in excess of 8p per unit. Wind energy costs vary, depending upon location, but are in the range 8p to 21p per unit. In the most favourable locations wind competes with coal and gas generation.

The reliability of a source has also to be considered, as well as how easily production can be started up and shut down as demand for electricity varies. Natural gas power stations have a short start-up time, while coal and then oil power stations take successively longer to start up; nuclear power stations take longest. They are all reliable in that they can produce electricity at any time of day and in any season of the year as long as fuel is available. Hydroelectric power stations are also very reliable and have a very short start-up time which means they can be switched on when the demand for electricity peaks. The electricity output of a tidal power station, although predictable, is not as reliable because it depends on the height of the tide which varies over daily, monthly and seasonal time scales. The wind and the Sun are even less reliable sources of energy since the output of a wind turbine changes with the strength of the wind and that of a solar cell with the intensity of light falling on it; the output may not be able to match the demand for electricity at a particular time.

Renewable sources are still only being used on a small scale globally. The contribution of the main energy sources to the world's total energy consumption at present is given in Table 15.1. (The use of biofuels is not well documented.) The pattern in the UK is similar but France generates nearly three-quarters of its electricity from nuclear plants; for Japan and Taiwan the proportion is one-third, and it is in the developing economies of East Asia where interest in nuclear energy is growing most dramatically. However, the great dependence on fossil fuels worldwide is evident. It is clear the world has an energy problem (Figure 15.10).

Table 15.1 World use of energy sources

Oil	Coal	Gas	Nuclear	Hydroelectric
36%	29%	23%	6%	6%

Consumption varies from one country to another; North America and Europe are responsible for about 42% of the world's energy consumption each year. Table 15.2 shows approximate values for the annual consumption per head of population for different areas.

These figures include the 'hidden' consumption in the manufacturing and transporting of goods. The world average consumption is 69×10^9 J per head per year.

Table 15.2 Energy consumption per head per year/J $\times 10^9$

N. America	UK	Japan	S. America	China	Africa
335	156	172	60	55	20

Figure 15.10 An energy supply crisis in California forces a blackout on stock exchange traders.

Questions

1 The pie chart in Figure 15.11 shows the percentages of the main energy sources used by a certain country.
   a What percentage is supplied by water power?
   b Which of the sources is/are renewable?
   c What is meant by 'renewable'?
   d Name two other renewable sources.
   e Why, if energy is always conserved, is it important to develop renewable sources?

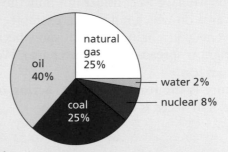

Figure 15.11

2 List six properties which you think the ideal energy source should have for generating electricity in a power station.
3 a List six social everyday benefits for which electrical energy is responsible.
   b Draw up two lists of suggestions for saving energy
      (i) in the home, and
      (ii) globally.

Checklist

After studying this chapter you should be able to

- distinguish between renewable and non-renewable energy sources,
- give some advantages and some disadvantages of using non-renewable fuels,
- describe the different ways of harnessing solar, wind, wave, tidal, hydroelectric, geothermal and biomass energy,
- describe the energy transfer processes in a thermal and a hydroelectric power station,
- compare and contrast the advantages and disadvantages of using different energy sources to generate electricity,
- discuss the environmental and economic issues of electricity production and consumption.

16 Pressure and liquid pressure

- Pressure
- Liquid pressure
- Water supply system
- Hydraulic machines
- Expression for liquid pressure
- Pressure gauges

● Pressure

To make sense of some effects in which a force acts on a body we have to consider not only the force but also the area on which it acts. For example, wearing skis prevents you sinking into soft snow because your weight is spread over a greater area. We say the **pressure** is less.

Pressure is the force (or thrust) **acting on unit area** (i.e. $1\,\text{m}^2$) and is calculated from

$$\text{pressure} = \frac{\text{force}}{\text{area}}$$

The unit of pressure is the **pascal** (Pa); it equals 1 newton per square metre (N/m^2) and is quite a small pressure. An apple in your hand exerts about 1000 Pa.

The greater the area over which a force acts, the less the pressure. Figure 16.1 shows the pressure exerted on the floor by the same box standing on end (Figure 16.1a) and lying flat (Figure 16.1b). This is why a tractor with wide wheels can move over soft ground. The pressure is large when the area is small and this is why nails are made with sharp points. Walnuts can be broken in the hand by squeezing two together but not one. Why?

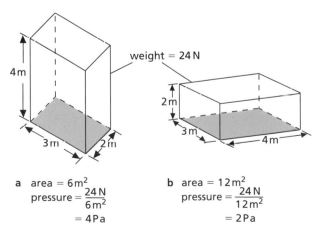

weight = 24 N

a area = $6\,\text{m}^2$
pressure = $\dfrac{24\,\text{N}}{6\,\text{m}^2}$
= 4 Pa

b area = $12\,\text{m}^2$
pressure = $\dfrac{24\,\text{N}}{12\,\text{m}^2}$
= 2 Pa

Figure 16.1

● Liquid pressure

1 **Pressure in a liquid increases with depth** because the further down you go, the greater the weight of liquid above. In Figure 16.2a water spurts out fastest and furthest from the lowest hole.

2 **Pressure at one depth acts equally in all directions.** The can of water in Figure 16.2b has similar holes all round it at the same level. Water comes out equally fast and spurts equally far from each hole. Hence the pressure exerted by the water at this depth is the same in all directions.

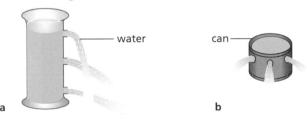

water

can

a

b

Figure 16.2

3 **A liquid finds its own level.** In the U-tube of Figure 16.3a the liquid pressure at the foot of P is greater than at the foot of Q because the left-hand column is higher than the right-hand one. When the clip is opened the liquid flows from P to Q until the pressure and the levels are the same, i.e. the liquid 'finds its own level'. Although the weight of liquid in Q is now greater than in P, it acts over a greater area because tube Q is wider.
In Figure 16.3b the liquid is at the same level in each tube and confirms that the pressure at the foot of a liquid column depends only on the *vertical* depth of the liquid and not on the tube width or shape.

4 **Pressure depends on the density of the liquid.** The denser the liquid, the greater the pressure at any given depth.

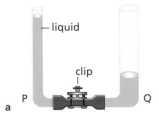

liquid

clip

P

Q

a

b

Figure 16.3

Water supply system

A town's water supply often comes from a reservoir on high ground. Water flows from it through pipes to any tap or storage tank that is below the level of water in the reservoir (Figure 16.4). The lower the place supplied, the greater the water pressure. In very tall buildings it may be necessary first to pump the water to a large tank in the roof.

Reservoirs for water supply or for hydroelectric power stations are often made in mountainous regions by building a dam at one end of a valley. The dam must be thicker at the bottom than at the top due to the large water pressure at the bottom.

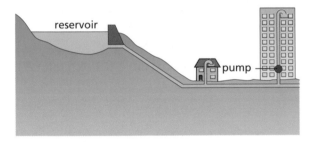

Figure 16.4 Water supply system. Why is the pump needed in the high-rise building?

Hydraulic machines

Liquids are almost incompressible (i.e. their volume cannot be reduced by squeezing) and they 'pass on' any pressure applied to them. Use is made of these facts in **hydraulic** machines. Figure 16.5 shows the principle on which they work.

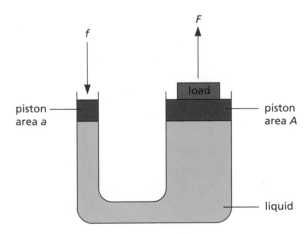

Figure 16.5 The hydraulic principle

Suppose a downward force f acts on a piston of area a. The pressure transmitted through the liquid is

$$\text{pressure} = \frac{\text{force}}{\text{area}} = \frac{f}{A}$$

This pressure acts on a second piston of larger area A, producing an upward force, $F = \text{pressure} \times \text{area}$:

$$F = \frac{f}{a} \times A$$

or

$$F = f \times \frac{A}{a}$$

Since A is larger than a, F must be larger than f and the hydraulic system is a force multiplier; the **multiplying factor** is A/a.

For example, if $f = 1\,\text{N}$, $a = \frac{1}{100}\,\text{m}^2$ and $A = \frac{1}{2}\,\text{m}^2$ then

$$F = 1\,\text{N} \times \frac{\frac{1}{2}\,\text{m}^2}{\frac{1}{100}\,\text{m}^2}$$

$$= 50\,\text{N}$$

A force of 1 N could lift a load of 50 N; the hydraulic system multiplies the force 50 times.

A **hydraulic jack** (Figure 16.6) has a platform on top of piston B and is used in garages to lift cars. Both valves open only to the right and they allow B to be raised a long way when A moves up and down repeatedly. When steel is forged using a hydraulic press there is a fixed plate above piston B and the sheets of steel are placed between B and the plate.

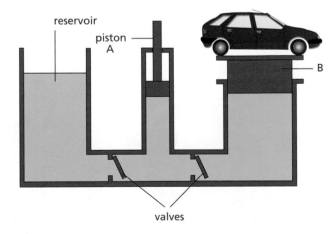

Figure 16.6 A hydraulic jack

Hydraulic fork-lift trucks and similar machines such as loaders (Figure 16.7) work in the same way.

Figure 16.7 A hydraulic machine in action

Hydraulic car brakes are shown in Figure 16.8. When the brake pedal is pushed, the piston in the master cylinder exerts a force on the brake fluid and the resulting pressure is transmitted equally to eight other pistons (four are shown). These force the brake shoes or pads against the wheels and stop the car.

● Expression for liquid pressure

In designing a dam an engineer has to calculate the pressure at various depths below the water surface. The pressure increases with depth and density.

An expression for the pressure at a depth h in a liquid of density ρ can be found by considering a horizontal area A (Figure 16.9). The force acting vertically downwards on A equals the weight of a liquid column of height h and cross-sectional area A above it. Then

$$\text{volume of liquid column} = hA$$

Since mass = volume × density we can say

$$\text{mass of liquid column} = hA\rho$$

Taking a mass of 1 kg to have weight 10 N,

$$\text{weight of liquid column} = 10hA\rho$$

$$\therefore \quad \text{force on area A} = 10hA\rho$$

As

$$\text{pressure} = \text{force}/\text{area} = 10hA\rho/A$$

then

$$\text{pressure} = 10h\rho$$

This is usually written as

$$\text{pressure} = \text{depth} \times \text{density} \times g$$

$$= h\rho g$$

where g is the acceleration of free fall (Chapter 4).

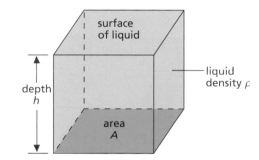

Figure 16.9

This pressure acts equally in all directions at depth h and depends only on h and ρ. Its value will be in Pa if h is in m and ρ in kg/m³.

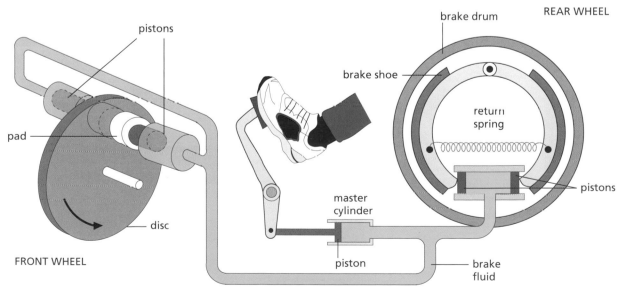

Figure 16.8 Hydraulic car brakes

Pressure gauges

These measure the pressure exerted by a fluid, in other words by a liquid or a gas.

a) Bourdon gauge

This works like the toy in Figure 16.10. The harder you blow into the paper tube, the more it uncurls. In a **Bourdon gauge** (Figure 16.11), when a fluid pressure is applied, the curved metal tube tries to straighten out and rotates a pointer over a scale. Car oil-pressure gauges and the gauges on gas cylinders are of this type.

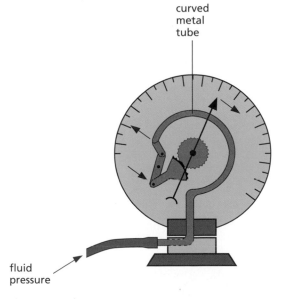

Figure 16.10 The harder you blow, the greater the pressure and the more it uncurls.

Figure 16.11 A Bourdon gauge

b) U-tube manometer

In Figure 16.12a each surface of the liquid is acted on equally by atmospheric pressure and the levels are the same. If one side is connected to, for example, the gas supply (Figure 16.12b), the gas exerts a pressure on surface A and level B rises until

pressure of gas = atmospheric pressure
+ pressure due to liquid column BC

The pressure of the liquid column BC therefore equals the amount by which the gas pressure *exceeds* atmospheric pressure. It equals $h\rho g$ (in Pa) where h is the vertical height of BC (in m) and ρ is the density of the liquid (in kg/m³). The height h is called the **head of liquid** and sometimes, instead of stating a pressure in Pa, we say that it is so many cm of water (or mercury for higher pressures).

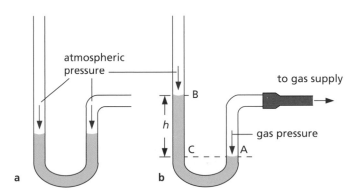

Figure 16.12 A U-tube manometer

c) Mercury barometer

A **barometer** is a manometer which measures atmospheric pressure. A simple barometer is shown in Figure 16.13. The pressure at X due to the weight of the column of mercury XY equals the atmospheric pressure on the surface of the mercury in the bowl. The height XY measures the atmospheric pressure in mm of mercury (mmHg).

The *vertical* height of the column is unchanged if the tube is tilted. Would it be different with a wider tube? The space above the mercury in the tube is a vacuum (except for a little mercury vapour).

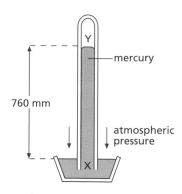

Figure 16.13 Mercury barometer

Questions

1 a What is the pressure on a surface when a force of 50 N acts on an area of
   (i) 2.0 m²,
   (ii) 100 m²,
   (iii) 0.50 m²?
  b A pressure of 10 Pa acts on an area of 3.0 m². What is the force acting on the area?
2 In a hydraulic press a force of 20 N is applied to a piston of area 0.20 m². The area of the other piston is 2.0 m². What is
  a the pressure transmitted through the liquid,
  b the force on the other piston?
3 a Why must a liquid and not a gas be used as the 'fluid' in a hydraulic machine?
  b On what other important property of a liquid do hydraulic machines depend?
4 What is the pressure 100 m below the surface of sea water of density 1150 kg/m³?
5 Figure 16.14 shows a simple barometer.
  a What is the region A?
  b What keeps the mercury in the tube?
  c What is the value of the atmospheric pressure being shown by the barometer?
  d What would happen to this reading if the barometer were taken up a high mountain? Give a reason.

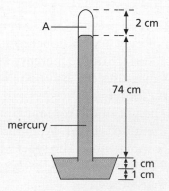

Figure 16.14

6 Which of the following will damage a wood-block floor that can withstand a pressure of 2000 kPa (2000 kN/m²)?
  1 A block weighing 2000 kN standing on an area of 2 m².
  2 An elephant weighing 200 kN standing on an area of 0.2 m².
  3 A girl of weight 0.5 kN wearing stiletto-heeled shoes standing on an area of 0.0002 m².
  Use the answer code:
  A 1, 2, 3
  B 1, 2
  C 2, 3
  D 1
  E 3
7 The pressure at a point in a liquid
  1 increases as the depth increases
  2 increases if the density of the liquid increases
  3 is greater vertically than horizontally.
  Which statement(s) is (are) correct?
  A 1, 2, 3
  B 1, 2
  C 2, 3
  D 1
  E 3

Checklist

After studying this chapter you should be able to

- relate pressure to force and area and give examples,
- define pressure and recall its unit,
- connect the pressure in a fluid with its depth and density,
- recall that pressure is transmitted through a liquid and use it to explain the hydraulic jack and hydraulic car brakes,

- use pressure = $h\rho g$ to solve problems,

- describe how a U-tube manometer may be used to measure fluid pressure,
- describe and use a simple mercury barometer.

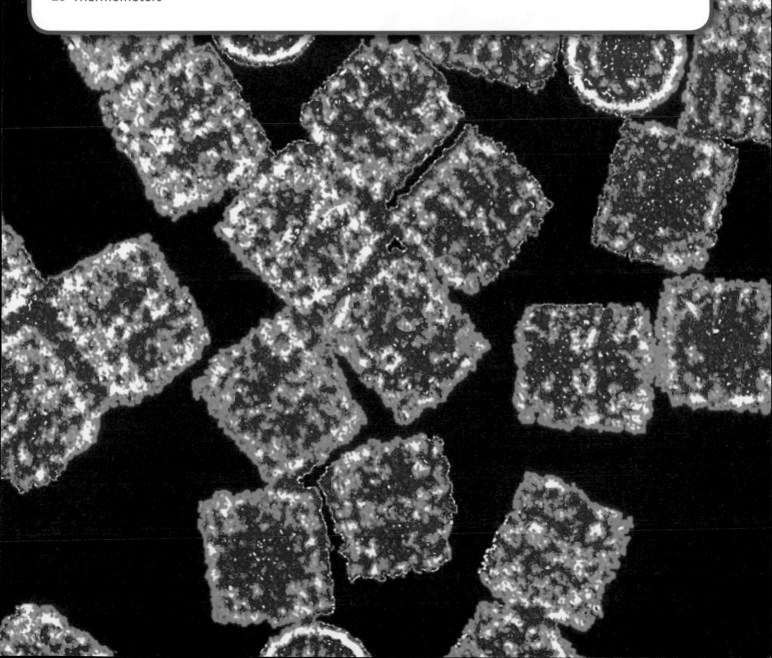

Thermal physics

Chapters

Simple kinetic molecular model of matter
17 Molecules
18 The gas laws

Thermal properties and temperature
19 Expansion of solids, liquids and gases
20 Thermometers

21 Specific heat capacity
22 Specific latent heat

Thermal processes
23 Conduction and convection
24 Radiation

(17) Molecules

- **Kinetic theory of matter**
- **Crystals**
- **Diffusion**
- **Practical work: Brownian motion**

Matter is made up of tiny particles or **molecules** which are too small for us to see directly. But they can be 'seen' by scientific 'eyes'. One of these is the **electron microscope**. Figure 17.1 is a photograph taken with such an instrument showing molecules of a protein. Molecules consist of even smaller particles called **atoms** and are in continuous motion.

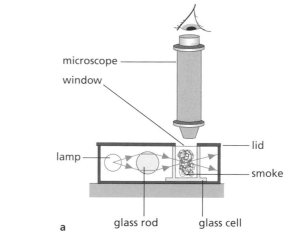

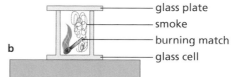

Figure 17.2

Carefully adjust the microscope until you see bright specks dancing around haphazardly (Figure 17.2c). The specks are smoke particles seen by reflected light; their random motion is called **Brownian motion**. It is due to collisions with fast-moving air molecules in the cell. A smoke particle is massive compared with an air molecule but if there are more high-speed molecules striking one side of it than the other at a given instant, the particle will move in the direction in which there is a net force. The imbalance, and hence the direction of the net force, changes rapidly in a random manner.

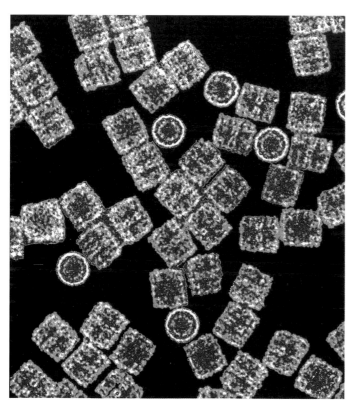

Figure 17.1 Protein molecules

Practical work

Brownian motion

The apparatus is shown in Figure 17.2a. First fill the glass cell with smoke using a match (Figure 17.2b). Replace the lid on the apparatus and set it on the microscope platform. Connect the lamp to a 12 V supply; the glass rod acts as a lens and focuses light on the smoke.

● Kinetic theory of matter

As well as being in continuous motion, molecules also exert strong electric forces on one another when they are close together. The forces are both attractive and repulsive. The former hold molecules together and the latter cause matter to resist compression.

The **kinetic theory** can explain the existence of the solid, liquid and gaseous states.

a) Solids

The theory states that in solids the molecules are close together and the attractive and repulsive forces between neighbouring molecules balance. Also each molecule vibrates to and fro about a fixed position.

It is just as if springs, representing the electric forces between molecules, hold the molecules together (Figure 17.3). This enables the solid to keep a definite shape and volume, while still allowing the individual molecules to vibrate backwards and forwards. The theory shows that the molecules in a solid could be arranged in a regular, repeating pattern like those formed by crystalline substances.

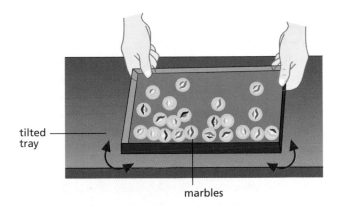

Figure 17.4 A model of molecular behaviour in a liquid

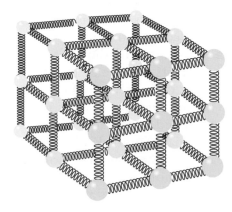

Figure 17.3 The electric forces between molecules in a solid can be represented by springs.

b) Liquids

The theory considers that in liquids the molecules are slightly further apart than in solids but still close enough together to have a definite volume. As well as vibrating, they can at the same time move rapidly over short distances, slipping past each other in all directions. They are never near another molecule long enough to get trapped in a regular pattern which would stop them from flowing and from taking the shape of the vessel containing them.

A model to represent the liquid state can be made by covering about a third of a tilted tray with marbles ('molecules') (Figure 17.4). It is then shaken to and fro and the motion of the marbles observed. They are able to move around but most stay in the lower half of the tray, so the liquid has a fairly definite volume. A few energetic ones 'escape' from the 'liquid' into the space above. They represent molecules that have 'evaporated' from the 'liquid' surface and become 'gas' or 'vapour' molecules. The thinning out of the marbles near the 'liquid' surface can also be seen.

c) Gases

The molecules in gases are much further apart than in solids or liquids (about ten times) and so gases are much less dense and can be squeezed (compressed) into a smaller space. The molecules dash around at very high speed (about 500 m/s for air molecules at 0 °C) in all the space available. It is only during the brief spells when they collide with other molecules or with the walls of the container that the molecular forces act.

A model of a gas is shown in Figure 17.5. The faster the vibrator works, the more often the ball-bearings have collisions with the lid, the tube and with each other, representing a gas at a higher temperature. Adding more ball-bearings is like pumping more air into a tyre; it increases the pressure. If a polystyrene ball (1 cm diameter) is dropped into the tube, its irregular motion represents Brownian motion.

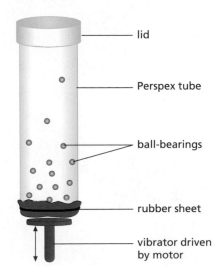

Figure 17.5 A model of molecular behaviour in a gas

● Crystals

Crystals have hard, flat sides and straight edges. Whatever their size, crystals of the same substance have the same shape. This can be seen by observing, through a microscope, very small cubic salt crystals growing as water evaporates from salt solution on a glass slide (Figure 17.6).

Figure 17.6 Salt crystals viewed under a microscope with polarised light

A calcite crystal will split cleanly if a trimming knife, held exactly parallel to one side of the crystal, is struck by a hammer (Figure 17.7).

Figure 17.7 Splitting a calcite crystal

These facts suggest that crystals are made of small particles (e.g. atoms) arranged in an orderly way in planes. Metals have crystalline structures, but many other common solids such as glass, plastics and wood do not.

● Diffusion

Smells, pleasant or otherwise, travel quickly and are caused by rapidly moving molecules. The spreading of a substance of its own accord is called **diffusion** and is due to molecular motion.

Diffusion of gases can be shown if some brown nitrogen dioxide gas is made by pouring a mixture of equal volumes of concentrated nitric acid and water onto copper turnings in a gas jar. When the reaction has stopped, a gas jar of air is inverted over the bottom jar (Figure 17.8). The brown colour spreads into the upper jar showing that nitrogen dioxide molecules diffuse upwards against gravity. Air molecules also diffuse into the lower jar.

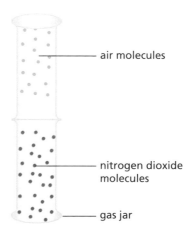

air molecules

nitrogen dioxide molecules

gas jar

Figure 17.8 Demonstrating diffusion of a gas

The speed of diffusion of a gas depends on the speed of its molecules and is greater for light molecules. The apparatus of Figure 17.9 shows this. When hydrogen surrounds the porous pot, the liquid in the U-tube moves in the direction of the arrows. This is because the lighter, faster molecules of hydrogen diffuse into the pot faster than the heavier, slower molecules of air diffuse out. The opposite happens when carbon dioxide surrounds the pot. Why?

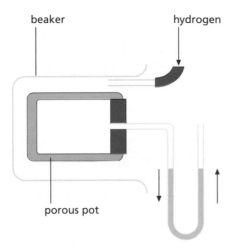

Figure 17.9 The speed of diffusion is greater for lighter molecules

Questions

1 Which one of the following statements is *not* true?
   A The molecules in a solid vibrate about a fixed position.
   B The molecules in a liquid are arranged in a regular pattern.
   C The molecules in a gas exert negligibly small forces on each other, except during collisions.
   D The densities of most liquids are about 1000 times greater than those of gases because liquid molecules are much closer together than gas molecules.
   E The molecules of a gas occupy all the space available.
2 Using what you know about the compressibility (squeezability) of the different states of matter, explain why
   a air is used to inflate tyres,
   b steel is used to make railway lines.

Checklist

After studying this chapter you should be able to

- describe and explain an experiment to show Brownian motion,
- use the kinetic theory to explain the physical properties of solids, liquids and gases.

(18) The gas laws

- Pressure of a gas
- Absolute zero
- The gas laws
- Gases and the kinetic theory

- Practical work: Effect on volume of temperature; Effect on pressure of temperature; Effect on volume of pressure

● Pressure of a gas

The air forming the Earth's atmosphere stretches upwards a long way. Air has weight; the air in a normal room weighs about the same as you do, about 500 N. Because of its weight the atmosphere exerts a large pressure at sea level, about $100\,000\,N/m^2 = 10^5\,Pa$ (or $100\,kPa$). This pressure acts equally in all directions.

A gas in a container exerts a pressure on the walls of the container. If air is removed from a can by a vacuum pump (Figure 18.1), the can collapses because the air pressure outside is greater than that inside. A space from which all the air has been removed is a **vacuum**. Alternatively the pressure in a container can be increased, for example by pumping more gas into the can; a Bourdon gauge (p. 69) is used for measuring fluid pressures.

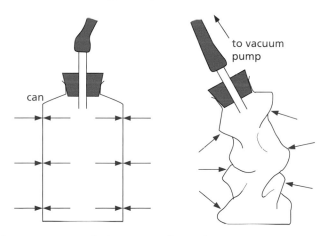

Figure 18.1 Atmospheric pressure collapses the evacuated can.

When a gas is heated, as air is in a jet engine, its pressure as well as its volume may change. To study the effect of temperature on these two quantities we must keep one fixed while the other is changed.

Practical work

Effect on volume of temperature (pressure constant) – Charles' law

Arrange the apparatus as in Figure 18.2. The index of concentrated sulfuric acid traps the air column to be investigated and also dries it. Adjust the capillary tube so that the bottom of the air column is opposite a convenient mark on the ruler.

Note the length of the air column (between the lower end of the index and the sealed end of the capillary tube) at different temperatures but, before taking a reading, stop heating and stir well to make sure that the air has reached the temperature of the water. Put the results in a table.

Plot a graph of volume (in cm, since the length of the air column is a measure of it) on the *y*-axis and temperature (in °C) on the *x*-axis.

The pressure of (and on) the air column is constant and equals atmospheric pressure plus the pressure of the acid index.

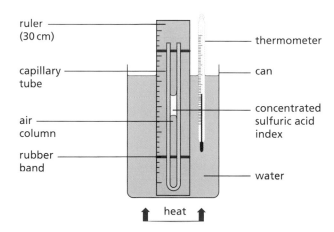

Figure 18.2

Practical work

Effect on pressure of temperature (volume constant) – the Pressure law

The apparatus is shown in Figure 18.3. The rubber tubing from the flask to the pressure gauge should be as short as possible. The flask must be in water almost to the top of its neck and be securely clamped to keep it off the bottom of the can.

Record the pressure over a wide range of temperatures, but before taking a reading, stop heating, stir and allow time for the gauge reading to become steady; the air in the flask will then be at the temperature of the water. Tabulate the results.

Plot a graph of pressure on the *y*-axis and temperature on the *x*-axis.

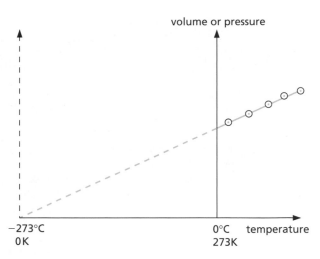

Figure 18.4

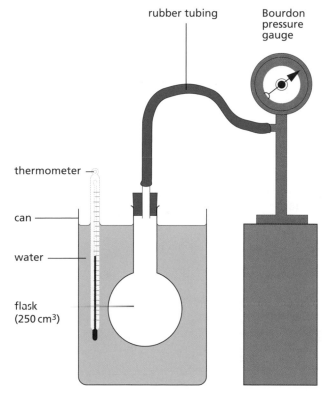

Figure 18.3

● Absolute zero

The volume–temperature and pressure–temperature graphs for a gas are straight lines (Figure 18.4). They show that gases expand **linearly** with temperature as measured on a mercury thermometer, i.e. equal temperature increases cause equal volume or pressure increases.

The graphs do not pass through the Celsius temperature origin (0 °C). If they are produced backwards they cut the temperature axis at about –273 °C. This temperature is called **absolute zero** because we believe it is the lowest temperature possible. It is the zero of the **absolute** or **Kelvin scale of temperature**. At absolute zero molecular motion ceases and a substance has no internal energy.

Degrees on this scale are called **kelvins** and are denoted by K. They are exactly the same size as Celsius degrees. Since –273 °C = 0 K, conversions from °C to K are made by adding 273. For example

$$0\,°\text{C} = 273\,\text{K}$$
$$15\,°\text{C} = 273 + 15 = 288\,\text{K}$$
$$100\,°\text{C} = 273 + 100 = 373\,\text{K}$$

Kelvin or absolute temperatures are represented by the letter *T*, and if θ (Greek letter 'theta') stands for a Celsius scale temperature then, in general,

$$T = 273 + \theta$$

Near absolute zero strange things occur. Liquid helium becomes a **superfluid**. It cannot be kept in an open vessel because it flows up the inside of the vessel, over the edge and down the outside. Some metals and compounds become **superconductors** of electricity and a current once started in them flows forever, without a battery. Figure 18.5 shows research equipment that is being used to create materials that are superconductors at very much higher temperatures, such as –23 °C. ▶▶

Figure 18.5 This equipment is being used to make films of complex composite materials that are superconducting at temperatures far above absolute zero.

Practical work

Effect on volume of pressure (temperature constant) – Boyle's law

Changes in the volume of a gas due to pressure changes can be studied using the apparatus in Figure 18.6. The volume V of air trapped in the glass tube is read off on the scale behind. The pressure is altered by pumping air from a foot pump into the space above the oil reservoir. This forces more oil into the glass tube and increases the pressure p on the air in it; p is measured by the Bourdon gauge.

If a graph of pressure against volume is plotted using the results, a curve like that in Figure 18.7a is obtained. Close examination of the graph shows that if p is doubled, V is halved. That is, **p is inversely proportional to V**. In symbols

$$p \propto \frac{1}{V} \quad \text{or} \quad p = \text{constant} \times \frac{1}{V}$$

$$\therefore \qquad pV = \text{constant}$$

If several pairs of readings, p_1 and V_1, p_2 and V_2, etc. are taken, then it can be confirmed that $p_1V_1 = p_2V_2 = \text{constant}$. This is **Boyle's law**, which is stated as follows:

> The pressure of a fixed mass of gas is inversely proportional to its volume if its temperature is kept constant.

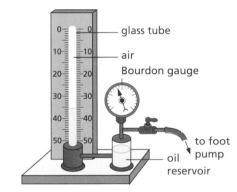

Figure 18.6

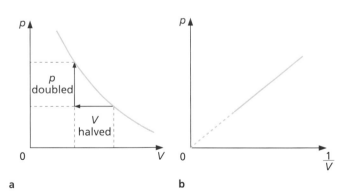

a

b

Figure 18.7

Since p is inversely proportional to V, then p is directly proportional to 1/V. A graph of p against 1/V is therefore a straight line through the origin (Figure 18.7b).

● The gas laws

Using absolute temperatures, the gas laws can be stated in a convenient form for calculations.

a) Charles' law

In Figure 18.4 (p. 77) the volume–temperature graph passes through the origin if temperatures are measured on the Kelvin scale, that is, if we take 0 K as the origin. We can then say that the volume V is directly proportional to the absolute temperature T, i.e. doubling T doubles V, etc. Therefore

$$V \propto T \text{ or } V = \text{constant} \times T$$

or

$$\frac{V}{T} = \text{constant} \qquad (1)$$

Charles' law may be stated as follows.

> The volume of a fixed mass of gas is directly proportional to its absolute temperature if the pressure is kept constant.

b) Pressure law

From Figure 18.4 we can say similarly for the pressure p that

$$p \propto T \text{ or } p = \text{constant} \times T$$

or

$$\frac{p}{T} = \text{constant} \qquad (2)$$

The Pressure law may be stated as follows.

> The pressure of a fixed mass of gas is directly proportional to its absolute temperature if the volume is kept constant.

c) Boyle's law

For a fixed mass of gas at constant temperature

$$pV = \text{constant} \qquad (3)$$

d) Combining the laws

The three equations can be combined giving

$$\frac{pV}{T} = \text{constant}$$

For cases in which p, V and T all change from, say, p_1, V_1 and T_1 to p_2, V_2 and T_2, then

$$\frac{p_1 V_1}{T_1} = \frac{p_2 V_2}{T_2} \qquad (4)$$

● Worked example

A bicycle pump contains 50 cm³ of air at 17 °C and at 1.0 atmosphere pressure. Find the pressure when the air is compressed to 10 cm³ and its temperature rises to 27 °C.

We have

$p_1 = 1.0$ atm	$p_2 = ?$
$V_1 = 50$ cm³	$V_2 = 10$ cm³
$T_1 = 273 + 17 = 290$ K	$T_2 = 273 + 27 = 300$ K

From equation (4) we get

$$p_2 = p_1 \times \frac{V_1}{V_2} \times \frac{T_2}{T_1} = 1 \times \frac{50}{10} \times \frac{300}{290} = 5.2 \text{ atm}$$

Notes

1 All temperatures must be in K.
2 Any units can be used for p and V provided the same units are used on both sides of the equation.
3 In some calculations the volume of the gas has to be found at standard temperature and pressure, or 's.t.p.'. This is temperature 0 °C and pressure 1 atmosphere (1 atm = 10^5 Pa).

● Gases and the kinetic theory

The kinetic theory can explain the behaviour of gases.

a) Cause of gas pressure

All the molecules in a gas are in rapid random motion, with a wide range of speeds, and repeatedly hit and rebound from the walls of the container in huge numbers per second. At each rebound, a gas molecule undergoes a change of momentum which produces a force on the walls of the container (see Chapter 12). The average force and hence the pressure they exert on the walls is constant since pressure is force on unit area.

b) Boyle's law

If the volume of a fixed mass of gas is halved by halving the volume of the container (Figure 18.8), the number of molecules per cm³ will be doubled. There will be twice as many collisions per second with the walls, i.e. the pressure is doubled. This is Boyle's law.

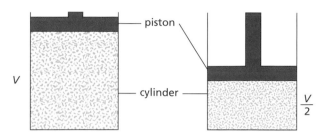

Figure 18.8 Halving the volume doubles the pressure.

c) Temperature

When a gas is heated and its temperature rises, the average speed of its molecules increases. If the volume of the gas stays constant, its pressure increases because there are more frequent and more violent collisions of the molecules with the walls. If the pressure of the gas is to remain constant, the volume must increase so that the frequency of collisions does not go up.

- Uses of expansion
- Precautions against expansion
- Bimetallic strip

- Linear expansivity
- Unusual expansion of water

In general, when matter is heated it expands and when cooled it contracts. If the changes are resisted large forces are created which are sometimes useful but at other times are a nuisance.

According to the kinetic theory (Chapter 17) the molecules of solids and liquids are in constant vibration. When heated they vibrate faster and force each other a little further apart. Expansion results, and this is greater for liquids than for solids; gases expand even more. The linear (length) expansion of solids is small and for the effect to be noticed, the solid must be long and/or the temperature change must be large.

● Uses of expansion

In Figure 19.1 the axles have been shrunk by cooling in liquid nitrogen at −196 °C until the gear wheels can be slipped on to them. On regaining normal temperature the axles expand to give a very tight fit.

Figure 19.1 'Shrink-fitting' of axles into gear wheels

In the kitchen, a tight metal lid can be removed from a glass jar by immersing the lid in hot water so that it expands.

● Precautions against expansion

Gaps used to be left between lengths of railway lines to allow for expansion in summer. They caused a familiar 'clickety-click' sound as the train passed over them. These days rails are welded into lengths of about 1 km and are held by concrete 'sleepers' that can withstand the large forces created without buckling. Also, at the joints the ends are tapered and overlap (Figure 19.2a). This gives a smoother journey and allows some expansion near the ends of each length of rail.

For similar reasons slight gaps are left between lengths of aluminium guttering. In central heating pipes 'expansion joints' are used to join lengths of pipe (Figure 19.2b); these allow the copper pipes to expand in length inside the joints when carrying very hot water.

Figure 19.2a Tapered overlap of rails

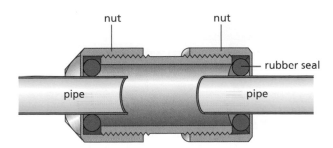

Figure 19.2b Expansion joint

● Bimetallic strip

If equal lengths of two different metals, such as copper and iron, are riveted together so that they cannot move separately, they form a **bimetallic strip** (Figure 19.3a). When heated, copper expands more than iron and to allow this the strip bends with copper on the outside (Figure 19.3b). If they had expanded equally, the strip would have stayed straight.

Bimetallic strips have many uses.

a

b

Figure 19.3 A bimetallic strip: **a** before heating; **b** after heating

a) Fire alarm

Heat from the fire makes the bimetallic strip bend and complete the electrical circuit, so ringing the alarm bell (Figure 19.4a).

A bimetallic strip is also used in this way to work the flashing direction indicator lamps in a car, being warmed by an electric heating coil wound round it.

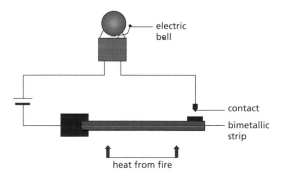

a

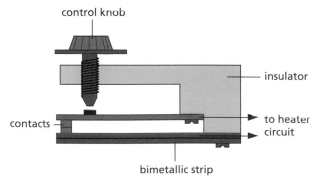

b

Figure 19.4 Uses of a bimetallic strip: **a** fire alarm; **b** a thermostat in an iron

b) Thermostat

A **thermostat** keeps the temperature of a room or an appliance constant. The one in Figure 19.4b uses a bimetallic strip in the electrical heating circuit of, for example, an iron.

When the iron reaches the required temperature the strip bends down, breaks the circuit at the contacts and switches off the heater. After cooling a little the strip remakes contact and turns the heater on again. A near-steady temperature results.

If the control knob is screwed down, the strip has to bend more to break the heating circuit and this needs a higher temperature.

● Linear expansivity

An engineer has to allow for the **linear expansion** of a bridge when designing it. The expansion can be calculated if all the following are known:

(i) the length of the bridge,
(ii) the range of temperature it will experience, and
(iii) the **linear expansivity** of the material to be used.

> The linear expansivity α of a substance is the increase in length of 1 m for a 1 °C rise in temperature.

The linear expansivity of a material is found by experiment. For steel it is 0.000 012 per °C. This means that 1 m will become 1.000 012 m for a temperature rise of 1 °C. A steel bridge 100 m long will expand by 0.000 012 × 100 m for each 1 °C rise in temperature. If the maximum **temperature change** expected is 60 °C (e.g. from −15 °C to +45 °C), the expansion will be 0.000 012 per °C × 100 m × 60 °C = 0.072 m, or 7.2 cm. In general,

> expansion = linear expansivity × original length
> × temperature rise

(The Greek letter delta, Δ, is often used to mean the 'difference' or the change in a quantity. So in the above calculation, the change in temperature, $\Delta\theta$, is 60 °C and the change in length, Δl, is 7.2 cm.)

Values of expansivity for liquids are typically about 5 times higher than that for steel; gases have expansivity values about 100 times that of steel. These figures indicate that gases expand much more readily than liquids, and liquids expand more readily than solids.

● Unusual expansion of water

As water is cooled to 4 °C it contracts, as we would expect. However between 4 °C and 0 °C it expands, surprisingly. **Water has a maximum density at 4 °C** (Figure 19.5).

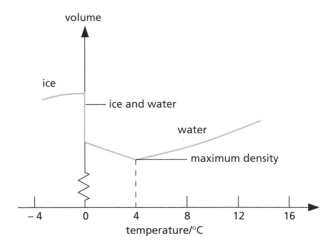

Figure 19.5 Water expands on cooling below 4 °C.

At 0 °C, when it freezes, a considerable volume expansion occurs and every 100 cm³ of water becomes 109 cm³ of ice. This accounts for the bursting of unlagged water pipes in very cold weather and for the fact that ice is less dense than cold water and so floats. Figure 19.6 shows a bottle of frozen milk, the main constituent of which is water.

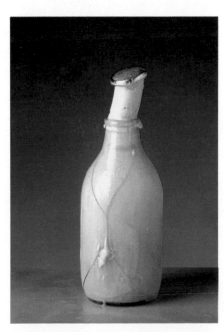

Figure 19.6 Result of the expansion of water on freezing

The unusual expansion of water between 4 °C and 0 °C explains why fish survive in a frozen pond. The water at the top of the pond cools first, contracts and being denser sinks to the bottom. Warmer, less dense water rises to the surface to be cooled. When all the water is at 4 °C the circulation stops. If the temperature of the surface water falls below 4 °C, it becomes less dense and remains at the top, eventually forming a layer of ice at 0 °C. Temperatures in the pond are then as in Figure 19.7.

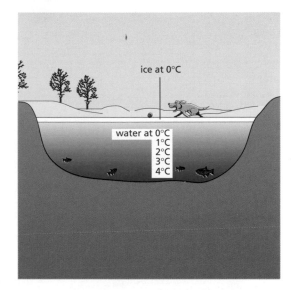

Figure 19.7 Fish can survive in a frozen pond.

The volume expansion of water between 4 °C and 0 °C is due to the breaking up of the groups that water molecules form above 4 °C. The new arrangement requires a larger volume and more than cancels out the contraction due to the fall in temperature.

Questions

1 Explain why
   a the metal lid on a glass jam jar can be unscrewed easily if the jar is inverted for a few seconds with the *lid* in very hot water,
   b furniture may creak at night after a warm day,
   c concrete roads are laid in sections with pitch between them.
2 A bimetallic strip is made from aluminium and copper. When heated it bends in the direction shown in Figure 19.8.
   Which metal expands more for the same rise in temperature, aluminium or copper?
   Draw a diagram to show how the bimetallic strip would appear if it were cooled to below room temperature.

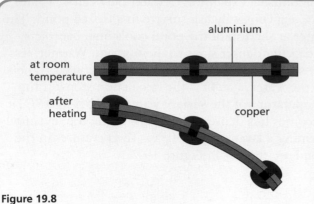

Figure 19.8

3 When a metal bar is heated the increase in length is greater if
1 the bar is long
2 the temperature rise is large
3 the bar has a large diameter.
Which statement(s) is (are) correct?
A 1, 2, 3   B 1, 2   C 2, 3   D 1   E 3

4 A bimetallic thermostat for use in an iron is shown in Figure 19.9.

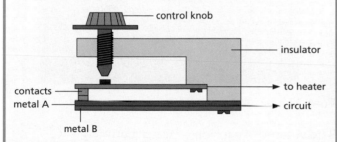

Figure 19.9

1 It operates by the bimetallic strip bending away from the contact.
2 Metal A has a greater expansivity than metal B.
3 Screwing in the control knob raises the temperature at which the contacts open.
Which statement(s) is (are) correct?
A 1, 2, 3   B 1, 2   C 2, 3   D 1   E 3

(20) Thermometers

- Liquid-in-glass thermometer
- Scale of temperature
- Clinical thermometer
- Thermocouple thermometer
- Other thermometers
- Heat and temperature

The **temperature** of a body tells us how hot the body is. It is measured using a thermometer and usually in **degrees Celsius** (°C). The kinetic theory (Chapter 17) regards temperature as a measure of the average kinetic energy (k.e.) of the molecules of the body. The greater this is, the faster the molecules move and the higher the temperature of the body.

There are different kinds of thermometer, each type being more suitable than another for a certain job. In each type the physical property used must vary continuously over a wide range of temperature. It must be accurately measurable with simple apparatus and vary in a similar way to other physical properties. Figure 20.1 shows the temperature of a lava flow being measured.

Figure 20.1 Use of a thermocouple probe thermometer to measure a temperature of about 1160 °C in lava

Liquid-in-glass thermometer

In this type the liquid in a glass bulb expands up a capillary tube when the bulb is heated. The liquid must be easily seen and must expand (or contract) rapidly and by a large amount over a wide range of temperature. It must not stick to the inside of the tube or the reading will be too high when the temperature is falling.

Mercury and coloured alcohol are in common use. Mercury freezes at −39 °C and boils at 357 °C.

Alcohol freezes at −115 °C and boils at 78 °C and is therefore more suitable for low temperatures.

Scale of temperature

A **scale** and unit of temperature are obtained by choosing two temperatures, called the **fixed points**, and dividing the range between them into a number of equal divisions or **degrees**.

On the Celsius scale (named after the Swedish scientist who suggested it), **the lower fixed point is the temperature of pure melting ice** and is taken as 0 °C. **The upper fixed point is the temperature of the steam above water boiling at normal atmospheric pressure**, 10^5 Pa (or N/m²), and is taken as 100 °C.

When the fixed points have been marked on the thermometer, the distance between them is divided into 100 equal degrees (Figure 20.2). The thermometer now has a linear scale, in other words it has been **calibrated** or **graduated**.

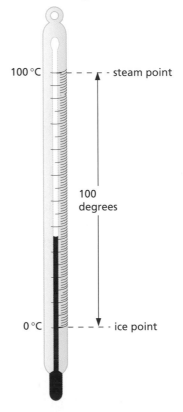

Figure 20.2 A temperature scale in degrees Celsius

● Clinical thermometer

A **clinical thermometer** is a special type of mercury-in-glass thermometer used by doctors and nurses. Its scale only extends over a few degrees on either side of the normal body temperature of 37 °C (Figure 20.3), i.e. it has a small **range**. Because of the very narrow capillary tube, temperatures can be measured very accurately, in other words, the thermometer has a high **sensitivity**.

The tube has a constriction (a narrower part) just beyond the bulb. When the thermometer is placed under the tongue the mercury expands, forcing its way past the constriction. When the thermometer is removed (after 1 minute) from the mouth, the mercury in the bulb cools and contracts, breaking the mercury thread at the constriction. The mercury beyond the constriction stays in the tube and shows the body temperature. After use the mercury is returned to the bulb by a flick of the wrist. Since mercury is a toxic material, digital thermometers are now replacing mercury thermometers for clinical use.

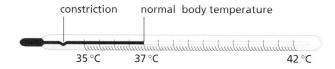

Figure 20.3 A clinical thermometer

● Thermocouple thermometer

A **thermocouple** consists of wires of two different materials, such as copper and iron, joined together (Figure 20.4). When one junction is at a higher temperature than the other, an electric current flows and produces a reading on a sensitive meter which depends on the temperature difference.

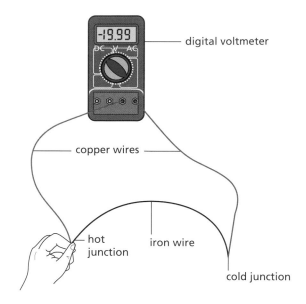

Figure 20.4 A simple thermocouple thermometer

Thermocouples are used in industry to measure a wide range of temperatures from −250 °C up to about 1500 °C, especially rapidly changing temperatures and those of small objects.

● Other thermometers

One type of **resistance thermometer** uses the fact that the electrical resistance (Chapter 38) of a platinum wire increases with temperature.

A resistance thermometer can measure temperatures accurately in the range −200 °C to 1200 °C but it is bulky and best used for steady temperatures. A **thermistor** can also be used but over a small range, such as −5 °C to 70 °C; its resistance decreases with temperature.

The **constant-volume gas thermometer** uses the change in pressure of a gas to measure temperatures over a wide range. It is an accurate but bulky instrument, basically similar to the apparatus of Figure 18.3 (p. 77).

Thermochromic liquids which change colour with temperature have a limited range around room temperatures.

● Heat and temperature

It is important not to confuse the temperature of a body with the heat energy that can be obtained from it. For example, a red-hot spark from a fire is at a higher temperature than the boiling water in a saucepan. In the boiling water the average k.e. of the molecules is lower than in the spark; but since there are many more water molecules, their total energy is greater, and therefore more heat energy can be supplied by the water than by the spark.

Heat passes from a body at a higher temperature to one at a lower temperature. This is because the average k.e. (and speed) of the molecules in the 'hot' body falls as a result of the collisions with molecules of the 'cold' body whose average k.e., and therefore temperature, increases. When the average k.e. of the molecules is the same in both bodies, they are at the same temperature. For example, if the red-hot spark landed in the boiling water, heat would pass from it to the water even though much more heat energy could be obtained from the water.

Heat is also called **thermal** or **internal energy**; it is the energy a body has because of the kinetic energy *and* the potential energy (p.e.) of its molecules. Increasing the temperature of a body increases its heat energy because the k.e. of its molecules increases. But as we will see later (Chapter 22), the internal energy of a body can also be increased by increasing the p.e. of its molecules.

3 a How must a property behave to measure temperature?
 b Name three properties that qualify.
 c Name a suitable thermometer for measuring
  (i) a steady temperature of 1000 °C,
  (ii) the changing temperature of a small object,
  (iii) a winter temperature at the North Pole.
4 Describe the main features of a clinical thermometer.

Checklist

After studying this chapter you should be able to

* define the fixed points on the Celsius scale,
* recall the properties of mercury and alcohol as liquids suitable for use in thermometers,
* describe clinical and thermocouple thermometers,

* understand the meaning of range, sensitivity and linearity in relation to thermometers,

* describe the structure and use of a thermocouple thermometer,

* recall some other types of thermometer and the physical properties on which they depend,
* distinguish between heat and temperature and recall that temperature decides the direction of heat flow,
* relate a rise in the temperature of a body to an increase in internal energy.

Questions

1 1530 °C 120 °C 55 °C 37 °C 19 °C 0 °C −12 °C −50 °C
From the above list of temperatures choose the most likely value for *each* of the following:
 a the melting point of iron,
 b the temperature of a room that is comfortably warm,
 c the melting point of pure ice at normal pressure,
 d the lowest outdoor temperature recorded in London in winter,
 e the normal body temperature of a healthy person.
2 In order to make a mercury thermometer that will measure small changes in temperature accurately, would you
 A decrease the volume of the mercury bulb
 B put the degree markings further apart
 C decrease the diameter of the capillary tube
 D put the degree markings closer together
 E leave the capillary tube open to the air?

(21) Specific heat capacity

- The heat equation
- Thermal capacity
- Importance of the high specific heat capacity of water

- Practical work: Finding specific heat capacities: water, aluminium

If 1 kg of water and 1 kg of paraffin are heated in turn for the same time by the same heater, the temperature rise of the paraffin is about *twice* that of the water. Since the heater gives equal amounts of heat energy to each liquid, it seems that different substances require different amounts of heat to cause the same temperature rise in the same mass, say 1 °C in 1 kg.

The 'thirst' of a substance for heat is measured by its **specific heat capacity** (symbol c).

The specific heat capacity of a substance is the heat required to produce a 1 °C rise in 1 kg.

Heat, like other forms of energy, is measured in joules (J) and the unit of specific heat capacity is the **joule per kilogram per °C**, i.e. J/(kg °C).

In physics, the word 'specific' means that 'unit mass' is being considered.

● The heat equation

If a substance has a specific heat capacity of 1000 J/(kg °C) then

1000 J raises the temperature of 1 kg by 1 °C

∴ 2 × 1000 J raises the temperature of 2 kg by 1 °C

∴ 3 × 2 × 1000 J raises the temperature of 2 kg by 3 °C

That is, 6000 J will raise the temperature of 2 kg of this substance by 3 °C. We have obtained this answer by multiplying together:

(i) the **mass** in kg,
(ii) the **temperature rise** in °C, and
(iii) the **specific heat capacity** in J/(kg °C).

If the temperature of the substance fell by 3 °C, the heat given out would also be 6000 J. In general, we can write the **heat equation** as

heat received or given out
= mass × temperature change × specific heat capacity
In symbols
$$Q = m \times \Delta\theta \times c$$

For example, if the temperature of a 5 kg mass of material of specific heat capacity 400 J/(kg °C) rises from 15 °C to 25 °C, the heat received, Q, is

$$Q = 5\,kg \times (25-15)\,°C \times 400\,J/(kg\,°C)$$
$$= 5\,kg \times 10\,°C \times 400\,J/(kg\,°C)$$
$$= 20\,000\,J$$

● Thermal capacity

The **thermal capacity** of a body is the quantity of heat needed to raise the temperature of the whole body by 1 °C.

For a temperature rise of 1 °C the heat equation becomes:

heat received = mass × 1 × specific heat capacity

so that

thermal capacity = mass × specific heat capacity
$$= m \times c$$

Thermal capacity is measured in **joules per °C**, i.e. J/°C.

For a copper block of mass 0.1 kg and specific heat capacity 390 J/(kg °C),

thermal capacity $= m \times c$
$$= 0.1\,kg \times 390\,J/(kg\,°C)$$
$$= 39\,J/°C$$

Practical work

Finding specific heat capacities

You need to know the power of the 12 V electric immersion heater to be used. (Precaution: Do not use one with a cracked seal.) A 40 W heater converts 40 joules of electrical energy into heat energy per second. If the power is not marked on the heater, ask about it.[1]

a) Water

Weigh out 1 kg of water into a container, such as an aluminium saucepan. Note the temperature of the water, insert the heater (Figure 21.1), switch on the 12 V supply and start timing. Stir the water and after 5 minutes switch off, but continue stirring and note the *highest* temperature reached.

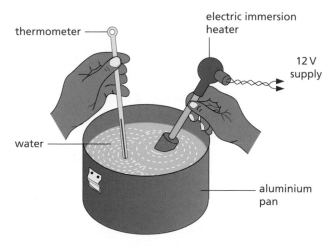

Figure 21.1

Assuming that the heat supplied by the heater equals the heat received by the water, work out the specific heat capacity of water in J/(kg °C), as shown below:

heat received by water (J)
= power of heater (J/s) × time heater on (s)

Rearranging the 'heat equation' we get

$$\frac{\text{specific heat}}{\text{capacity of water}} = \frac{\text{heat received by water (J)}}{\text{mass (kg)} \times \text{temp. rise (°C)}}$$

Suggest causes of error in this experiment.

b) Aluminium

An aluminium cylinder weighing 1 kg and having two holes drilled in it is used. Place the immersion heater in the central hole and a thermometer in the other hole (Figure 21.2).

Note the temperature, connect the heater to a 12 V supply and

---

[1]The power is found by immersing the heater in water, connecting it to a 12 V d.c. supply and measuring the current taken (usually 3–4 amperes). Then power in watts = volts × amperes.

switch it on for 5 minutes. When the temperature stops rising record its highest value.

Calculate the specific heat capacity as before.

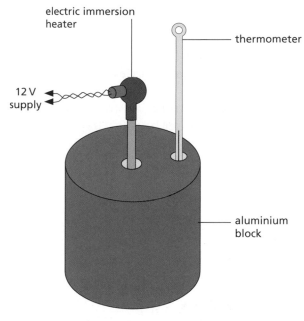

Figure 21.2

● Importance of the high specific heat capacity of water

The specific heat capacity of water is 4200 J/(kg °C) and that of soil is about 800 J/(kg °C). As a result, the temperature of the sea rises and falls more slowly than that of the land. A certain mass of water needs five times more heat than the same mass of soil for its temperature to rise by 1 °C. Water also has to give out more heat to fall 1 °C. Since islands are surrounded by water they experience much smaller changes of temperature from summer to winter than large land masses such as Central Asia.

The high specific heat capacity of water (as well as its cheapness and availability) accounts for its use in cooling engines and in the radiators of central heating systems.

● Worked examples

1 A tank holding 60 kg of water is heated by a 3 kW electric immersion heater. If the specific heat capacity of water is 4200 J/(kg °C), estimate the time for the temperature to rise from 10 °C to 60 °C.

A 3 kW (3000 W) heater supplies 3000 J of heat energy per second.

Let t = time taken in seconds to raise the temperature of the water by (60−10) = 50 °C,

$$\therefore \text{ heat supplied to water in time } t \text{ seconds} = (3000 \times t)\,\text{J}$$

From the heat equation, we can say

$$\text{heat received by water} = 60\,\text{kg} \times 4200\,\text{J}/(\text{kg °C}) \times 50\,°\text{C}$$

Assuming heat supplied = heat received

$$3000\,\text{J/s} \times t = (60 \times 4200 \times 50)\,\text{J}$$

$$\therefore \quad t = \frac{(60 \times 4200 \times 50)\,\text{J}}{3000\,\text{J/s}} = 4200\,\text{s} \ (70\,\text{min})$$

2 A piece of aluminium of mass 0.5 kg is heated to 100 °C and then placed in 0.4 kg of water at 10 °C. If the resulting temperature of the mixture is 30 °C, what is the specific heat capacity of aluminium if that of water is 4200 J/(kg °C)?

When two substances at different temperatures are mixed, heat flows from the one at the higher temperature to the one at the lower temperature until both are at the same temperature – the temperature of the mixture. If there is no loss of heat, then in this case:

heat given out by aluminium = heat taken in by water

Using the heat equation and letting c be the specific heat capacity of aluminium in J/(kg °C), we have

$$\text{heat given out} = 0.5\,\text{kg} \times c \times (100 - 30)\,°\text{C}$$

$$\text{heat taken in} = 0.4\,\text{kg} \times 4200\,\text{J}/(\text{kg °C}) \times (30 - 10)\,°\text{C}$$

$$\therefore 0.5\,\text{kg} \times c \times 70\,°\text{C} = 0.4\,\text{kg} \times 4200\,\text{J}/(\text{kg °C}) \times 20\,°\text{C}$$

$$c = \frac{(4200 \times 8)\,\text{J}}{35\,\text{kg °C}} = 960\,\text{J}/(\text{kg °C})$$

Questions

1 How much heat is needed to raise the temperature by 10°C of 5 kg of a substance of specific heat capacity 300 J/(kg°C)? What is the thermal capacity of the substance?

2 The same quantity of heat was given to different masses of three substances A, B and C. The temperature rise in each case is shown in the table. Calculate the specific heat capacities of A, B and C.

Material	Mass/kg	Heat given/J	Temp. rise/°C
A	1.0	2000	1.0
B	2.0	2000	5.0
C	0.5	2000	4.0

3 The jam in a hot pop tart always seems hotter than the pastry. Why?

Checklist

After studying this chapter you should be able to

- define specific heat capacity, c,
- define thermal capacity,
- solve problems on specific heat capacity using the heat equation $Q = m \times \Delta\theta \times c$,
- describe experiments to measure the specific heat capacity of metals and liquids by electrical heating,
- explain the importance of the high specific heat capacity of water.

Specific latent heat

- Specific latent heat of fusion
- Specific latent heat of vaporisation
- Latent heat and the kinetic theory
- Evaporation and boiling
- Condensation and solidification

- Cooling by evaporation
- Liquefaction of gases and vapours
- Practical work: Cooling curve of ethanamide; Specific latent heat of fusion for ice; Specific latent heat of vaporisation for steam

When a solid is heated, it may melt and **change its state** from solid to liquid. If ice is heated it becomes water. The opposite process, freezing, occurs when a liquid solidifies.

A pure substance melts at a definite temperature, called the **melting point**; it solidifies at the same temperature – sometimes then called the **freezing point**.

Practical work

Cooling curve of ethanamide

Half fill a test tube with ethanamide (acetamide) and place it in a beaker of water (Figure 22.1a). Heat the water until all the ethanamide has melted and its temperature reaches about 90°C.

Remove the test tube and arrange it as in Figure 22.1b, with a thermometer in the liquid ethanamide. Record the temperature every minute until it has fallen to 70°C.

Plot a cooling curve of temperature against time. What is the freezing (melting) point of ethanamide?

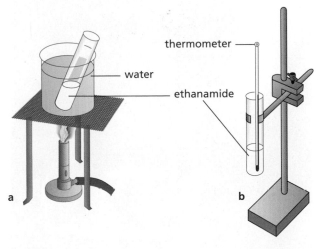

Figure 22.1

● Specific latent heat of fusion

The previous experiment shows that the temperature of liquid ethanamide falls until it starts to solidify (at 82°C) and remains constant until it has all solidified. The cooling curve in Figure 22.2 is for a pure substance; the flat part AB occurs at the melting point when the substance is solidifying.

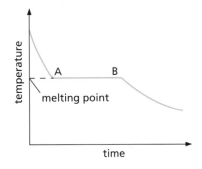

Figure 22.2 Cooling curve

During solidification a substance loses heat to its surroundings but its temperature does not fall. Conversely when a solid is melting, the heat supplied does not cause a temperature rise; heat is added but the substance does not get hotter. For example, the temperature of a well-stirred ice–water mixture remains at 0°C until all the ice is melted.

Heat that is **absorbed** by a solid during melting or given out by a liquid during solidification is called **latent heat of fusion**. 'Latent' means hidden and 'fusion' means melting. Latent heat does not cause a temperature change; it seems to disappear.

The **specific latent heat of fusion** (l_f) of a substance is the quantity of heat needed to change *unit mass* from solid to liquid without temperature change.

Specific latent heat is measured in J/kg or J/g. In general, the quantity of heat Q to change a mass m from solid to liquid is given by

$$Q = m \times l_f$$

Practical work

Specific latent heat of fusion for ice

Through measurement of the mass of water m produced when energy Q is transferred to melting ice, the specific latent heat of fusion for ice can be calculated.

Insert a 12 V electric immersion heater of known power P into a funnel, and pack crushed ice around it as shown in Figure 22.3.

To correct for heat transferred from the surroundings, collect the melted ice in a beaker for time t (e.g. 4 minutes); weigh the beaker plus the melted ice, m_1. Empty the beaker, switch on the heater, and collect the melted ice for the same time t; re-weigh the beaker plus the melted ice, m_2. The mass of ice melted by the heater is then

$$m = m_2 - m_1$$

The electrical energy supplied by the heater is given by $Q = P \times t$, where P is in J/s and t is in seconds; Q will be in joules. Alternatively, a joulemeter can be used to record Q directly.

Calculate the specific latent heat of fusion, l_f, for ice using

$$Q = m \times l_f$$

How does it compare with the accepted value of 340 J/g? How could the experiment be improved?

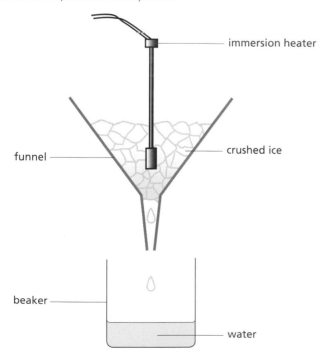

funnel

immersion heater

crushed ice

beaker

water

Figure 22.3

● Specific latent heat of vaporisation

Latent heat is also needed to change a liquid into a vapour. The reading of a thermometer placed in water that is boiling remains constant at 100 °C even though heat, called **latent heat of vaporisation**, is still being absorbed by the water from whatever is heating it. When steam condenses to form water, latent heat is given out.

> The **specific latent heat of vaporisation** (l_v) of a substance is the quantity of heat needed to change unit mass from liquid to vapour without change of temperature.

Again, the specific latent heat is measured in J/kg or J/g. In general, the quantity of heat Q to change a mass m from liquid to vapour is given by

$$Q = m \times l_v$$

Figure 22.4 Why is a scald from steam often more serious than one from boiling water?

Specific latent heat of vaporisation for steam

Through measurement of the mass of vapour *m* produced when energy *Q* is transferred to boiling water, the specific latent heat of vaporisation for steam can be calculated.

Water in the flask (Figure 22.5) is heated to boiling point by an immersion heater of power *P*. Steam passes out through the holes in the top of the flask, down the outside of the flask and into the inner tube of a condenser, where it changes back to liquid (because cold water is flowing through the outer tube), and is collected in a beaker.

After the water has been boiling for some time, it becomes enclosed by a 'jacket' of vapour at the boiling point, which helps to reduce loss of heat to the surroundings. The rate of vaporisation becomes equal to the rate of condensation, and the electrical energy is only being used to transfer latent heat to the water (not to raise its temperature).

The electrical energy *Q* supplied by the heater is given by

$$Q = P \times t = ItV$$

where *I* is the steady current through the heater and *V* is the p.d. across it. *Q* is in joules if *P* is in J/s and *t* is in seconds. Alternatively, a joulemeter can be used to record *Q* directly.

If a mass of water *m* is collected in time *t*, then the specific latent heat of vaporisation l_v can be calculated using

$$Q = m \times l_v$$

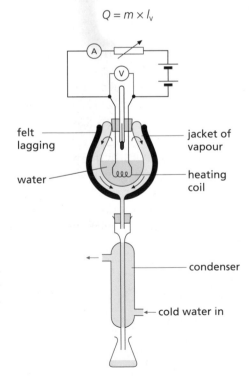

felt lagging

jacket of vapour

water

heating coil

condenser

cold water in

Figure 22.5

● Latent heat and the kinetic theory

a) Fusion

The kinetic theory explains latent heat of fusion as being the energy that enables the molecules of a solid to overcome the intermolecular forces that hold them in place, and when it exceeds a certain value they break free. Their vibratory motion about fixed positions changes to the slightly greater range of movement they have as liquid molecules, and the solid melts.

The energy input is used to increase the potential energy (p.e.) of the molecules, but not their average kinetic energy (k.e.) as happens when the heat causes a temperature rise.

b) Vaporisation

If liquid molecules are to overcome the forces holding them together and gain the freedom to move around independently as gas molecules, they need a large amount of energy. They receive this as latent heat of vaporisation which, like latent heat of fusion, increases the potential energy of the molecules but not their kinetic energy. It also gives the molecules the energy required to push back the surrounding atmosphere in the large expansion that occurs when a liquid vaporises.

To change 1 kg of water at 100 °C to steam at 100 °C needs over *five* times as much heat as is needed to raise the temperature of 1 kg of water at 0 °C to water at 100 °C (see *Worked example 1*, p. 95).

● Evaporation and boiling

a) Evaporation

A few energetic molecules close to the surface of a liquid may escape and become gas molecules. This process occurs at all temperatures and is called **evaporation**. It happens more rapidly when

(i) the **temperature is higher**, since then more molecules in the liquid are moving fast enough to escape from the surface,

(ii) the **surface area of the liquid is large**, so giving more molecules a chance to escape because more are near the surface, and

(iii) a **wind** or **draught** is blowing over the surface carrying vapour molecules away from the surface, thus stopping them from returning to the liquid and making it easier for more liquid molecules to break free. (Evaporation into a vacuum occurs much more rapidly than into a region where there are gas molecules.)

b) Boiling

For a pure liquid boiling occurs at a definite temperature called its **boiling point** and is accompanied by bubbles that form within the liquid, containing the gaseous or vapour form of the particular substance.

Latent heat is needed in both evaporation and boiling and is stored in the vapour, from which it is released when the vapour is cooled or compressed and changes to liquid again.

● Condensation and solidification

In **condensation**, a gas changes to a liquid state and latent heat of vaporisation is released. In **solidification**, a liquid changes to a solid and latent heat of fusion is given out. In each case the potential energy of the molecules decreases. Condensation of steam is easily achieved by contact with a cold surface, for example a cold windowpane. In Figure 22.5, the latent heat released when the steam condenses to water is transferred to the cold water flowing through the condenser.

● Cooling by evaporation

In evaporation, latent heat is obtained by the liquid from its surroundings, as may be shown by the following demonstration, **done in a fume cupboard**.

a) Demonstration

Dichloromethane is a **volatile** liquid, i.e. it has a low boiling point and evaporates readily at room temperature, especially when air is blown through it (Figure 22.6). Latent heat is taken first from the liquid itself and then from the water below the can. The water soon freezes causing the block and can to stick together.

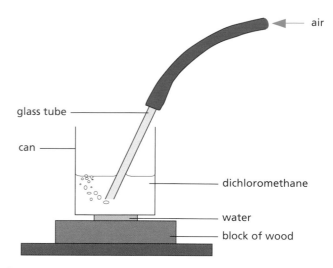

Figure 22.6 Demonstrating cooling by evaporation

b) Explanation

Evaporation occurs when faster-moving molecules escape from the surface of the liquid. The average speed and therefore the average kinetic energy of the molecules left behind decreases, i.e. the temperature of the liquid falls. Any body in contact with an evaporating liquid will be cooled by the evaporation.

c) Uses

Water evaporates from the skin when we sweat. This is the body's way of losing unwanted heat and keeping a constant temperature. After vigorous exercise there is a risk of the body being overcooled, especially in a draught; it is then less able to resist infection.

Ether acts as a local anaesthetic by chilling (as well as cleaning) your arm when you are having an injection. Refrigerators, freezers and air-conditioning systems use cooling by evaporation on a large scale.

Volatile liquids are used in perfumes.

● Liquefaction of gases and vapours

A vapour can be liquefied if it is compressed enough. However, a gas must be cooled below a certain critical temperature T_c before liquefaction by pressure can occur. The critical temperatures for some gases are given in Table 22.1.

Table 22.1 Critical temperatures of some gases

	Carbon dioxide	Oxygen	Air	Nitrogen	Hydrogen	Helium
T_c/K	304	154	132	126	33.3	5.3
T_c/°C	+31	−119	−141	−147	−239.7	−267.7

Low-temperature liquids have many uses. Liquid hydrogen and oxygen are used as the fuel and oxidant respectively in space rockets. Liquid nitrogen is used in industry as a coolant in, for example, shrink-fitting (see Figure 19.1, p. 81). Materials that behave as superconductors (see Chapter 18) when cooled by liquid nitrogen are increasingly used in electrical power engineering and electronics.

● Worked examples

The values in Table 22.2 are required.

Table 22.2

	Water	Ice	Aluminium
Specific heat capacity/J/(g °C)	4.2	2.0	0.90
Specific latent heat/J/g	2300	340	

1 How much heat is needed to change 20 g of ice at 0 °C to steam at 100 °C?

There are three stages in the change.

Heat to change 20 g **ice at 0 °C** to **water at 0 °C**

> = mass of ice × specific latent heat of ice
> = 20 × 340 J/g = 6800 J

Heat to change 20 g **water at 0 °C** to **water at 100 °C**

> = mass of water × specific heat capacity of water
> × temperature rise
> = 20 g × 4.2 J/(g °C) × 100 °C = 8400 J

Heat to change 20 g **water at 100 °C** to **steam at 100 °C**

> = mass of water × specific latent heat of steam
> = 20 g × 2300 J/g = 46 000 J

∴ Total heat supplied

> = 6800 + 8400 + 46 000 = 61 200 J

2 An aluminium can of mass 100 g contains 200 g of water. Both, initially at 15 °C, are placed in a freezer at −5.0 °C. Calculate the quantity of heat that has to be removed from the water and the can for their temperatures to fall to −5.0 °C.

Heat lost by **can** in falling from 15 °C to −5.0 °C

> = mass of can × specific heat capacity of aluminium × temperature fall
> = 100 g × 0.90 J/(g °C) × (15 − [−5]) °C
> = 100 g × 0.90 J/(g °C) × 20 °C
> = 1800 J

Heat lost by **water** in falling from 15 °C to 0 °C

> = mass of water × specific heat capacity of water × temperature fall
> = 200 g × 4.2 J/(g °C) × 15 °C
> = 12 600 J

Heat lost by **water** at 0 °C freezing to ice at 0 °C

> = mass of water × specific latent heat of ice
> = 200 g × 340 J/g
> = 68 000 J

Heat lost by **ice** in falling from 0 °C to −5.0 °C

> = mass of ice × specific heat capacity of ice × temperature fall
> = 200 g × 2.0 J/(g °C) × 5.0 °C
> = 2000 J

∴ Total heat removed

> = 1800 + 12 600 + 68 000 + 2000 = 84 400 J

Questions

Use values given in Table 22.2.
1 a How much heat will change 10 g of ice at 0 °C to water at 0 °C?
 b What quantity of heat must be removed from 20 g of water at 0 °C to change it to ice at 0 °C?
2 a How much heat is needed to change 5 g of ice at 0 °C to water at 50 °C?
 b If a freezer cools 200 g of water from 20 °C to its freezing point in 10 minutes, how much heat is removed per minute from the water?
3 How long will it take a 50 W heater to melt 100 g of ice at 0 °C?
4 Some small aluminium rivets of total mass 170 g and at 100 °C are emptied into a hole in a large block of ice at 0 °C.
 a What will be the final temperature of the rivets?
 b How much ice will melt?

 ▶▶

5 a How much heat is needed to change 4 g of water at 100 °C to steam at 100 °C?

  b Find the heat given out when 10 g of steam at 100 °C condenses and cools to water at 50 °C.

6 A 3 kW electric kettle is left on for 2 minutes after the water starts to boil. What mass of water is boiled off in this time?

7 a Why is ice good for cooling drinks?

  b Why do engineers often use superheated steam (steam above 100 °C) to transfer heat?

8 Some water is stored in a bag of porous material, such as canvas, which is hung where it is exposed to a draught of air. Explain why the temperature of the water is lower than that of the air.

9 Explain why a bottle of milk keeps better when it stands in water in a porous pot in a draught.

10 A certain liquid has a specific heat capacity of 4.0 J/(g °C). How much heat must be supplied to raise the temperature of 10 g of the liquid from 20 °C to 50 °C?

Checklist

After studying this chapter you should be able to

• describe an experiment to show that during a change of state the temperature stays constant,
• state the meaning of melting point and boiling point,
• describe condensation and solidification,

• define specific latent heat of fusion, l_f,

• define specific latent heat of vaporisation, l_v,

• explain latent heat using the kinetic theory,

• solve problems on latent heat, using $Q = ml$,

• distinguish between evaporation and boiling,

• describe an experiment to measure specific latent heats for ice and steam,

• explain cooling by evaporation using the kinetic theory.

Conduction and convection

- Conduction
- Uses of conductors
- Conduction and the kinetic theory
- Convection in liquids

- Convection in air
- Natural convection currents
- Energy losses from buildings
- Ventilation

To keep a building or a house at a comfortable temperature in winter and in summer, if it is to be done economically and efficiently, requires a knowledge of how heat travels.

● Conduction

The handle of a metal spoon held in a hot drink soon gets warm. Heat passes along the spoon by **conduction**.

> Conduction is the flow of thermal energy (heat) through matter from places of higher temperature to places of lower temperature without movement of the matter as a whole.

A simple demonstration of the different conducting powers of various metals is shown in Figure 23.1. A match is fixed to one end of each rod using a little melted wax. The other ends of the rods are heated by a burner. When the temperatures of the far ends reach the melting point of wax, the matches drop off. The match on copper falls first, showing it is the best **conductor**, followed by aluminium, brass and then iron.

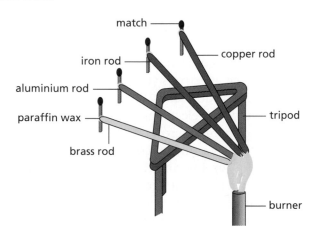

Figure 23.1 Comparing conducting powers

Heat is conducted faster through a rod if it has a large cross-sectional area, is short and has a large temperature difference between its ends.

Most metals are good conductors of heat; materials such as wood, glass, cork, plastics and fabrics are bad conductors. The arrangement in Figure 23.2 can be used to show the difference between brass and wood. If the rod is passed through a flame several times, the paper over the wood scorches but not the paper over the brass. The brass conducts the heat away from the paper quickly, preventing the paper from reaching the temperature at which it burns. The wood conducts the heat away only very slowly.

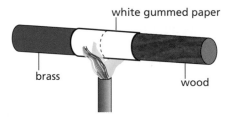

Figure 23.2 The paper over the brass does not burn.

Metal objects below body temperature *feel* colder than those made of bad conductors – even if all the objects are at exactly the same temperature – because they carry heat away faster from the hand.

Liquids and gases also conduct heat but only very slowly. Water is a very poor conductor, as shown in Figure 23.3. The water at the top of the tube can be boiled before the ice at the bottom melts.

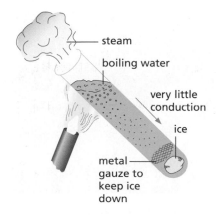

Figure 23.3 Water is a poor conductor of heat.

● Uses of conductors

a) Good conductors

These are used whenever heat is required to travel quickly through something. Saucepans, boilers and radiators are made of metals such as aluminium, iron and copper.

b) Bad conductors (insulators)

The handles of some saucepans are made of wood or plastic. Cork is used for table mats.

Air is one of the worst conductors and so one of the best **insulators**. This is why houses with cavity walls (two layers of bricks separated by an air space) and double-glazed windows keep warmer in winter and cooler in summer.

Materials that trap air, such as wool, felt, fur, feathers, polystyrene foam, fibreglass, are also very bad conductors. Some of these materials are used as 'lagging' to insulate water pipes, hot water cylinders, ovens, refrigerators and the walls and roofs of houses (Figures 23.4a and 23.4b). Others are used to make warm winter clothes like 'fleece' jackets (Figure 23.4c).

Figure 23.4a Lagging in a cavity wall provides extra insulation.

Figure 23.4b Laying lagging in a house loft

Figure 23.4c Fleece jackets help you to retain your body warmth.

'Wet suits' are worn by divers and water skiers to keep them warm. The suit gets wet and a layer of water gathers between the person's body and the suit. The water is warmed by body heat and stays warm because the suit is made of an insulating fabric, such as neoprene, a synthetic rubber.

● Conduction and the kinetic theory

Two processes occur in metals. Metals have a large number of 'free' electrons (Chapter 36) which wander about inside them. When one part of a metal is heated, the electrons there move faster (their kinetic energy increases) and further. As a result they 'jostle' atoms in cooler parts, so passing on their energy and raising the temperature of these parts. This process occurs quickly.

The second process is much slower. The atoms themselves at the hot part make 'colder' neighbouring atoms vibrate more vigorously. This is less important in metals but is the only way conduction occurs in non-metals since these do not have 'free' electrons; hence non-metals are poor conductors of heat.

● Convection in liquids

Convection is the usual method by which thermal energy (heat) travels through fluids such as liquids and gases. It can be shown in water by dropping a few crystals of potassium permanganate down a tube to the bottom of a beaker or flask of water. When the tube is removed and the beaker heated just below the crystals by a *small* flame (Figure 23.5a), purple streaks of water rise upwards and fan outwards.

Figure 23.5a Convection currents shown by potassium permanganate in water

Streams of warm moving fluids are called **convection currents**. They arise when a fluid is heated because it expands, becomes less dense and is forced upwards by surrounding cooler, denser fluid which moves under it. We say 'hot water (or hot air) rises'. Warm fluid behaves like a cork released under water: being less dense it bobs up. Lava lamps (Figure 23.5b) use this principle.

> Convection is the flow of heat through a fluid from places of higher temperature to places of lower temperature by movement of the fluid itself.

Figure 23.5b Lava lamps make use of convection.

● Convection in air

Black marks often appear on the wall or ceiling above a lamp or a radiator. They are caused by dust being carried upwards in air convection currents produced by the hot lamp or radiator.

A laboratory demonstration of convection currents in air can be given using the apparatus of Figure 23.6. The direction of the convection current created by the candle is made visible by the smoke from the touch paper (made by soaking brown paper in strong potassium nitrate solution and drying it).

Convection currents set up by electric, gas and oil heaters help to warm our homes. Many so-called 'radiators' are really convector heaters.

Where should the input and extraction ducts for cold/hot air be located in a room?

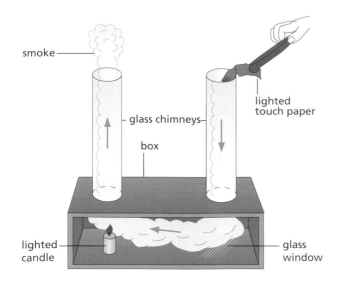

Figure 23.6 Demonstrating convection in air

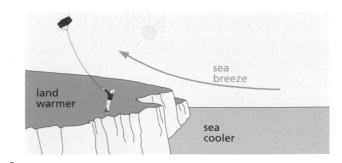

a

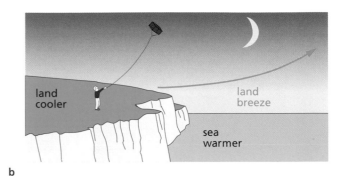

b

Figure 23.7 Coastal breezes are due to convection: **a** day; **b** night

● Natural convection currents

a) Coastal breezes

During the day the temperature of the land increases more quickly than that of the sea (because the specific heat capacity of the land is much smaller; see Chapter 21). The hot air above the land rises and is replaced by colder air from the sea. A breeze from the sea results (Figure 23.7a).

At night the opposite happens. The sea has more heat to lose and cools more slowly. The air above the sea is warmer than that over the land and a breeze blows from the land (Figure 23.7b).

b) Gliding

Gliders, including 'hang-gliders' (Figure 23.8), depend on hot air currents, called **thermals**.

● Energy losses from buildings

The inside of a building can only be kept at a steady temperature above that outside by heating it at a rate which equals the rate at which it is losing energy. The loss occurs mainly by conduction through the walls, roof, floors and windows. For a typical house in the UK where no special precautions have been taken,

Figure 23.8 Once airborne, a hang-glider pilot can stay aloft for several hours by flying from one thermal to another.

the contribution each of these makes to the total loss is shown in Table 23.1a.

As fuels (and electricity) become more expensive and the burning of fuels becomes of greater environmental concern (Chapter 15), more people are considering it worthwhile to reduce heat losses from their homes. The substantial reduction of this loss which can be achieved, especially by wall and roof insulation, is shown in Table 23.1b.

Table 23.1 Energy losses from a typical house

a

Percentage of total energy loss due to				
walls	roof	floors	windows	draughts
35	25	15	10	15

b

Percentage of each loss saved by				
insulating walls	insulating roof	carpets on floors	double glazing	draught excluders
65	80	≈30	50	≈60
Percentage of total loss saved = 60				

● Ventilation

In addition to supplying heat to compensate for the energy losses from a building, a heating system has also to warm the ventilated cold air, needed for comfort, which comes in to replace stale air.

If the rate of heat loss is, say, 6000 J/s, or 6 kW, and the warming of ventilated air requires 2 kW, then the total power needed to maintain a certain temperature (e.g. 20 °C) in the building is 8 kW. Some of this is supplied by each person's 'body heat', estimated to be roughly equal to a 100 W heater.

d A vacuum is an even better heat insulator than air. Suggest one (scientific) reason why the double glazing should not have a vacuum between the sheets of glass.

e The manufacturers of roof lagging suggest that two layers of fibreglass are more effective than one. Describe how you might set up an experiment in the laboratory to test whether this is true.

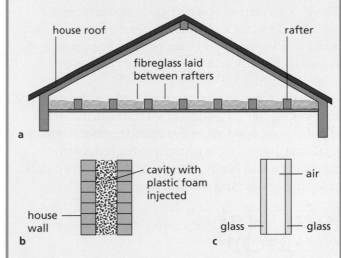

Figure 23.9 a Roof insulation; **b** cavity wall insulation; **c** double glazing

3 What is the advantage of placing an electric immersion heater
   a near the top,
   b near the bottom,
   of a tank of water?
4 Explain why on a cold day the metal handlebars of a bicycle feel colder than the rubber grips.

Questions

1 Explain why
   a newspaper wrapping keeps hot things hot, e.g. fish and chips, and cold things cold, e.g. ice cream,
   b fur coats would keep their owners warmer if they were worn inside out,
   c a string vest helps to keep a person warm even though it is a collection of holes bounded by string.
2 Figure 23.9 illustrates three ways of reducing heat losses from a house.
   a As far as you can, explain how each of the three methods reduces heat losses. Draw diagrams where they will help your explanations.
   b Why are fibreglass and plastic foam good substances to use?
   c Air is one of the worst conductors of heat. What is the point of replacing it by plastic foam as shown in the Figure 23.9b?

Checklist

After studying this chapter you should be able to

● describe experiments to show the different conducting powers of various substances,
● name good and bad conductors and state uses for each,
● explain conduction using the kinetic theory,
● describe experiments to show convection in fluids (liquids and gases),
● relate convection to phenomena such as land and sea breezes,
● explain the importance of insulating a building.

(24) Radiation

- Good and bad absorbers
- Good and bad emitters
- Vacuum flask

- The greenhouse
- Rate of cooling of an object

Radiation is a third way in which heat can travel, but whereas conduction and convection both need matter to be present, radiation can occur in a vacuum; particles of matter are not involved. Radiation is the way heat reaches us from the Sun.

Radiation has all the properties of electromagnetic waves (Chapter 32), such as it travels at the speed of radio waves and gives interference effects. When it falls on an object, it is partly reflected, partly transmitted and partly absorbed; the absorbed part raises the temperature of the object.

Radiation is the flow of heat from one place to another by means of electromagnetic waves.

Radiation is emitted by all bodies above absolute zero and consists mostly of **infrared radiation** (Chapter 32) but light and **ultraviolet** are also present if the body is very hot (e.g. the Sun).

Figure 24.1 Why are buildings in hot countries often painted white?

● Good and bad absorbers

Some surfaces absorb radiation better than others, as may be shown using the apparatus in Figure 24.2. The inside surface of one lid is shiny and of the other dull black. The coins are stuck on the outside of each lid with candle wax. If the heater is midway between the lids they each receive the same amount of radiation. After a few minutes the wax on the black lid melts and the coin falls off. The shiny lid stays cool and the wax unmelted.

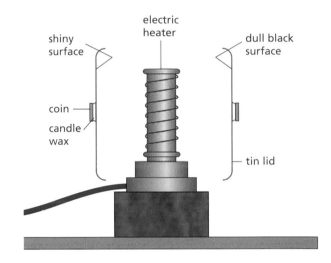

Figure 24.2 Comparing absorbers of radiation

Dull black surfaces are better absorbers of radiation than white shiny surfaces – the latter are good **reflectors** of radiation. Reflectors on electric fires are made of polished metal because of its good reflecting properties.

● Good and bad emitters

Some surfaces also emit radiation better than others when they are hot. If you hold the backs of your hands on either side of a hot copper sheet that has one side polished and the other side blackened (Figure 24.3), it will be found that the **dull black surface is a better emitter of radiation than the shiny one**.

The cooling fins on the heat exchangers at the back of a refrigerator are painted black so that they lose heat more quickly. By contrast, saucepans that are polished are poor emitters and keep their heat longer.

In general, surfaces that are good absorbers of radiation are good emitters when hot.

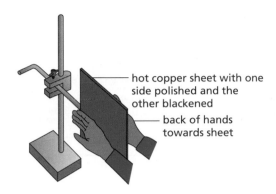

Figure 24.3 Comparing emitters of radiation

● Vacuum flask

A vacuum or Thermos flask keeps hot liquids hot or cold liquids cold. It is very difficult for heat to travel into or out of the flask.

Transfer by conduction and convection is minimised by making the flask a double-walled glass vessel with a vacuum between the walls (Figure 24.4). Radiation is reduced by silvering both walls on the vacuum side. Then if, for example, a hot liquid is stored, the small amount of radiation from the hot inside wall is reflected back across the vacuum by the silvering on the outer wall. The slight heat loss that does occur is by conduction up the thin glass walls and through the stopper.

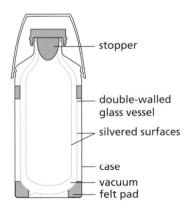

Figure 24.4 A vacuum flask

● The greenhouse

The warmth from the Sun is not cut off by a sheet of glass but the warmth from a red-hot fire can be blocked by glass. The radiation from very hot bodies like the Sun is mostly in the form of light and short-wavelength infrared. The radiation from less hot objects, like a fire, is largely long-wavelength infrared which, unlike light and short-wavelength infrared, cannot pass through glass.

Light and short-wavelength infrared from the Sun penetrate the glass of a greenhouse and are absorbed by the soil, plants, etc., raising their temperature. These in turn emit infrared but, because of their relatively low temperature, this has a long wavelength and is not transmitted by the glass. The greenhouse thus acts as a 'heat-trap' and its temperature rises.

Carbon dioxide and other gases such as methane in the Earth's atmosphere act in a similar way to the glass of a greenhouse in trapping heat; this has serious implications for the global climate.

● Rate of cooling of an object

The rate at which an object cools, i.e. at which its temperature falls, can be shown to be proportional to the ratio of its surface area A to its volume V.

For a cube of side l

$$\frac{A_1}{V_1} = 6 \times \frac{l^2}{l^3} = \frac{6}{l}$$

For a cube of side $2l$

$$\frac{A_2}{V_2} = 6 \times \frac{4l^2}{8l^3} = \frac{3}{l}$$

$$= \frac{1}{2} \times \frac{6}{l} = \frac{1}{2}\frac{A_1}{V_1}$$

The larger cube has the smaller A/V ratio and so cools more slowly.

▶▶

You could investigate this using two aluminium cubes, one having twice the length of side of the other. Each needs holes for a thermometer and an electric heater to raise them to the same starting temperature. Graphs of temperature against time for both blocks can then be obtained. It is important that the blocks are at the same starting temperature because the higher the temperature a body is above its surroundings, the greater the amount of radiation it emits per second and the faster it cools. This can be seen from its cooling curve since the gradient of the graph is steeper at high temperatures than it is at low temperatures (see Figure 22.2 where the ethanamide is cooling).

Checklist

After studying this chapter you should be able to

- describe the effect of surface colour and texture on the emission, absorption and reflection of radiation,
- describe experiments to study factors affecting the absorption and emission of radiation,
- recall that good absorbers are also good emitters,
- explain how a knowledge of heat transfer affects the design of a vacuum flask,
- explain how a greenhouse acts as a 'heat-trap'.

Questions

1 The door canopy in Figure 24.5 shows in a striking way the difference between white and black surfaces when radiation falls on them. Explain why.

Figure 24.5

2 a The Earth has been warmed by the radiation from the Sun for millions of years yet we think its average temperature has remained fairly steady. Why is this?
  b Why is frost less likely on a cloudy night than a clear one?

Section 3

Properties of waves

Chapters

General wave properties
25 Mechanical waves

Light
26 Light rays
27 Reflection of light

28 Plane mirrors
29 Refraction of light
30 Total internal reflection
31 Lenses
32 Electromagnetic radiation

Sound
33 Sound waves

- Types of wave
- Describing waves
- The wave equation
- Wavefronts and rays
- Reflection
- Refraction
- Diffraction
- Wave theory
- Interference
- Polarisation
- Practical work: The ripple tank

● Types of wave

Several kinds of wave occur in physics. **Mechanical waves** are produced by a disturbance, such as a vibrating object, in a material medium and are transmitted by the particles of the medium vibrating to and fro. Such waves can be seen or felt and include waves on a rope or spring, water waves and sound waves in air or in other materials.

A **progressive** or travelling wave is a disturbance which carries energy from one place to another without transferring matter. There are two types, **transverse** and **longitudinal**. Longitudinal waves are dealt with in Chapter 33.

In a transverse wave, the direction of the disturbance is at **right angles** to the direction of travel of the wave. A transverse wave can be sent along a rope (or a spring) by fixing one end and moving the other rapidly up and down (Figure 25.1). The disturbance generated by the hand is passed on from one part of the rope to the next which performs the same motion but slightly later. The humps and hollows of the wave travel along the rope as each part of the rope vibrates transversely about its undisturbed position.

Water waves are transverse waves.

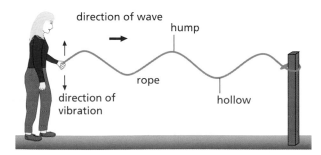

Figure 25.1 A transverse wave

● Describing waves

Terms used to describe waves can be explained with the aid of a **displacement–distance graph** (Figure 25.2). It shows, at a certain instant of time, the distance moved (sideways from their undisturbed positions) by the parts of the medium vibrating at different distances from the cause of the wave.

a) Wavelength

The **wavelength** of a wave, represented by the Greek letter λ ('lambda'), is the distance between successive crests.

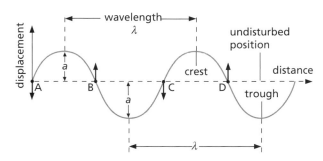

Figure 25.2 Displacement–distance graph for a wave at a particular instant

b) Frequency

The **frequency** f is the number of complete waves generated per second. If the end of a rope is moved up and down twice in a second, two waves are produced in this time. The frequency of the wave is 2 vibrations per second or 2 hertz (2 Hz; the **hertz** being the unit of frequency), which is the same as the frequency of the movement of the end of the rope. That is, the frequencies of the wave and its source are equal.

The frequency of a wave is also the number of crests passing a chosen point per second.

c) Speed

The **speed** v of the wave is the distance moved in the direction of travel of the wave by a crest or any point on the wave in 1 second.

d) Amplitude

The **amplitude** a is the height of a crest or the depth of a trough measured from the undisturbed position of what is carrying the wave, such as a rope.

e) Phase

The short arrows at A, B, C, D on Figure 25.2 show the directions of vibration of the parts of the rope at these points. The parts at A and C have the same speed in the same direction and are **in phase**. At B and D the parts are also in phase with each other but they are **out of phase** with those at A and C because their directions of vibration are opposite.

● The wave equation

The faster the end of a rope is vibrated, the shorter the wavelength of the wave produced. That is, the higher the frequency of a wave, the smaller its wavelength. There is a useful connection between f, λ and v, which is true for all types of wave.

Suppose waves of wavelength $\lambda = 20$ cm travel on a long rope and three crests pass a certain point every second. The frequency $f = 3$ Hz. If Figure 25.3 represents this wave motion then, if crest A is at P at a particular time, 1 second later it will be at Q, a distance from P of three wavelengths, i.e. $3 \times 20 = 60$ cm. The speed of the wave is $v = 60$ cm per second (60 cm/s), obtained by multiplying f by λ. Hence

speed of wave = frequency × wavelength

or

$$v = f\lambda$$

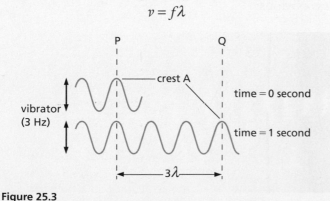

Figure 25.3

The ripple tank

The behaviour of water waves can be studied in a ripple tank. It consists of a transparent tray containing water, having a light source above and a white screen below to receive the wave images (Figure 25.4).

Pulses (i.e. short bursts) of ripples are obtained by dipping a finger in the water for circular ripples and a ruler for straight ripples. **Continuous ripples** are generated using an electric motor and a bar. The bar gives straight ripples if it just touches the water or circular ripples if it is raised and has a small ball fitted to it.

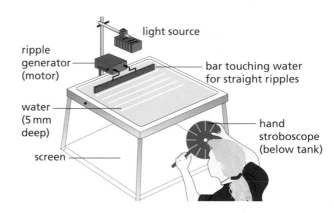

Figure 25.4 A ripple tank

Continuous ripples are studied more easily if they are *apparently* stopped ('frozen') by viewing the screen through a disc with equally spaced slits, which can be spun by hand, i.e. a **stroboscope**. If the disc speed is such that the waves have advanced one wavelength each time a slit passes your eye, they appear at rest.

● Wavefronts and rays

In two dimensions, a **wavefront** is a line on which the disturbance has the same phase at all points; the **crests of waves** in a ripple tank can be thought of as wavefronts. A vibrating source produces a succession of wavefronts, all of the same shape. In a ripple tank, straight wavefronts are produced by a vibrating bar (a line source) and circular wavefronts are produced by a vibrating ball (a point source). A line drawn at right angles to a wavefront, which shows its direction of travel, is called a **ray**. Straight wavefronts and the corresponding rays are shown in Figure 25.5; circular wavefronts can be seen in Figure 25.13.

● Reflection

In Figure 25.5 **straight** water waves are falling on a metal strip placed in a ripple tank at an angle of 60°, i.e. the angle *i* between the direction of travel of the waves and the normal to the strip is 60°, as is the angle between the wavefront and the strip. (The perpendicular to the strip at the point where the incident ray strikes is called the **normal**.) The wavefronts are represented by straight lines and can be thought of as the crests of the waves. They are at right angles to the direction of travel, i.e. to the rays. The angle of **reflection** *r* is 60°. Incidence at other angles shows that the **angle of incidence and angle of reflection are always equal.**

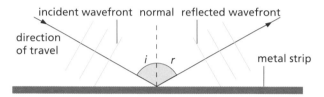

Figure 25.5 Reflection of waves

● Refraction

If a glass plate is placed in a ripple tank so that the water over the glass plate is about 1 mm deep but is 5 mm deep elsewhere, continuous straight waves in the shallow region are found to have a shorter wavelength than those in the deeper parts, i.e. the wavefronts are closer together (Figure 25.6). Both sets of waves have the frequency of the vibrating bar and, since $v = f\lambda$, if λ has decreased so has v, since f is fixed. Hence **waves travel more slowly in shallow water.**

Figure 25.6 Waves in shallower water have a shorter wavelength.

When the plate is at an angle to the waves (Figure 25.7a), their direction of travel in the shallow region is bent towards the normal (Figure 25.7b).

The change in the direction of travel of the waves, which occurs when their speed and hence wavelength changes, is termed **refraction**.

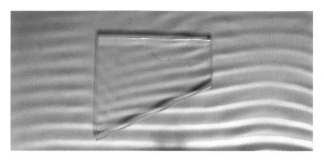

Figure 25.7a Waves are refracted at the boundary between deep and shallow regions.

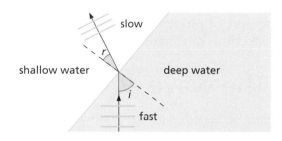

Figure 25.7b The direction of travel is bent towards the normal in the shallow region.

● Diffraction

In Figures 25.8a and 25.8b, straight water waves in a ripple tank are meeting gaps formed by obstacles. In Figure 25.8a the gap width is about the same as the wavelength of the waves (1 cm); the wavefronts that pass through become circular and spread out in all directions. In Figure 25.8b the gap is wide (10 cm) compared with the wavelength and the waves continue straight on; some spreading occurs but it is less obvious.

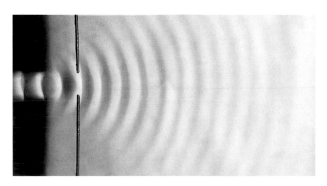

Figure 25.8a Spreading of waves after passing through a narrow gap

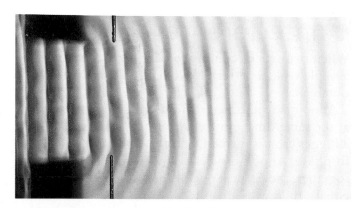

Figure 25.8b Spreading of waves after passing through a wide gap

Figure 25.9 Model of a harbour used to study wave behaviour

The spreading of waves at the edges of obstacles is called **diffraction**; when designing harbours, engineers use models like that in Figure 25.9 to study it.

● Wave theory

If the position of a wavefront is known at one instant, its position at a later time can be found using **Huygens' construction**. Each point on the wavefront is considered to be a source of **secondary** spherical wavelets (Figure 25.10) which spread out at the wave speed; the new wavefront is the surface that touches all the wavelets (in the forward direction). In Figure 25.10; the straight wavefront AB is travelling from left to right with speed v. At a time t later, the spherical wavelets from AB will be a distance vt from the secondary sources and the new surface which touches all the wavelets is the straight wavefront CD.

Wave theory can be used to explain reflection, refraction and diffraction effects.

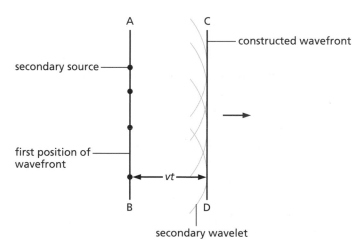

Figure 25.10 Huygens' construction for a straight wavefront

a) Reflection and wave theory

Figure 25.11 shows a straight wavefront AB incident at an angle i on a reflecting surface; the

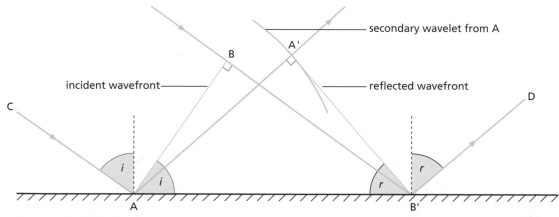

Figure 25.11 Reflection of a straight wavefront

wavefront has just reached the surface at A. The position of the wavefront a little later, when B reaches the reflecting surface, can be found using Huygens' construction. A circle of radius BB′ is drawn about A; the reflected wavefront is then A′B′, the tangent to the wavelet from B′. Measurements of the angle of incidence i and the angle of reflection r show that they are equal.

b) Refraction and wave theory

Huygens' construction can also be used to find the position of a wavefront when it enters a second medium in which the speed of travel of the wave is different from in the first (see Figure 25.7b).

In Figure 25.12, point A, on the straight wavefront AB, has just reached the boundary between two media. When B reaches the boundary the secondary wavelet from A will have moved on to A′. If the wave travels more slowly in the second medium the distance AA′ is shorter than BB′. The new wavefront is then A′B′, the tangent to the wavelet from B′; it is clear that the direction of travel of the wave has changed – refraction has occurred.

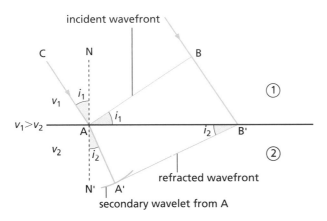

Figure 25.12 Refraction of a straight wavefront

c) Diffraction and wave theory

Diffraction effects, such as those shown in Figure 25.8, can be explained by describing what happens to secondary wavelets arising from point sources on the unrestricted part of the wavefronts in the gaps.

● Interference

When two sets of continuous circular waves cross in a ripple tank, a pattern like that in Figure 25.13 is obtained.

At points where a crest from one source, S_1, arrives at the same time as a crest from the other source, S_2, a bigger crest is formed, and the waves are said to be **in phase**. At points where a crest and a trough arrive together, they cancel out (if their amplitudes are equal); the waves are exactly **out of phase** (because they have travelled different distances from S_1 and S_2) and the water is undisturbed; in Figure 25.13 the blurred lines radiating from between S_1 and S_2 join such points.

Figure 25.13 Interference of circular waves

Interference or **superposition** is the combination of waves to give a larger or a smaller wave (Figure 25.14).

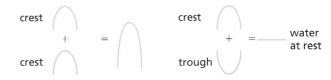

Figure 25.14

Study this effect with two ball 'dippers' about 3 cm apart on the bar of a ripple tank. Also observe the effect of changing **(i)** the frequency and **(ii)** the separation of the dippers; use a stroboscope when necessary. You will find that if the frequency is increased, i.e. the wavelength decreased, the blurred lines are closer together. Increasing the separation has the same effect.

Similar patterns are obtained if straight waves fall on two small gaps: interference occurs between the sets of emerging (circular) diffracted waves.

● Polarisation

This effect occurs only with transverse waves. It can be shown by fixing a rope at one end, D in Figure 25.15, and passing it through two slits, B and C. If end A is vibrated in all directions (as shown by the short arrowed lines), vibrations of the rope occur in every plane and transverse waves travel towards B.

At B only waves due to vibrations in a vertical plane can emerge from the vertical slit. The wave between B and C is said to be **plane polarised** (in the vertical plane containing the slit at B). By contrast the waves between A and B are unpolarised. If the slit at C is vertical, the wave travels on, but if it is horizontal as shown, the wave is stopped and the slits are said to be 'crossed'.

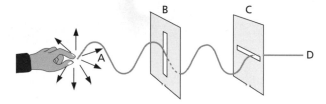

Figure 25.15 Polarising waves on a rope

Questions

1 The lines in Figure 25.16 are crests of straight ripples.
   a What is the wavelength of the ripples?
   b If 5 seconds ago ripple A occupied the position now occupied by ripple F, what is the frequency of the ripples?
   c What is the speed of the ripples?

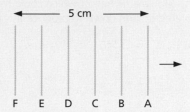

Figure 25.16

2 During the refraction of a wave which two of the following properties change?
   A the speed
   B the frequency
   C the wavelength

3 One side of a ripple tank ABCD is raised slightly (Figure 25.17), and a ripple started at P by a finger. After 1 second the shape of the ripple is as shown.
   a Why is it not circular?
   b Which side of the tank has been raised?

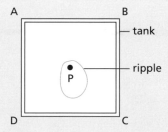

Figure 25.17

▶▶

4 Figure 25.18 gives a full-scale representation of the water in a ripple tank 1 second after the vibrator was started. The coloured lines represent crests.

a What is represented at A at this instant?

b Estimate
  (i) the wavelength,
  (ii) the speed of the waves, and
  (iii) the frequency of the vibrator.

c Sketch a suitable attachment which could have been vibrated up and down to produce this wave pattern.

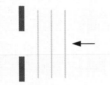

Figure 25.18

5 Copy Figure 25.19 and show on it what happens to the waves as they pass through the gap, if the water is much shallower on the right-hand side than on the left.

Figure 25.19

Checklist

After studying this chapter you should be able to

- describe the production of pulses and progressive transverse waves on ropes, springs and ripple tanks,
- recall the meaning of wavelength, frequency, speed and amplitude,
- represent a transverse wave on a displacement–distance graph and extract information from it,
- recall the wave equation $v = f\lambda$ and use it to solve problems,

- use the term wavefront,
- recall that the angle of reflection equals the angle of incidence and draw a diagram for the reflection of straight wavefronts at a plane surface,
- describe experiments to show reflection and refraction of waves,
- recall that refraction at a straight boundary is due to change of wave speed but not of frequency,
- draw a diagram for the refraction of straight wavefronts at a straight boundary,
- explain the term diffraction,
- describe experiments to show diffraction of waves,
- draw diagrams for the diffraction of straight wavefronts at single slits of different widths,

- predict the effect of changing the wavelength or the size of the gap on diffraction of waves at a single slit.

26 Light rays

- Sources of light
- Rays and beams
- Shadows

- Speed of light
- Practical work: The pinhole camera

● Sources of light

You can see an object only if light from it enters your eyes. Some objects such as the Sun, electric lamps and candles make their own light. We call these **luminous** sources.

Most things you see do not make their own light but reflect it from a luminous source. They are **non-luminous** objects. This page, you and the Moon are examples. Figure 26.1 shows some others.

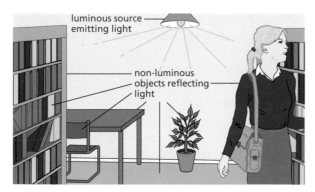

Figure 26.1 Luminous and non-luminous objects

Luminous sources radiate light when their atoms become 'excited' as a result of receiving energy. In a light bulb, for example, the energy comes from electricity. The 'excited' atoms give off their light haphazardly in most luminous sources.

A light source that works differently is the **laser**, invented in 1960. In laser light sources the excited atoms act together and emit a narrow, very bright beam of light. The laser has a host of applications. It is used in industry to cut through plate metal, in scanners to read bar codes at shop-and library check-outs, in CD players, in optical fibre telecommunication systems, in delicate medical operations on the eye or inner ear (for example Figure 26.2), in printing and in surveying and range-finding.

Figure 26.2 Laser surgery in the inner ear

● Rays and beams

Sunbeams streaming through trees (Figure 26.3) and light from a cinema projector on its way to the screen both suggest that light travels in straight lines. The beams are visible because dust particles in the air reflect light into our eyes.

The direction of the path in which light is travelling is called a **ray** and is represented in diagrams by a straight line with an arrow on it. A beam is a stream of light and is shown by a number of rays, as in Figure 26.4. A beam may be **parallel**, **diverging** (spreading out) or **converging** (getting narrower).

Figure 26.3 Light travels in straight lines

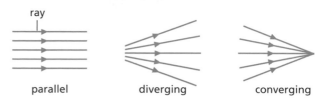

ray

parallel   diverging   converging

Figure 26.4 Beams of light

Practical work

The pinhole camera

A simple pinhole camera is shown in Figure 26.5a. Make a small pinhole in the centre of the black paper. Half darken the room. Hold the box at arm's length so that the pinhole end is nearer to and about 1 metre from a luminous object, such as a carbon filament lamp or a candle. Look at the **image** on the screen (an image is a likeness of an object and need not be an exact copy).

Can you see *three* ways in which the image differs from the object? What is the effect of moving the camera closer to the object?

Make the pinhole larger. What happens to the

(i) brightness,
(ii) sharpness,
(iii) size of the image?

Make several small pinholes round the large hole (Figure 26.5b), and view the image again.

The formation of an image is shown in Figure 26.6.

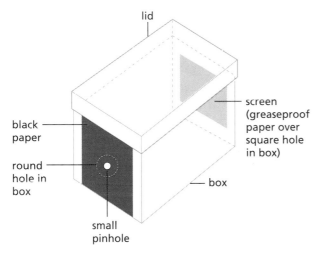

a **A pinhole camera**

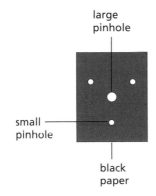

b

Figure 26.5

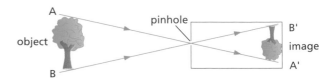

Figure 26.6 Forming an image in a pinhole camera

● Shadows

Shadows are formed for two reasons. First, because some objects, which are said to be **opaque**, do not allow light to pass through them. Secondly, light travels in straight lines.

The sharpness of the shadow depends on the size of the light source. A very small source of light, called a **point source**, gives a sharp shadow which is equally dark all over. This may be shown as in Figure 26.7a where the small hole in the card acts as a point source.

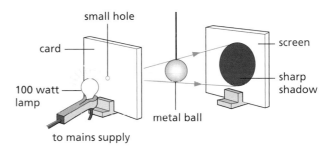

a **With a point source**

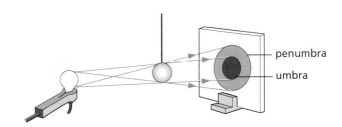

b **With an extended source**

Figure 26.7 Forming a shadow

If the card is removed the lamp acts as a large or **extended source** (Figure 26.7b). The shadow is then larger and has a central dark region, the **umbra**, surrounded by a ring of partial shadow, the **penumbra**. You can see by the rays that some light reaches the penumbra but none reaches the umbra.

● Speed of light

Proof that light travels very much faster than sound is provided by a thunderstorm. The flash of lightning is seen before the thunder is heard. The length of the time lapse is greater the further the observer is from the storm.

The speed of light has a definite value; light does not travel instantaneously from one point to another but takes a certain, very small time. Its speed is about 1 million times greater than that of sound.

Questions

1 How would the size and brightness of the image formed by a pinhole camera change if the camera were made longer?
2 What changes would occur in the image if the single pinhole in a camera were replaced by
a four pinholes close together,
b a hole 1 cm wide?
3 In Figure 26.8 the completely dark region on the screen is
A PQ   B PR   C QR   D QS   E RS

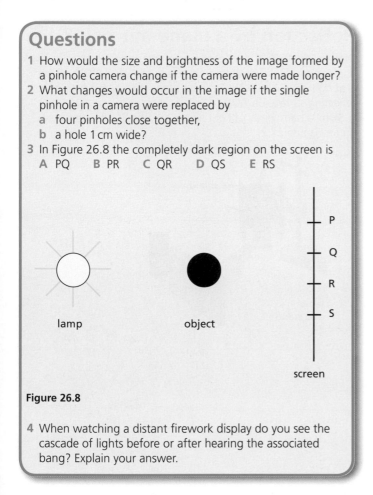

Figure 26.8

4 When watching a distant firework display do you see the cascade of lights before or after hearing the associated bang? Explain your answer.

- ● Law of reflection
- ● Periscope

- ● Regular and diffuse reflection
- ● Practical work: Reflection by a plane mirror

If we know how light behaves when it is reflected, we can use a mirror to change the direction in which the light is travelling. This happens when a mirror is placed at the entrance of a concealed drive to give warning of approaching traffic.

An ordinary mirror is made by depositing a thin layer of silver on one side of a piece of glass and protecting it with paint. The silver – at the *back* of the glass – acts as the reflecting surface.

● Law of reflection

Terms used in connection with reflection are shown in Figure 27.1. The perpendicular to the mirror at the point where the incident ray strikes it is called the **normal**. Note that the **angle of incidence** i is the angle between the incident ray and the normal; similarly the **angle of reflection** r is the angle between the reflected ray and the normal.

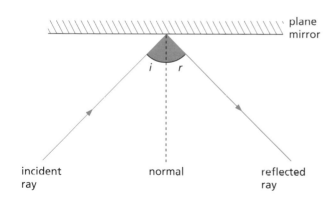

Figure 27.1 Reflection of light by a plane mirror

The law of reflection states:

The angle of incidence equals the angle of reflection.

The incident ray, the reflected ray and the normal all lie in the same plane. (This means that they could all be drawn on a flat sheet of paper.)

Practical work

Reflection by a plane mirror

Draw a line AOB on a sheet of paper and using a protractor mark angles on it. Measure them from the perpendicular ON, which is at right angles to AOB. Set up a plane (flat) mirror with its reflecting surface on AOB.

Shine a narrow ray of light along, say, the 30° line onto the mirror (Figure 27.2).

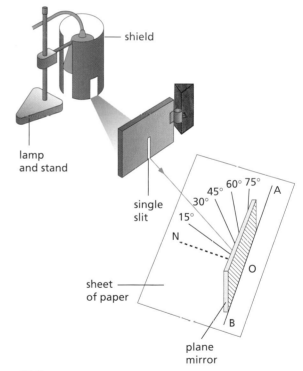

Figure 27.2

Mark the position of the reflected ray, remove the mirror and measure the angle between the reflected ray and ON. Repeat for rays at other angles. What can you conclude?

● Periscope

A simple **periscope** consists of a tube containing two plane mirrors, fixed parallel to and facing each other. Each makes an angle of 45° with the line joining them (Figure 27.3). Light from the object is turned through 90° at each reflection and an observer is able to see over a crowd, for example (Figure 27.4), or over the top of an obstacle.

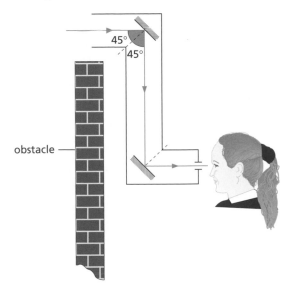

Figure 27.3 Action of a simple periscope

Figure 27.4 Periscopes being used by people in a crowd.

In more elaborate periscopes like those used in submarines, prisms replace mirrors (see Chapter 30).

Make your own periscope from a long, narrow cardboard box measuring about 40 cm × 5 cm × 5 cm (Such as one in which aluminium cooking foil or clingfilm is sold), two plane mirrors (7.5 cm × 5 cm) and sticky tape. When you have got it to work, make modifications that turn it into a 'see-back-o-scope', which lets you see what is behind you.

● Regular and diffuse reflection

If a parallel beam of light falls on a plane mirror it is reflected as a parallel beam (Figure 27.5a) and **regular reflection** occurs. Most surfaces, however, reflect light irregularly and the rays in an incident parallel beam are reflected in many directions (Figure 27.5b).

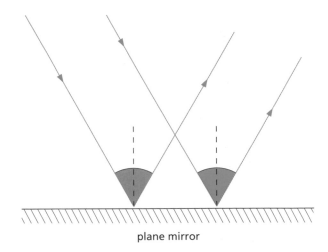

a Regular reflection

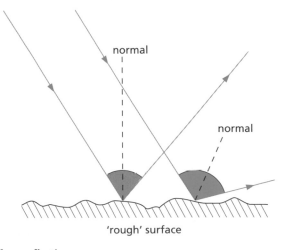

b Diffuse reflection

Figure 27.5

Irregular or **diffuse reflection** happens because, unlike a mirror, the surface of an object is not perfectly smooth. At each point on the surface the laws of reflection are obeyed but the angle of incidence, and so the angle of reflection, varies from point to point. The reflected rays are scattered haphazardly. Most objects, being rough, are seen by diffuse reflection.

Checklist

After studying this chapter you should be able to

- state the law of reflection and use it to solve problems,
- describe an experiment to show that the angle of incidence equals the angle of reflection,
- draw a ray diagram to show how a periscope works.

Questions

1 Figure 27.6 shows a ray of light PQ striking a mirror AB. The mirror AB and the mirror CD are at right angles to each other. QN is a normal to the mirror AB.

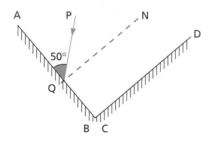

Figure 27.6

  a What is the value of the angle of incidence of the ray PQ on the mirror AB?
  b Copy the diagram, and continue the ray PQ to show the path it takes after reflection at both mirrors.
  c What are the values of the angle of reflection at AB, the angle of incidence at CD and the angle of reflection at CD?
  d What do you notice about the path of the ray PQ and the final reflected ray?
2 A ray of light strikes a plane mirror at an angle of incidence of 60°, is reflected from the mirror and then strikes a second plane mirror placed so that the angle between the mirrors is 45°. The angle of reflection at the second mirror, in degrees, is
    A 15   B 25   C 45   D 65   E 75
3 A person stands in front of a mirror (Figure 27.7). How much of the mirror is used to see from eye to toes?

Figure 27.7

Plane mirrors

● Real and virtual images
● Lateral inversion
● Properties of the image

● Kaleidoscope
● Practical work: Position of the image

When you look into a plane mirror on the wall of a room you see an image of the room behind the mirror; it is as if there were another room. Restaurants sometimes have a large mirror on one wall just to make them look larger. You may be able to say how much larger after the next experiment.

The position of the image formed by a mirror depends on the position of the object.

Practical work

Position of the image

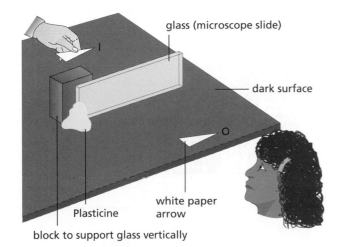

Figure 28.1

Support a piece of thin glass on the bench, as in Figure 28.1. It must be *vertical* (at 90° to the bench). Place a small paper arrow, O, about 10 cm from the glass. The glass acts as a poor mirror and an image of O will be seen in it; the darker the bench top, the brighter the image will be.

Lay another identical arrow, I, on the bench behind the glass; move it until it coincides with the image of O. How do the sizes of O and its image compare? Imagine a line joining them. What can you say about it? Measure the distances of the points of O and I from the glass along the line joining them. How do they compare? Try placing O at other distances.

● Real and virtual images

A **real image** is one which can be produced on a screen (as in a pinhole camera) and is formed by rays that actually pass through it.

A **virtual image** cannot be formed on a screen and is produced by rays which seem to come from it but do not pass through it. The image in a plane mirror is virtual. Rays from a point on an object are reflected at the mirror and appear to our eyes to come from a point behind the mirror where the rays would intersect when produced backwards (Figure 28.2). IA and IB are construction lines and are shown as broken lines.

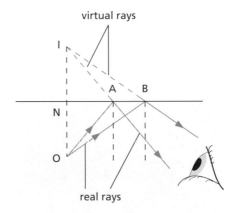

Figure 28.2 A plane mirror forms a virtual image.

● Lateral inversion

If you close your left eye, your image in a plane mirror seems to close the right eye. In a mirror image, left and right are interchanged and the image appears to be **laterally inverted**. The effect occurs whenever an image is formed by one reflection and is very evident if print is viewed in a mirror (Figure 28.3). What happens if two reflections occur, as in a periscope?

Figure 28.3 The image in a plane mirror is laterally inverted.

● Properties of the image

The image in a plane mirror is

(i) as far behind the mirror as the object is in front, with the line joining the object and image being perpendicular to the mirror,
(ii) the same size as the object,
(iii) virtual,
(iv) laterally inverted.

● Kaleidoscope

Figure 28.4 A kaleidoscope produces patterns using the images formed by two plane mirrors – the patterns change as the object moves.

To see how a kaleidoscope works, draw on a sheet of paper two lines at right angles to one another. Using different coloured pens or pencils, draw a design between them (Figure 28.5a). Place a small mirror along each line and look into the mirrors (Figure 28.5b). You will see three reflections which join up to give a circular pattern. If you make the angle between the mirrors smaller, more reflections appear but you always get a complete design.

In a kaleidoscope the two mirrors are usually kept at the same angle (about 60°) and different designs are made by hundreds of tiny coloured beads which can be moved around between the mirrors.

Now make a kaleidoscope using a cardboard tube (from half a kitchen roll), some thin card, grease-proof paper, clear sticky tape, small pieces of different coloured cellophane and two mirrors (10 cm × 3 cm) or a single plastic mirror (10 cm × 6 cm) bent to form two mirrors at 60° to each other, as shown in Figure 28.5c.

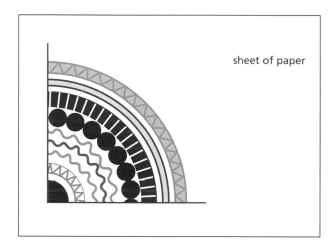

Figure 28.5a

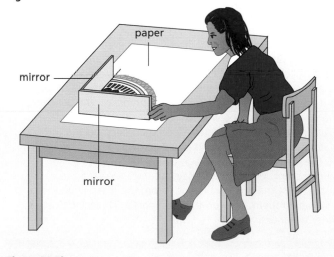

Figure 28.5b

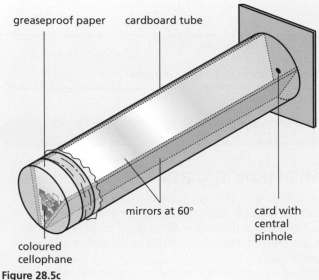

greaseproof paper   cardboard tube

mirrors at 60°

coloured
cellophane

card with
central
pinhole

Figure 28.5c

Questions

1 In Figure 28.6 at which of the points A to E will the
observer see the image in the plane mirror of the object?

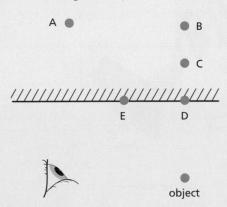

A ●   ● B

● C

E   D

object

Figure 28.6

2 Figure 28.7 shows the image in a plane mirror of a clock.
The correct time is
A 2.25   B 2.35   C 6.45   D 9.25

Figure 28.7

3 A girl stands 5 m away from a large plane mirror. How far
must she walk to be 2 m away from her image?
4 The image in a plane mirror is
A upright, real and larger
B upright, virtual and the same size
C inverted, real and smaller
D inverted, virtual and the same size
E inverted, real and larger.

Checklist

After studying this chapter you should be able to

• describe an experiment to show that the image in a plane
mirror is as far behind the mirror as the object is in front,
and that the line joining the object and image is at right
angles to the mirror,
• draw a diagram to explain the formation of a virtual image
by a plane mirror,
• explain the apparent lateral inversion of an image in a plane
mirror,
• explain how a kaleidoscope works.

(29) Refraction of light

- Facts about refraction
- Real and apparent depth
- Refractive index

- Refraction by a prism
- Dispersion
- Practical work: Refraction in glass

If you place a coin in an empty dish and move back until you *just* cannot see it, the result is surprising if someone *gently* pours in water. Try it.

Although light travels in straight lines in a transparent material, such as air, if it passes into a different material, such as water, it changes direction at the boundary between the two, i.e. it is bent. The **bending of light** when it passes from one material (called a **medium**) to another is called **refraction**. It causes effects such as the coin trick.

● Facts about refraction

(i) A ray of light is bent **towards** the normal when it enters an optically denser medium at an angle, for example from air to glass as in Figure 29.1a. The angle of refraction r is less than the angle of incidence i.

(ii) A ray of light is bent **away from** the normal when it enters an optically less dense medium, for example from glass to air.

(iii) A ray emerging from a parallel-sided block is **parallel** to the ray entering, but is **displaced sideways**, like the ray in Figure 29.1a.

(iv) A ray travelling along the normal direction at a boundary is **not refracted** (Figure 29.1b).

Note 'Optically denser' means having a greater refraction effect; the actual density may or may not be greater.

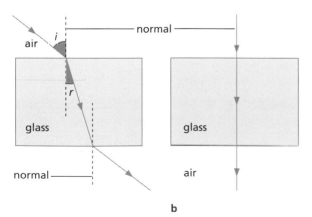

a   b

Figure 29.1 Refraction of light in glass

Practical work

Refraction in glass

Shine a ray of light at an angle on to a glass block (which has its lower face painted white or frosted), as in Figure 29.2. Draw the outline ABCD of the block on the sheet of paper under it. Mark the positions of the various rays in air and in glass.

Remove the block and draw the normals on the paper at the points where the ray enters side AB (see Figure 29.2) and where it leaves side CD.

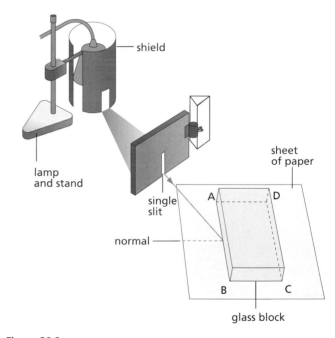

Figure 29.2

What *two* things happen to the light falling on AB? When the ray enters the glass at AB, is it bent towards or away from the part of the normal in the block? How is it bent at CD? What can you say about the direction of the ray falling on AB and the direction of the ray leaving CD?

What happens if the ray hits AB at right angles?

● Real and apparent depth

Rays of light from a point O on the bottom of a pool are refracted away from the normal at the water surface because they are passing into an optically less dense medium, i.e. air (Figure 29.3). On entering the eye they appear to come from a point I that is *above* O; I is the virtual image of O formed by refraction. The apparent depth of the pool is less than its real depth.

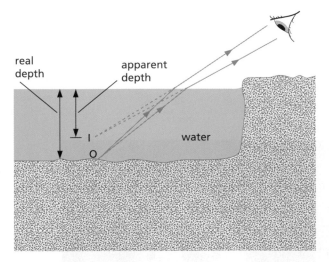

Figure 29.3 A pool of water appears shallower than it is.

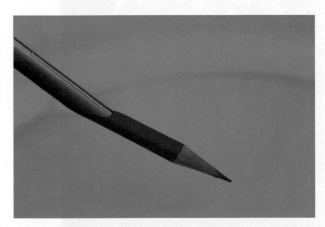

Figure 29.4 A pencil placed in water seems to bend at the water surface. Why?

● Refractive index

Light is refracted because its speed changes when it enters another medium. An analogy helps to explain why.

Suppose three people A, B, C are marching in line, with hands linked, on a good road surface. If they approach marshy ground at an angle (Figure 29.5a), person A is slowed down first, followed by B and then C. This causes the whole line to swing round and change its direction of motion.

In air (and a vacuum) light travels at 300 000 km/s (3×10^8 m/s); in glass its speed falls to 200 000 km/s (2×10^8 m/s) (Figure 29.5b). The **refractive index**, n, of a medium, in this case glass, is defined by the equation

$$\text{refractive index, } n = \frac{\text{speed of light in air (or a vacuum)}}{\text{speed of light in medium}}$$

$$\text{for glass,} \quad n = \frac{300\,000 \text{ km/s}}{200\,000 \text{ km/s}} = \frac{3}{2}$$

Experiments also show that

$$n = \frac{\text{sine of angle between ray in air and normal}}{\text{sine of angle between ray in glass and normal}}$$

$$= \frac{\sin i}{\sin r} \quad \text{(see Figure 29.1a)}$$

The more light is slowed down when it enters a medium from air, the greater is the refractive index of the medium and the more the light is bent.

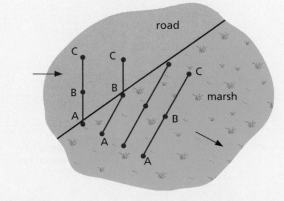

Figure 29.5a

123

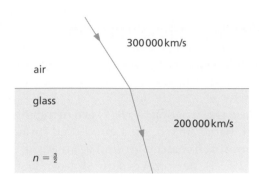

Figure 29.5b

We saw earlier (Chapter 25) that water waves are refracted when their speed changes. The change in the direction of travel of a light ray when its speed changes on entering another medium suggests that light may also be a type of wave motion.

● Refraction by a prism

In a triangular glass prism (Figure 29.6a), the bending of a ray due to refraction at the first surface is added to the bending of the ray at the second surface (Figure 29.6b); the overall change in direction of the ray is called the **deviation**.

The bendings of the ray do not cancel out as they do in a parallel-sided block where the emergent ray, although displaced, is parallel to the incident ray.

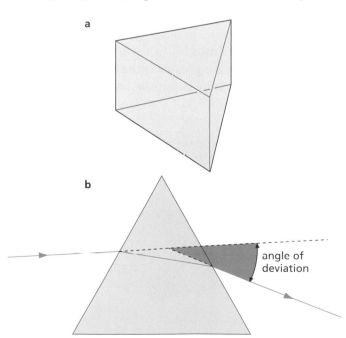

Figure 29.6

● Dispersion

When sunlight (white light) falls on a triangular glass prism (Figure 29.7a), a band of colours called a **spectrum** is obtained (Figure 29.7b). The effect is termed **dispersion**. It arises because white light is a mixture of many colours; the prism separates the colours because the refractive index of glass is different for each colour (it is greatest for violet light).

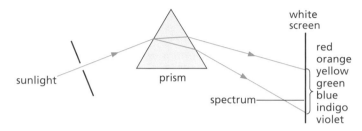

Figure 29.7a Forming a spectrum with a prism

Figure 29.7b White light shining through cut crystal can produce several spectra.

Questions

1 Figure 29.8 shows a ray of light entering a rectangular block of glass.
   a Copy the diagram and draw the normal at the point of entry.
   b Sketch the approximate path of the ray through the block and out of the other side.

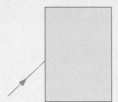

Figure 29.8

2 Draw two rays from a point on a fish in a stream to show where someone on the bank will see the fish. Where must the person aim to spear the fish?
3 What is the speed of light in a medium of refractive index 6/5 if its speed in air is 300 000 km/s?
4 Figure 29.9 shows a ray of light OP striking a glass prism and then passing through it. Which of the rays A to D is the correct representation of the emerging ray?

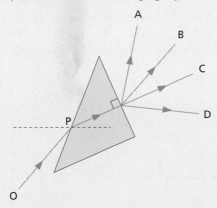

Figure 29.9

5 A beam of white light strikes the face of a prism. Copy Figure 29.10 and draw the path taken by red and blue rays of light as they pass through the prism and on to the screen AB.

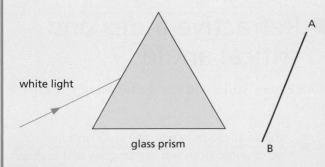

Figure 29.10

6 Which diagram in Figure 29.11 shows the ray of light refracted correctly?

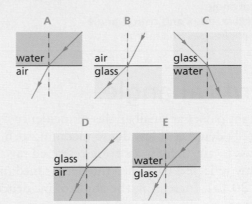

Figure 29.11

7 Which diagram in Figure 29.12 shows the correct path of the ray through the prism?

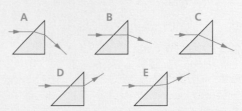

Figure 29.12

Checklist

After studying this chapter you should be able to

* state what the term refraction means,
* give examples of effects that show light can be refracted,
* describe an experiment to study refraction,
* draw diagrams of the passage of light rays through rectangular blocks and recall that lateral displacement occurs for a parallel-sided block,
* recall that light is refracted because it changes speed when it enters another medium,

* recall the definition of refractive index as $n =$ speed in air/speed in medium,

* recall and use the equation $n = \sin i/\sin r$

* draw a diagram for the passage of a light ray through a prism,
* explain the terms spectrum and dispersion,
* describe how a prism is used to produce a spectrum from white light.

30 Total internal reflection

- Critical angle
- Refractive index and critical angle
- Multiple images in a mirror

- Totally reflecting prisms
- Light pipes and optical fibres
- Practical work: Critical angle of glass

● Critical angle

When light passes at small angles of incidence from an optically dense to a less dense medium, such as from glass to air, there is a strong refracted ray and a weak ray reflected back into the denser medium (Figure 30.1a). Increasing the angle of incidence increases the angle of refraction.

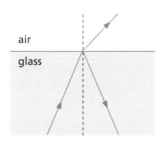

a

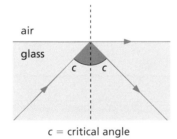

c = critical angle

b

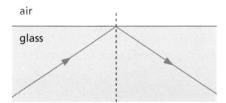

c

Figure 30.1

At a certain angle of incidence, called the **critical angle**, c, the angle of refraction is 90° (Figure 30.1b). For angles of incidence greater than c, the refracted ray disappears and all the incident light is reflected inside the denser medium (Figure 30.1c). The light does not cross the boundary and is said to undergo **total internal reflection**.

Critical angle of glass

Place a semicircular glass block on a sheet of paper (Figure 30.2), and draw the outline LOMN where O is the centre and ON the normal at O to LOM. Direct a narrow ray (at an angle of about 30° to the normal ON) along a radius towards O. The ray is not refracted at the curved surface. Why? Note the refracted ray emerging from LOM into the air and also the weak internally reflected ray in the glass.

Slowly rotate the paper so that the angle of incidence on LOM increases until total internal reflection *just* occurs. Mark the incident ray. Measure the angle of incidence; this is the critical angle.

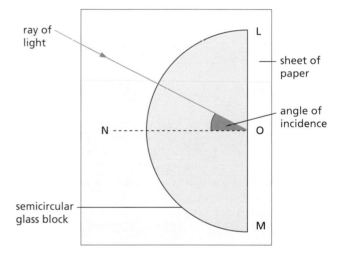

Figure 30.2

● Refractive index and critical angle

From Figure 30.1b and the definition of refractive index:

$$n = \frac{\text{sine of angle between ray in air and normal}}{\text{sine of angle between ray in glass and normal}}$$

126

$$= \frac{\sin 90°}{\sin c}$$

$$= \frac{1}{\sin c} \qquad \text{(because } \sin 90° = 1\text{)}$$

So, if $n = \frac{3}{2}$, then $\sin c = \frac{2}{3}$, and c must be 42°.

Worked example

If the critical angle for diamond is 24°, calculate its refractive index.

Critical angle, $c = 24°$
$$\sin 24° = 0.4$$

$$n = \frac{\sin 90°}{\sin c} = \frac{1}{\sin 24°}$$

$$= \frac{1}{0.4} = 2.5$$

Multiple images in a mirror

An ordinary mirror silvered at the back forms several images of one object, because of multiple reflections inside the glass (Figure 30.3a and Figure 30.3b). These blur the main image I (which is formed by one reflection at the silvering), especially if the glass is thick. The problem is absent in front-silvered mirrors but such mirrors are easily damaged.

Figure 30.3b The multiple images in a mirror cause blurring.

Totally reflecting prisms

The defects of mirrors are overcome if 45° right-angled glass prisms are used. The critical angle of ordinary glass is about 42° and a ray falling normally on face PQ of such a prism (Figure 30.4a) hits face PR at 45°. Total internal reflection occurs and the ray is turned through 90°. Totally reflecting prisms replace mirrors in good periscopes.

Light can also be reflected through 180° by a prism (Figure 30.4b); this happens in binoculars.

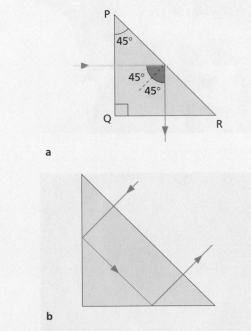

a

b

Figure 30.4 Reflection of light by a prism

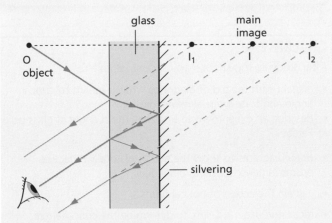

Figure 30.3a Multiple reflections in a mirror

● Light pipes and optical fibres

Light can be trapped by total internal reflection inside a bent glass rod and 'piped' along a curved path (Figure 30.5). A single, very thin glass fibre behaves in the same way.

Figure 30.5 Light travels through a curved glass rod or fibre by total internal reflection.

If several thousand such fibres are taped together, a flexible light pipe is obtained that can be used, for example, by doctors as an 'endoscope' (Figure 30.6a), to obtain an image from inside the body (Figure 30.6b), or by engineers to light up some awkward spot for inspection. The latest telephone 'cables' are optical (very pure glass) fibres carrying information as pulses of laser light.

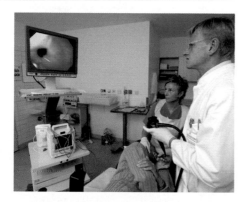

Figure 30.6a Endoscope in use

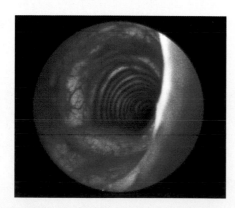

Figure 30.6b Trachea (windpipe) viewed by an endoscope

Questions

1 Figure 30.7 shows rays of light in a semicircular glass block.

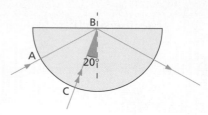

Figure 30.7

a Explain why the ray entering the glass at A is not bent as it enters.
b Explain why the ray AB is reflected at B and not refracted.
c Ray CB does not stop at B. Copy the diagram and draw its approximate path after it leaves B.
2 Copy Figures 30.8a and 30.8b and complete the paths of the rays through the glass prisms.

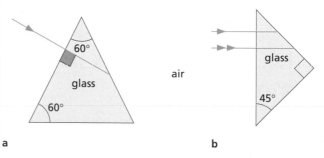

a   b

Figure 30.8

3 Name two instruments that use prisms to reflect light.
4 Light travels up through a pond of water of critical angle 49°. What happens at the surface if the angle of incidence is: **a** 30°; **b** 60°?
5 Calculate the critical angle for water if $n = \frac{4}{3}$.

Checklist

After studying this chapter you should be able to

- explain with the aid of diagrams what is meant by critical angle and total internal reflection,
- describe an experiment to find the critical angle of glass or Perspex,
- draw diagrams to show the action of totally reflecting prisms in periscopes and binoculars,
- explain the action of optical fibres,
- recall and use $n = 1/\sin c$ to determine the critical angle.

31 Lenses

- Converging and diverging lenses
- Principal focus
- Ray diagrams
- Magnification
- Power of a lens

- Magnifying glass
- Spectacles
- Practical work: Focal length, *f*, of a converging lens; Images formed by a converging lens

Converging and diverging lenses

Lenses are used in optical instruments such as cameras, spectacles, microscopes and telescopes; they often have spherical surfaces and there are two types. A **converging** (or convex) lens is thickest in the centre and bends light inwards (Figure 31.1a). You may have used one as a magnifying glass (Figure 31.2a) or as a burning glass. A **diverging** (or concave) lens is thinnest in the centre and spreads light out (Figure 31.1b); it always gives a diminished image (Figure 31.2b).

The centre of a lens is its **optical centre**, C; the line through C at right angles to the lens is the **principal axis**.

The action of a lens can be understood by treating it as a number of prisms (most with the tip removed), each of which bends the ray towards its base, as in Figure 31.1c and 31.1d. The centre acts as a parallel-sided block.

Figure 31.1d

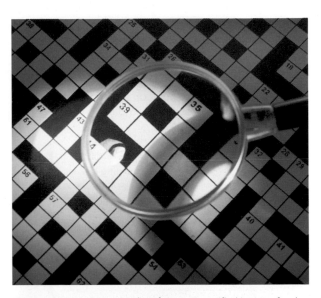

Figure 31.2a A converging lens forms a magnified image of a close object.

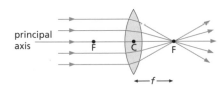

Figure 31.1a

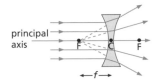

Figure 31.1b

Figure 31.1c

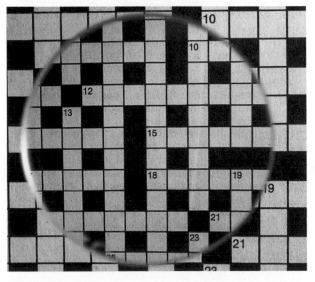

Figure 31.2b A diverging lens always forms a diminished image.

129

● Principal focus

When a beam of light parallel to the principal axis passes through a converging lens it is refracted so as to converge to a point on the axis called the **principal focus**, F. It is a real focus. A diverging lens has a virtual principal focus behind the lens, from which the refracted beam seems to diverge.

Since light can fall on both faces of a lens it has two principal foci, one on each side, equidistant from C. The distance CF is the **focal length** f of the lens (see Figure 31.1a); it is an important property of a lens. The more curved the lens faces are, the smaller is f and the more powerful is the lens.

Practical work

Focal length, f, of a converging lens

We use the fact that rays from a point on a very distant object, i.e. at infinity, are nearly parallel (Figure 31.3a).

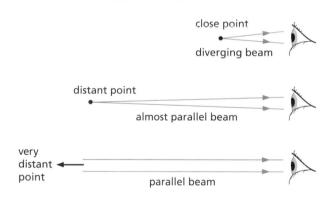

Figure 31.3a

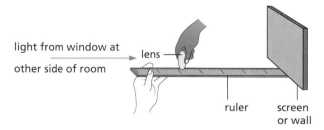

Figure 31.3b

Move the lens, arranged as in Figure 31.3b, until a **sharp** image of a window at the other side of the room is obtained on the screen. The distance between the lens and the screen is then f, roughly. Why?

Practical work

Images formed by a converging lens

In the formation of images by lenses, two important points on the principal axis are F and 2F; 2F is at a distance of twice the focal length from C.

First find the focal length of the lens by the 'distant object method' just described, then fix the lens upright with Plasticine at the centre of a metre rule. Place small pieces of Plasticine at the points F and 2F on both sides of the lens, as in Figure 31.4.

Place a small light source, such as a torch bulb, as the object supported on the rule beyond 2F and move a white card, on the other side of the lens from the light, until a sharp image is obtained on the card.

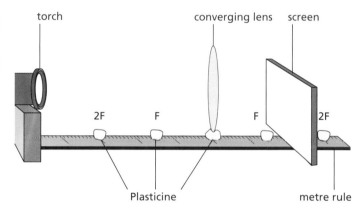

Figure 31.4

Note and record, in a table like the one below, the image position as 'beyond 2F', 'between 2F and F' or 'between F and lens'. Also note whether the image is 'larger' or 'smaller' than the actual bulb or 'same size' and if it is 'upright' or 'inverted'. Now repeat with the light at 2F, then between 2F and F.

Object position	Image position	Larger, smaller or same size?	Upright or inverted?
beyond 2F			
at 2F			
between 2F and F			
between F and lens			

So far all the images have been real since they can be obtained on a screen. When the light is between F and the lens, the image is **virtual** and is seen by *looking through the lens* at the light. Do this. Is the virtual image larger or smaller than the object? Is it upright or inverted? Record your findings in your table.

● Ray diagrams

Information about the images formed by a lens can be obtained by drawing two of the following rays.

1 A ray parallel to the principal axis which is refracted through the principal focus, F.
2 A ray through the optical centre, C, which is undeviated for a thin lens.
3 A ray through the principal focus, F, which is refracted parallel to the principal axis.

In diagrams a thin lens is represented by a straight line at which all the refraction is considered to occur.

In each ray diagram in Figure 31.5, two rays are drawn from the top A of an object OA. Where these rays intersect after refraction gives the top B of the image IB. The foot I of each image is on the axis since ray OC passes through the lens undeviated. In Figure 31.5d, the broken rays, and the image, are virtual. In all parts of Figure 31.5, the lens is a converging lens.

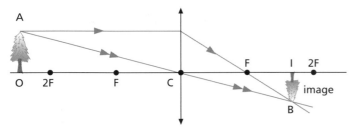

Image is between F and 2F, real, inverted, smaller

Figure 31.5a Object beyond 2F

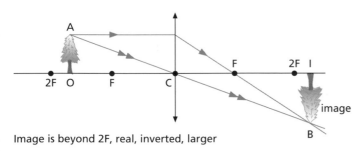

Image is beyond 2F, real, inverted, larger

Figure 31.5b Object at 2F

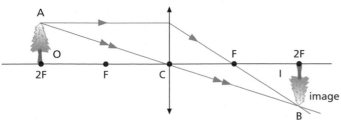

Image is at 2F, real, inverted, same size

Figure 31.5c Object between 2F and F

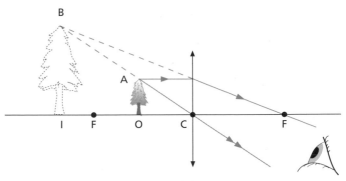

Image is behind object, virtual, erect, larger

Figure 31.5d Object between F and C

● Magnification

The **linear magnification**, m is given by

$$m = \frac{\text{height of image}}{\text{height of object}}$$

It can be shown that in all cases

$$m = \frac{\text{distance of image from lens}}{\text{distance of object from lens}}$$

● Power of a lens

The shorter the focal length of a lens, the stronger it is, i.e. the more it converges or diverges a beam of light. We define the **power**, P, of a lens to be 1/focal length of the lens, where the focal length is measured in metres:

$$P = \frac{1}{f}$$

● Magnifying glass

The apparent size of an object depends on its actual size and on its distance from the eye. The sleepers on a railway track are all the same length but those nearby seem longer. This is because they enclose a larger angle at your eye than more distant ones: their image on the retina is larger, so making them appear bigger.

▶▶

A converging lens gives an enlarged, upright virtual image of an object placed between the lens and its principal focus F (Figure 31.6a). It acts as a magnifying glass since the angle β made at the eye by the image, formed at the near point (see next section), is greater than the angle α made by the object when it is viewed directly at the near point without the magnifying glass (Figure 31.6b).

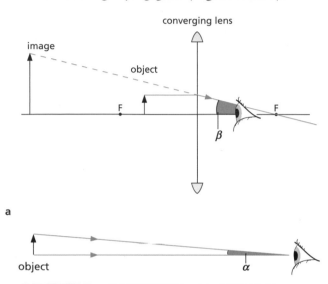

a

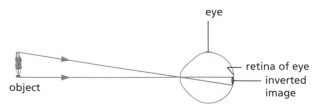

b

Figure 31.6 Magnification by a converging lens: angle β is larger than angle α

The fatter (more curved) a converging lens is, the shorter its focal length and the more it magnifies. Too much curvature, however, distorts the image.

● Spectacles

From the ray diagrams shown in Figure 31.5 (p. 131) we would expect that the converging lens in the eye will form a real **inverted** image on the retina as shown in Figure 31.7. Since an object normally appears upright, the brain must invert the image.

Figure 31.7 Inverted image on the retina

The average adult eye can focus objects comfortably from about 25 cm (the **near point**) to infinity (the **far point**). Your near point may be less than 25 cm; it gets further away with age.

a) Short sight

A **short-sighted** person sees near objects clearly but distant objects appear blurred. The image of a distant object is formed in front of the retina because the eyeball is too long or because the eye lens cannot be made thin enough (Figure 31.8a). The problem is corrected by a diverging spectacle lens (or contact lens) which diverges the light before it enters the eye, to give an image on the retina (Figure 31.8b).

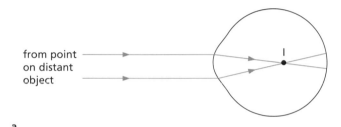

a

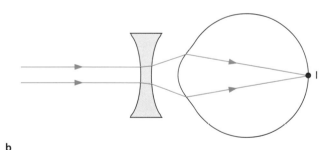

b

Figure 31.8 Short sight and its correction by a diverging lens

b) Long sight

A **long-sighted** person sees distant objects clearly but close objects appear blurred. The image of a near object is focused behind the retina because the eyeball is too short or because the eye lens cannot be made thick enough (Figure 31.9a). A converging spectacle lens (or contact lens) corrects the problem (Figure 31.9b).

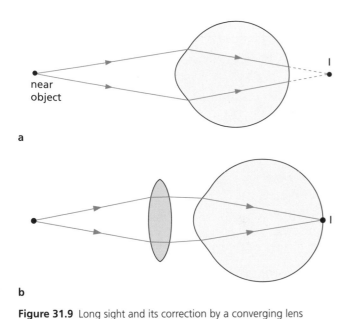

a

b

Figure 31.9 Long sight and its correction by a converging lens

Questions

1 A small torch bulb is placed at the focal point of a converging lens. When the bulb is switched on, does the lens produce a convergent, divergent or parallel beam of light?

2 a What kind of lens is shown in Figure 31.10?

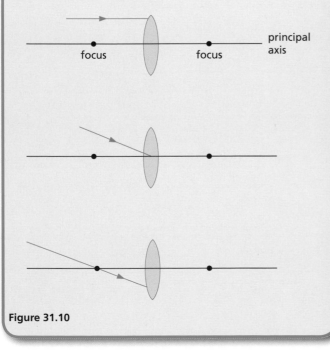

Figure 31.10

b Copy the diagrams and complete them to show the path of the light after passing through the lens.

c Figure 31.11 shows an object AB 6 cm high placed 18 cm in front of a lens of focal length 6 cm. Draw the diagram to scale and, by tracing the paths of rays from A, find the position and size of the image formed.

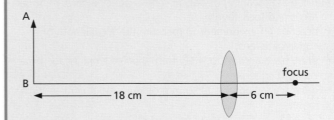

Figure 31.11

3 Where must the object be placed for the image formed by a converging lens to be
   a real, inverted and smaller than the object,
   b real, inverted and same size as the object,
   c real, inverted and larger than the object,
   d virtual, upright and larger than the object?

4 Figure 31.12 shows a camera focused on an object in the middle distance. Should the lens be moved towards or away from the film so that the image of a more distant object is in focus?

Figure 31.12

5 a Three converging lenses are available, having focal lengths of 4 cm, 40 cm and 4 m, respectively. Which one would you choose as a magnifying glass?

   b An object 2 cm high is viewed through a converging lens of focal length 8 cm. The object is 4 cm from the lens. By means of a ray diagram find the position, nature and magnification of the image.

6 An object is placed 10 cm in front of a lens, A; the details of the image are given below. The process is repeated for a different lens, B.
   Lens A Real, inverted, magnified and at a great distance.
   Lens B Real, inverted and same size as the object.
   Estimate the focal length of each lens and state whether it is converging or diverging.

Checklist

After studying this chapter you should be able to

- explain the action of a lens in terms of refraction by a number of small prisms,
- draw diagrams showing the effects of a converging lens on a beam of parallel rays,
- recall the meaning of optical centre, principal axis, principal focus and focal length,
- describe an experiment to measure the focal length of a converging lens,
- draw ray diagrams to show formation of a real image by a converging lens,
- draw scale diagrams to solve problems about converging lenses,

- draw ray diagrams to show formation of a virtual image by a single lens,

- show how a single lens is used as a magnifying glass.

32 Electromagnetic radiation

- Properties
- Light waves
- Infrared radiation
- Ultraviolet radiation

- Radio waves
- X-rays
- Practical work: Wave nature of microwaves

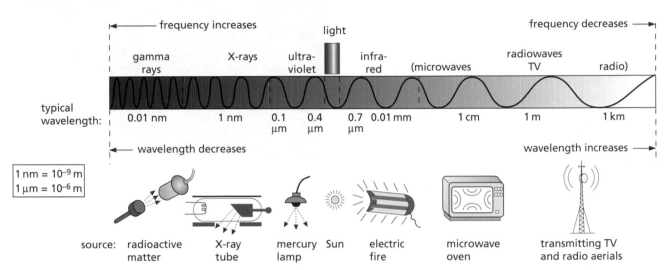

Figure 32.1 The electromagnetic spectrum and sources of each type of radiation

Light is one member of the family of electromagnetic radiation which forms a continuous spectrum beyond both ends of the visible (light) spectrum (Figure 32.1). While each type of radiation has a different source, all result from electrons in atoms undergoing an energy change and all have certain properties in common.

● Properties

1 All types of electromagnetic radiation **travel through a vacuum at 300 000 km/s** (3×10^8 m/s), i.e. with the speed of light.
2 They **exhibit interference, diffraction and polarisation,** which suggests they have a transverse wave nature.
3 They **obey the wave equation**, $v = f\lambda$, where v is the speed of light, f is the frequency of the waves and λ is the wavelength. Since v is constant for a particular medium, it follows that large f means small λ.
4 They **carry energy from one place to another and can be absorbed by matter to cause heating and other effects.** The higher the frequency and the smaller the wavelength of the radiation, the greater is the energy carried, i.e. gamma rays are more 'energetic' than radio waves. This is shown

by the **photoelectric effect** in which electrons are ejected from metal surfaces when electromagnetic waves fall on them. As the frequency of the waves increases so too does the speed (and energy) with which electrons are emitted.

Because of its electrical origin, its ability to travel in a vacuum (e.g. from the Sun to the Earth) and its wave-like properties (i.e. point 2 above), electromagnetic radiation is regarded as a **progressive transverse wave**. The wave is a combination of travelling electric and magnetic fields. The fields vary in value and are directed at right angles to each other and to the direction of travel of the wave, as shown by the representation in Figure 32.2.

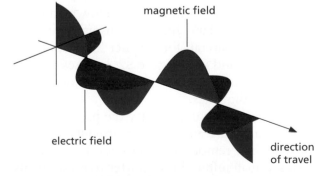

Figure 32.2 An electromagnetic wave

● Light waves

Red light has the longest wavelength, which is about 0.0007 mm (7×10^{-7} m = 0.7 μm), while violet light has the shortest wavelength of about 0.0004 mm (4×10^{-7} m = 0.4 μm). Colours between these in the spectrum of white light have intermediate values. Light of one colour and so of one wavelength is called **monochromatic** light.

Since $v = f\lambda$ for all waves including light, it follows that red light has a lower frequency, f, than violet light since (i) the wavelength, λ, of red light is greater, and (ii) all colours travel with the same speed, v, of 3×10^8 m/s in air (strictly, in a vacuum). It is the frequency of light which decides its colour, rather than its wavelength which is different in different media, as is the speed (Chapter 29).

Different frequencies of light travel at different speeds through a transparent medium and so are refracted by different amounts. This explains dispersion (Chapter 29), in other word why the refractive index of a material depends on the wavelength of the light.

The **amplitude** of a light (or any other) wave is greater the higher the **intensity** of the source; in the case of light the greater the intensity the brighter it is.

● Infrared radiation

Our bodies detect **infrared** radiation (IR) by its heating effect on the skin. It is sometimes called 'radiant heat' or 'heat radiation'.

Anything which is hot but not glowing, i.e. below 500 °C, emits IR alone. At about 500 °C a body becomes red hot and emits red light as well as IR – the heating element of an electric fire, a toaster or a grill are examples. At about 1500 °C, things such as lamp filaments are white hot and radiate IR and white light, i.e. all the colours of the visible spectrum.

Infrared is also detected by special temperature-sensitive photographic films which allow pictures to be taken in the dark. Infrared sensors are used on satellites and aircraft for weather forecasting, monitoring of land use (Figure 32.3), assessing heat loss from buildings, intruder alarms and locating victims of earthquakes.

Infrared lamps are used to dry the paint on cars during manufacture and in the treatment of muscular complaints. The remote control for an electronic device contains a small infrared transmitter to send signals to the device, such as a television or DVD player.

Figure 32.3 Infrared aerial photograph of Washington DC

● Ultraviolet radiation

Ultraviolet (UV) rays have shorter wavelengths than light. They cause sun tan and produce vitamins in the skin but can penetrate deeper, causing skin cancer. Dark skin is able to absorb more UV, so reducing the amount reaching deeper tissues. Exposure to the harmful UV rays present in sunlight can be reduced by wearing protective clothing such as a hat or by using sunscreen lotion.

Ultraviolet causes fluorescent paints and clothes washed in some detergents to fluoresce (Figure 32.4). They glow by re-radiating as light the energy they absorb as UV. This effect may be used to verify 'invisible' signatures on bank documents.

Figure 32.4 White clothes fluorescing in a club

A UV lamp used for scientific or medical purposes contains mercury vapour and this emits UV when an electric current passes through it. Fluorescent tubes also contain mercury vapour and their inner surfaces are coated with special powders called phosphors which radiate light.

● Radio waves

Radio waves have the longest wavelengths in the electromagnetic spectrum. They are radiated from aerials and used to 'carry' sound, pictures and other information over long distances.

a) Long, medium and short waves (wavelengths of 2 km to 10 m)

These diffract round obstacles so can be received even when hills are in their way (Figure 32.5a). They are also reflected by layers of electrically charged particles in the upper atmosphere (the **ionosphere**), which makes long-distance radio reception possible (Figure 32.5b).

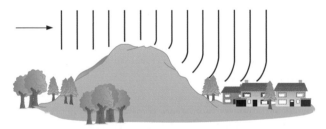

a Diffraction of radio waves

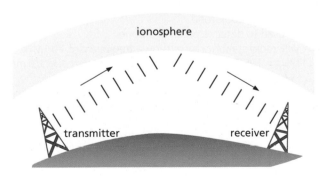

b Reflection of radio waves

Figure 32.5

b) VHF (very high frequency) and UHF (ultra high frequency) waves (wavelengths of 10 m to 10 cm)

These shorter wavelength radio waves need a clear, straight-line path to the receiver. They are not reflected by the ionosphere. They are used for local radio and for television.

c) Microwaves (wavelengths of a few cm)

These are used for international telecommunications and television relay via geostationary satellites and for mobile phone networks via microwave aerial towers and low-orbit satellites (Chapter 9). The microwave signals are transmitted through the ionosphere by dish aerials, amplified by the satellite and sent back to a dish aerial in another part of the world.

Microwaves are also used for **radar** detection of ships and aircraft, and in police speed traps.

Microwaves can be used for cooking since they cause water molecules in the moisture of the food to vibrate vigorously at the frequency of the microwaves. As a result, heating occurs inside the food which cooks itself.

Living cells can be damaged or killed by the heat produced when microwaves are absorbed by water in the cells. There is some debate at present as to whether their use in mobile phones is harmful; 'hands-free' mode, where separate earphones are used, may be safer.

Practical work

Wave nature of microwaves

The 3 cm microwave transmitter and receiver shown in Figure 32.6 can be used. The three metal plates can be set up for double-slit interference with 'slits' about 3 cm wide; the reading on the meter connected to the horn receiver rises and falls as it is moved across behind the plates.

Diffraction can be shown similarly using the two wide metal plates to form a single slit.

If the grid of vertical metal wires is placed in front of the transmitter, the signal is absorbed but transmission occurs when the wires are horizontal, showing that the microwaves are vertically polarised. This can also be shown by rotating the receiver through 90° in a vertical plane from the maximum signal position, when the signal decreases to a minimum.

▶▶

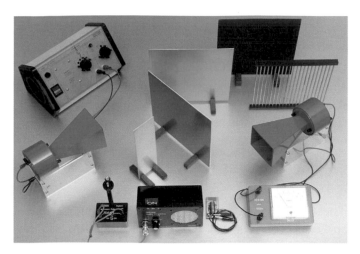

Figure 32.6

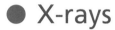

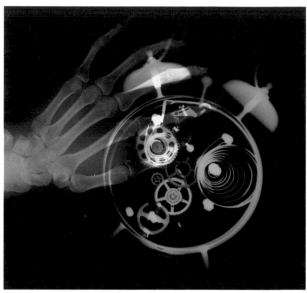

Figure 32.7 X-rays cannot penetrate bone and metal.

● X-rays

These are produced when high-speed electrons are stopped by a metal target in an X-ray tube. **X-rays** have smaller wavelengths than UV.

They are absorbed to some extent by living cells but can penetrate some solid objects and affect a photographic film. With materials like bones, teeth and metals which they do not pass through easily, shadow pictures can be taken, like that in Figure 32.7 of a hand on an alarm clock. They are widely used in dentistry and in medicine, for example to detect broken bones. X-rays are also used in security machines at airports for scanning luggage; some body scanners, now being introduced to screen passengers, use very low doses of X-rays. In industry X-ray photography is used to inspect welded joints.

X-ray machines need to be shielded with lead since normal body cells can be killed by high doses and made cancerous by lower doses.

Gamma rays (Chapter 49) are more penetrating and dangerous than X-rays. They are used to kill cancer cells and also harmful bacteria in food and on surgical instruments.

Questions

1 Give the approximate wavelength in micrometres (μm) of
   a red light,
   b violet light.
2 Which of the following types of radiation has
   a the longest wavelength,
   b the highest frequency?
      A UV
      B radio waves
      C light
      D X-rays
      E IR
3 Name one type of electromagnetic radiation which
   a causes sun tan,
   b is used for satellite communication,
   c is used to sterilise surgical instruments,
   d is used in a TV remote control,
   e is used to cook food,
   f is used to detect a break in a bone.
4 A VHF radio station transmits on a frequency of 100 MHz (1 MHz = 10^6 Hz). If the speed of radio waves is 3×10^8 m/s,
   a what is the wavelength of the waves,
   b how long does the transmission take to travel 60 km?

5 In the diagram in Figure 32.8 light waves are incident on an air–glass boundary. Some are reflected and some are refracted in the glass. One of the following is the same for the incident wave and the refracted wave inside the glass. Which?
  A speed
  B wavelength
  C direction
  D brightness
  E frequency

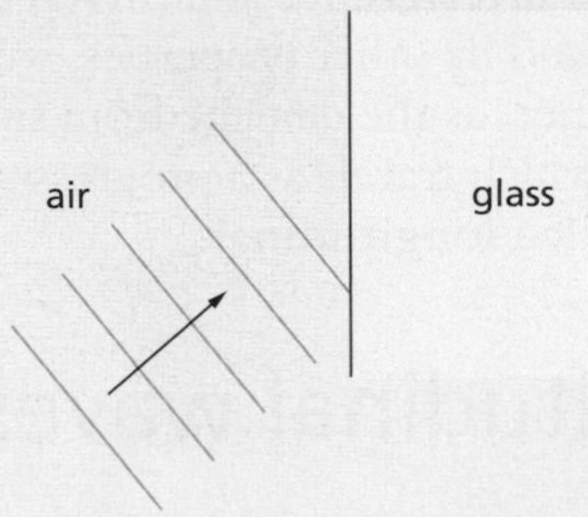

air   glass

Figure 32.8

Checklist

After studying this chapter you should be able to

- recall the types of electromagnetic radiation,
- recall that all electromagnetic waves have the same speed in space and are progressive transverse waves,
- recall that the colour of light depends on its frequency, that red light has a lower frequency (but longer wavelength) than blue light and that all colours travel at the same speed in air,

- use the term monochromatic,

- distinguish between infrared radiation, ultraviolet radiation, radio waves and X-rays in terms of their wavelengths, properties and uses,
- be aware of the harmful effects of different types of electromagnetic radiation and of how exposure to them can be reduced.

33 Sound waves

- Origin and transmission of sound
- Longitudinal waves
- Reflection and echoes
- Speed of sound
- Limits of audibility

- Musical notes
- Ultrasonics
- Seismic waves
- Practical work: Speed of sound in air

Origin and transmission of sound

Sources of sound all have some part that **vibrates**. A guitar has strings (Figure 33.1), a drum has a stretched skin and the human voice has vocal cords. The sound travels through the air to our ears and we hear it. That the air is necessary may be shown by pumping the air out of a glass jar containing a ringing electric bell (Figure 33.2); the sound disappears though the striker can still be seen hitting the gong. Evidently sound cannot travel in a vacuum as light can. Other materials, including solids and liquids, transmit sound.

Figure 33.1 A guitar string vibrating. The sound waves produced are amplified when they pass through the circular hole into the guitar's sound box.

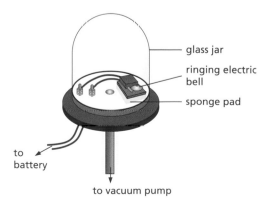

Figure 33.2 Sound cannot travel through a vacuum

Sound also gives **interference** and **diffraction** effects. Because of this and its other properties, we believe it is a form of energy (as the damage from supersonic booms shows) which travels as a progressive wave, but of a type called **longitudinal**.

Longitudinal waves

a) Waves on a spring

In a progressive longitudinal wave the particles of the transmitting medium vibrate to and fro along the same line as that in which the wave is travelling and not at right angles to it as in a transverse wave. A longitudinal wave can be sent along a spring, stretched out on the bench and fixed at one end, if the free end is repeatedly pushed and pulled sharply, as shown in Figure 33.3. **Compressions** C (where the coils are closer together) and **rarefactions** R (where the coils are further apart) travel along the spring.

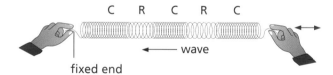

Figure 33.3 A longitudinal wave

b) Sound waves

A sound wave, produced for example by a loudspeaker, consists of a train of compressions ('squashes') and rarefactions ('stretches') in the air (Figure 33.4).

The speaker has a cone which is made to vibrate in and out by an electric current. When the cone moves out, the air in front is compressed; when it moves in, the air is rarefied (goes 'thinner'). The wave progresses through the air but the air as a whole does not move. The air particles (molecules) vibrate backwards and forwards a little as the wave passes. When the wave enters your ear the compressions and rarefactions cause small, rapid pressure changes on the eardrum and you experience the sensation of sound.

The number of compressions produced per second is the frequency f of the sound wave (and equals the frequency of the vibrating loudspeaker cone); the distance between successive compressions is the wavelength λ. As with transverse waves, the speed, v, $= f\lambda$.

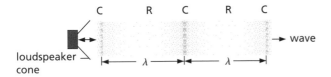

Figure 33.4 Sound travels as a longitudinal wave.

● Reflection and echoes

Sound waves are reflected well from hard, flat surfaces such as walls or cliffs and obey the same laws of reflection as light. The reflected sound forms an **echo**.

If the reflecting surface is nearer than 15 m from the source of sound, the echo joins up with the original sound which then seems to be prolonged. This is called **reverberation**. Some is desirable in a concert hall to stop it sounding 'dead', but too much causes 'confusion'. Modern concert halls are designed for the optimal amount of reverberation. Seats and some wall surfaces are covered with sound-absorbing material.

● Speed of sound

The **speed of sound** depends on the material through which it is passing. It is greater in solids than in liquids or gases because the molecules in a solid are closer together than in a liquid or a gas. Some values are given in Table 33.1.

Table 33.1 Speed of sound in different materials

Material	air (0 °C)	water	concrete	steel
Speed/m/s	330	1400	5000	6000

In air the speed **increases with temperature** and at high altitudes, where the temperature is lower, it is less than at sea level. Changes of atmospheric pressure do not affect it.

An estimate of the speed of sound can be made directly if you stand about 100 metres from a high wall or building and clap your hands. Echoes are produced. When the clapping rate is such that each clap coincides with the echo of the previous one, the sound has travelled to the wall and back in the time between two claps, i.e. one interval. By timing 30 intervals with a stopwatch, the time t for one interval can be found. Also, knowing the distance d to the wall, a rough value is obtained from

$$\text{speed of sound in air} = \frac{2d}{t}$$

The speed of sound in air can be found directly by measuring the time t taken for a sound to travel past two microphones separated by a distance d:

$$\text{speed of sound in air} = \frac{\text{distance travelled by the sound}}{\text{time taken}}$$
$$= \frac{d}{t}$$

● Limits of audibility

Humans hear only sounds with frequencies from about 20 Hz to 20 000 Hz. These are the limits of audibility; the upper limit decreases with age.

Practical work

Speed of sound in air

Set two microphones about a metre apart, and attach one to the 'start' terminal and the other to the 'stop' terminal of a digital timer, as shown in Figure 33.5. The timer should have millisecond accuracy. Measure and record the distance d between the centres of the microphones with a metre ruler. With the small hammer and metal plate to one side of the 'start' microphone, produce a sharp sound. When the sound reaches the 'start' microphone, the timer should start; when it reaches the 'stop' microphone, the timer should stop. The time displayed is then the time taken for the sound to travel the distance d. Record the time and then reset the timer; repeat the experiment a few times and work out an *average* value for t.

Calculate the speed of sound in air from d/t. How does your value compare with that given in Table 33.1?

▶▶

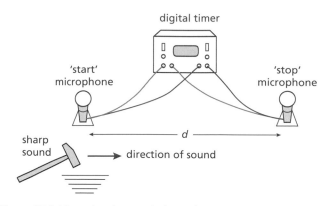

Figure 33.5 Measuring the speed of sound

A set of tuning forks with frequencies marked on them can also be used. A tuning fork (Figure 33.7), has two steel prongs which vibrate when struck; the prongs move in and out *together*, generating compressions and rarefactions.

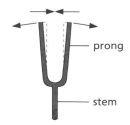

Figure 33.7 A tuning fork

● Musical notes

Irregular vibrations such as those of motor engines cause **noise**; regular vibrations such as occur in the instruments of a brass band (Figure 33.6), produce **musical notes** which have three properties – **pitch**, **loudness** and **quality**.

Figure 33.6 Musical instruments produce regular sound vibrations.

a) Pitch

The pitch of a note depends on the frequency of the sound wave reaching the ear, i.e. on the frequency of the source of sound. A high-pitched note has a high frequency and a short wavelength. The frequency of middle C is 256 vibrations per second or 256 Hz and that of upper C is 512 Hz. Notes are an **octave** apart if the frequency of one is twice that of the other. Pitch is like colour in light; both depend on the frequency.

Notes of known frequency can be produced in the laboratory by a signal generator supplying alternating electric current (a.c.) to a loudspeaker. The cone of the speaker vibrates at the frequency of the a.c. which can be varied and read off a scale on the generator.

b) Loudness

A note becomes louder when more sound energy enters our ears per second than before. This will happen when the source is vibrating with a larger amplitude. If a violin string is bowed more strongly, its amplitude of vibration increases as does that of the resulting sound wave and the note heard is louder because more energy has been used to produce it.

c) Quality

The same note on different instruments sounds different; we say the notes differ in **quality** or **timbre**. The difference arises because no instrument (except a tuning fork and a signal generator) emits a 'pure' note, i.e. of one frequency. Notes consist of a main or **fundamental** frequency mixed with others, called **overtones**, which are usually weaker and have frequencies that are exact multiples of the fundamental. The number and strength of the overtones decides the quality of a note. A violin has more and stronger higher overtones than a piano. Overtones of 256 Hz (middle C) are 512 Hz, 768 Hz and so on.

The **waveform** of a note played near a microphone connected to a cathode ray oscilloscope (CRO; see Chapter 48) can be displayed on the CRO screen. Those for the *same* note on three instruments are shown in Figure 33.8. Their different shapes show that while they have the same fundamental frequency, their quality differs. The 'pure' note of a tuning fork has a **sine** waveform and is the simplest kind of sound wave.

Note Although the waveform on the CRO screen is transverse, it represents a longitudinal sound wave.

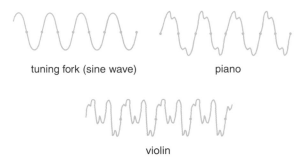

tuning fork (sine wave)   piano

violin

Figure 33.8 Notes of the same frequency (pitch) but different quality

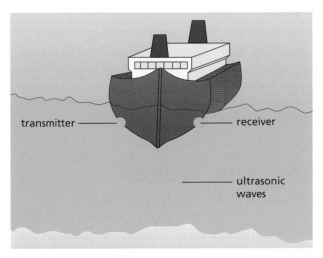

transmitter ———   ——— receiver

——— ultrasonic
waves

Figure 33.9 A ship using sonar

● Ultrasonics

Sound waves with frequencies above 20 kHz are called **ultrasonic** waves; their frequency is too high to be detected by the human ear but they can be detected electronically and displayed on a CRO.

a) Quartz crystal oscillators

Ultrasonic waves are produced by a quartz crystal which is made to vibrate electrically at the required frequency; they are emitted in a narrow beam in the direction in which the crystal oscillates. An ultrasonic receiver also consists of a quartz crystal but it works in reverse, i.e. when it is set into vibration by ultrasonic waves it generates an electrical signal which is then amplified. The same quartz crystal can act as both a transmitter and a receiver.

b) Ultrasonic echo techniques

Ultrasonic waves are partially or totally reflected from surfaces at which the density of the medium changes; this property is exploited in techniques such as the non-destructive testing of materials, sonar and medical ultrasound imaging. A bat emitting ultrasonic waves can judge the distance of an object from the time taken by the reflected wave or 'echo' to return.

Ships with **sonar** can determine the depth of a shoal of fish or the sea bed (Figure 33.9) in the same way; motion sensors (Chapter 2) also work on this principle.

In **medical ultrasound imaging**, used in antenatal clinics to monitor the health and sometimes to determine the sex of an unborn baby, an ultrasonic transmitter/receiver is scanned over the mother's abdomen and a detailed image of the fetus is built up (Figure 33.10). Reflection of the ultrasonic pulses occurs from boundaries of soft tissue, in addition to bone, so images can be obtained of internal organs that cannot be seen by using X-rays. Less detail of bone structure is seen than with X-rays, as the wavelength of ultrasonic waves is larger, typically about 1 mm, but ultrasound has no harmful effects on human tissue.

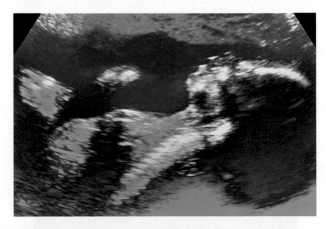

Figure 33.10 Checking the development of a fetus using ultrasound imaging

▶▶

c) Other uses

Ultrasound can also be used in ultrasonic drills to cut holes of any shape or size in hard materials such as glass and steel. Jewellery, or more mundane objects such as street lamp covers, can be cleaned by immersion in a tank of solvent which has an ultrasonic vibrator in the base.

● Seismic waves

Earthquakes produce both longitudinal waves (P-waves) and transverse waves (S-waves) that are known as **seismic** waves. These travel through the Earth at speeds of up to 13 000 m/s.

When seismic waves pass under buildings, severe structural damage may occur. If the earthquake occurs under the sea, the seismic energy can be transmitted to the water and produce **tsunami** waves that may travel for very large distances across the ocean. As a tsunami wave approaches shallow coastal waters, it slows down (see Chapter 25) and its amplitude increases, which can lead to massive coastal destruction. This happened in Sri Lanka (see Figure 33.11) and Thailand after the great 2004 Sumatra–Andaman earthquake. The time of arrival of a tsunami wave can be predicted if its speed of travel and the distance from the epicentre of the earthquake are known; it took about 2 hours for tsunami waves to cross the ocean to Sri Lanka from Indonesia. A similar time was needed for the tsunami waves to travel the shorter distance to Thailand. This was because the route was through shallower water and the waves travelled more slowly. If an early-warning system had been in place, many lives could have been saved.

Figure 33.11 This satellite image shows the tsunami that hit the south-western coast of Sri Lanka on 26 December 2004 as it pulled back out to sea, having caused utter devastation in coastal areas.

Questions

1 If 5 seconds elapse between a lightning flash and the clap of thunder, how far away is the storm? (Speed of sound = 330 m/s.)

2 a A girl stands 160 m away from a high wall and claps her hands at a steady rate so that each clap coincides with the echo of the one before. If her clapping rate is 60 per minute, what value does this give for the speed of sound?

 b If she moves 40 m closer to the wall she finds the clapping rate has to be 80 per minute. What value do these measurements give for the speed of sound?

 c If she moves again and finds the clapping rate becomes 30 per minute, how far is she from the wall if the speed of sound is the value you found in **a**?

3 a What properties of sound suggest it is a wave motion?

 b How does a progressive transverse wave differ from a longitudinal one? Which type of wave is a sound wave?

4 a Draw the waveform of
   (i) a loud, low-pitched note, and
   (ii) a soft, high-pitched note.

 b If the speed of sound is 340 m/s what is the wavelength of a note of frequency
   (i) 340 Hz,
   (ii) 170 Hz?

Checklist

After studying this chapter you should be able to

- recall that sound is produced by vibrations,
- describe an experiment to show that sound is not transmitted through a vacuum,
- describe how sound travels in a medium as progressive longitudinal waves,
- recall the limits of audibility (i.e. the range of frequencies) for the normal human ear,
- explain echoes and reverberation,
- describe an experiment to measure the speed of sound in air,
- solve problems using the speed of sound, e.g. distance away of thundercloud,

- state the order of magnitude of the speed of sound in air, liquid and solids,

- relate the loudness and pitch of sound waves to amplitude and frequency.

Section 4

Electricity and magnetism

Chapters

Simple phenomena of magnetism
34 Magnetic fields

Electrical quantities and circuits
35 Static electricity
36 Electric current
37 Potential difference
38 Resistance
39 Capacitors

40 Electric power
41 Electronic systems
42 Digital electronics

Electromagnetic effects
43 Generators
44 Transformers
45 Electromagnets
46 Electric motors
47 Electric meters
48 Electrons

34 Magnetic fields

- Properties of magnets
- Magnetisation of iron and steel
- Magnetic fields

- Earth's magnetic field
- Practical work: Plotting lines of force

● Properties of magnets

a) Magnetic materials

Magnets attract strongly only certain materials such as iron, steel, nickel and cobalt, which are called **ferro-magnetics**.

b) Magnetic poles

The **poles** are the places in a magnet to which magnetic materials, such as iron filings, are attracted. They are near the ends of a bar magnet and occur in pairs of equal strength.

c) North and south poles

If a magnet is supported so that it can swing in a horizontal plane it comes to rest with one pole, the **north-seeking** or N pole, always pointing roughly towards the Earth's north pole. A magnet can therefore be used as a **compass**.

d) Law of magnetic poles

If the N pole of a magnet is brought near the N pole of another magnet, repulsion occurs. Two S (south-seeking) poles also repel. By contrast, N and S poles always attract. The law of magnetic poles summarises these facts and states:

Like poles repel, unlike poles attract.
The force between magnetic poles decreases as their separation increases.

● Magnetisation of iron and steel

Chains of small iron nails and steel paper clips can be hung from a magnet (Figure 34.1). Each nail or clip magnetises the one below it and the unlike poles so formed attract.

If the iron chain is removed by pulling the top nail away from the magnet, the chain collapses, showing that **magnetism induced in iron is temporary**.

When the same is done with the steel chain, it does not collapse; **magnetism induced in steel is permanent**.

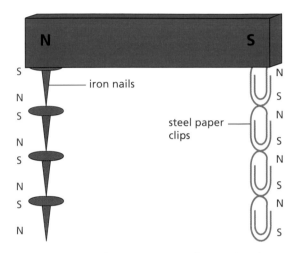

Figure 34.1 Investigating the magnetisation of iron and steel

Magnetic materials such as iron that magnetise easily but readily lose their magnetism (are easily demagnetised) are said to be **soft**. Those such as steel that are harder to magnetise than iron but stay magnetised are **hard**. Both types have their uses; very hard ones are used to make permanent magnets. Solenoids can be used to magnetise and demagnetise magnetic materials (p. 210); dropping or heating a magnet also causes demagnetisation. Hammering a magnetic material in a magnetic field causes magnetisation but in the absence of a field it causes demagnetisation. 'Stroking' a magnetic material several times in the same direction with one pole of a magnet will also cause it to become magnetised.

● Magnetic fields

The space surrounding a magnet where it produces a magnetic force is called a **magnetic field**. The force around a bar magnet can be detected and shown to vary in direction, using the apparatus in Figure 34.2. If the floating magnet is released near the N pole of the bar magnet, it is repelled to the S pole and moves along a curved path known as a **line of force** or a field line. It moves in the opposite direction if its south pole is uppermost.

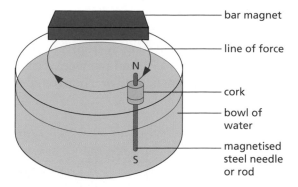

Figure 34.2 Detecting magnetic force

It is useful to consider that a magnetic field has a direction and to represent the field by lines of force. It has been decided that the **direction of the field at any point should be the direction of the force on a N pole**. To show the direction, arrows are put on the lines of force and point away from a N pole towards a S pole. The magnetic field is stronger in regions where the field lines are close together than where they are further apart.

The force between two magnets is a result of the interaction of their magnetic fields.

Practical work

Plotting lines of force

a) Plotting compass method
A plotting compass is a small pivoted magnet in a glass case with non-magnetic metal walls (Figure 34.3a).

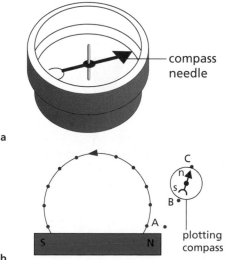

Figure 34.3

Lay a bar magnet on a sheet of paper. Place the plotting compass at a point such as A (Figure 34.3b), near one pole of the magnet. In Figure 34.3b it is the N pole. Mark the position of the poles (n, s) of the compass by pencil dots B, A. Move the compass so that pole s is exactly over B, mark the new position of n by dot C.

Continue this process until the other pole of the bar magnet is reached (in Figure 34.3b it is the S pole). Join the dots to give one line of force and show its direction by putting an arrow on it. Plot other lines by starting at different points round the magnet.

A typical field pattern is shown in Figure 34.4.

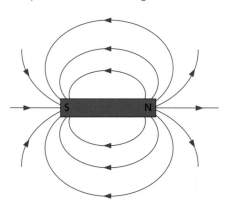

Figure 34.4 Magnetic field lines around a bar magnet

The combined field due to two neighbouring magnets can also be plotted to give patterns like those in Figures 34.5a, b. In part a, where two like poles are facing each other, the point X is called a **neutral point**. At X the field due to one magnet cancels out that due to the other and there are no lines of force.

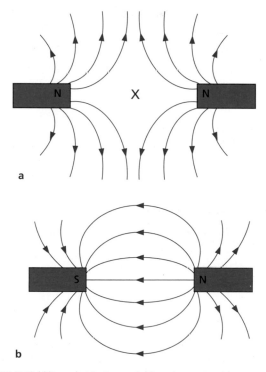

Figure 34.5 Field lines due to two neighbouring magnets

►►

b) Iron filings method

Place a sheet of paper *on top of* a bar magnet and sprinkle iron filings *thinly and evenly* on to the paper from a 'pepper pot'.

Tap the paper gently with a pencil and the filings should form patterns showing the lines of force. Each filing turns in the direction of the field when the paper is tapped.

This method is quick but no use for weak fields. Figures 34.6a, b show typical patterns with two magnets. Why are they different? What combination of poles would give the observed patterns?

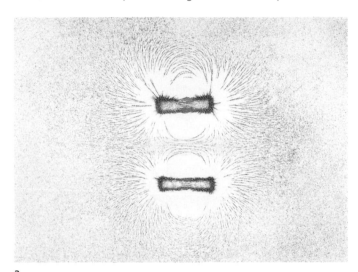

a

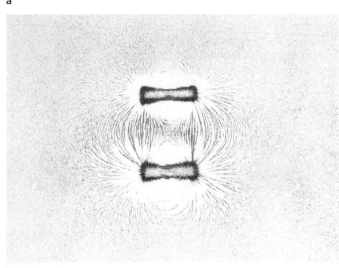

b

Figure 34.6 Field lines round two bar magnets shown by iron filings

● Earth's magnetic field

If lines of force are plotted on a sheet of paper with no magnets nearby, a set of parallel straight lines is obtained. They run roughly from S to N geographically (Figure 34.7), and represent a small part of the Earth's magnetic field in a horizontal plane.

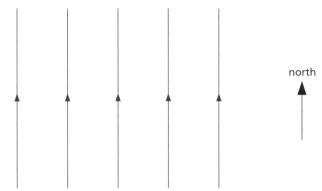

north

Figure 34.7 Lines of force due to the Earth's field

At most places on the Earth's surface a magnetic compass points slightly east or west of true north, i.e. the Earth's geographical and magnetic north poles do not coincide. The angle between magnetic north and true north is called the **declination** (Figure 34.8). In Hong Kong in 2014 it was 2° 35′ W of N and changing slowly.

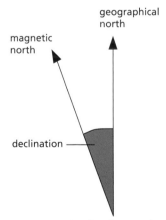

geographical north

magnetic north

declination

Figure 34.8 The Earth's geographical and magnetic poles do not coincide.

Questions

1 Which one of these statements is true?
A magnet attracts
   A plastics   B any metal   C iron and steel
   D aluminium   E carbon
2 Copy Figure 34.9 which shows a plotting compass and a magnet. Label the N pole of the magnet and draw the field line on which the compass lies.

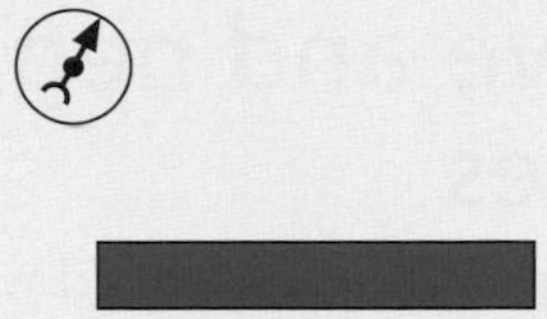

Figure 34.9

Checklist

After studying this chapter you should be able to

- state the properties of magnets,
- explain what is meant by soft and hard magnetic materials,
- recall that a magnetic field is the region round a magnet where a magnetic force is exerted and is represented by lines of force whose direction at any point is the direction of the force on a N pole,
- map magnetic fields (by the plotting compass and iron filings methods) round (a) one magnet, (b) two magnets,
- recall that at a neutral point the field due to one magnet cancels that due to another.

- Positive and negative charges
- Charges, atoms and electrons
- Electrons, insulators and conductors
- Electrostatic induction
- Attraction between uncharged and charged objects

- Dangers of static electricity
- Uses of static electricity
- van de Graaff generator
- Electric fields
- Practical work: Gold-leaf electroscope

Figure 35.1 A flash of lightning is nature's most spectacular static electricity effect.

Clothes containing nylon often crackle when they are taken off. We say they are 'charged with static electricity'; the crackles are caused by tiny electric sparks which can be seen in the dark. Pens and combs made of certain plastics become charged when rubbed on your sleeve and can then attract scraps of paper.

● Positive and negative charges

When a strip of polythene is rubbed with a cloth it becomes charged. If it is hung up and another rubbed polythene strip is brought near, repulsion occurs (Figure 35.2). Attraction occurs when a rubbed strip of cellulose acetate is brought near.

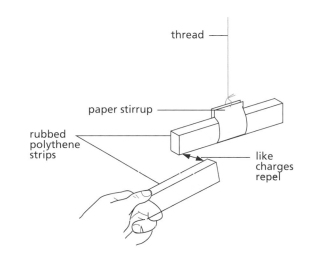

Figure 35.2 Investigating charges

This shows there are two kinds of electric charge. That on cellulose acetate is taken as **positive** (+) and that on polythene is **negative** (−). It also shows that:

Like charges (+ and +, or − and −) repel, while unlike charges (+ and −) attract.

The force between electric charges decreases as their separation increases.

● Charges, atoms and electrons

There is evidence (Chapter 50) that we can picture an atom as being made up of a small central nucleus containing positively charged particles called **protons**, surrounded by an equal number of negatively charged **electrons**. The charges on a proton and an electron are equal and opposite so an atom as a whole is normally electrically neutral, i.e. has no net charge.

Hydrogen is the simplest atom with one proton and one electron (Figure 35.3). A copper atom has 29 protons in the nucleus and 29 surrounding electrons. Every nucleus except hydrogen also contains uncharged particles called **neutrons**.

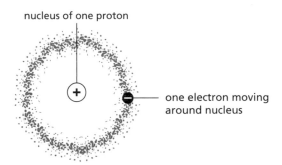

nucleus of one proton

one electron moving around nucleus

Figure 35.3 Hydrogen atom

The production of charges by rubbing can be explained by supposing that electrons are transferred from one material to the other. For example, when cellulose acetate is rubbed with a cloth, electrons go from the acetate to the cloth, leaving the acetate short of electrons, i.e. positively charged. The cloth now has more electrons than protons and becomes negatively charged. Note that it is only electrons which move; the protons remain fixed in the nucleus.

How does polythene become charged when rubbed?

Gold-leaf electroscope

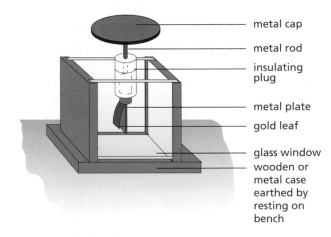

metal cap
metal rod
insulating plug
metal plate
gold leaf
glass window
wooden or metal case earthed by resting on bench

Figure 35.4 Gold-leaf electroscope

A gold-leaf electroscope consists of a metal cap on a metal rod at the foot of which is a metal plate with a leaf of gold foil attached (Figure 35.4). The rod is held by an insulating plastic plug in a case with glass sides to protect the leaf from draughts.

a) Detecting a charge

Bring a charged polythene strip towards the cap: the leaf rises away from the plate. When you remove the charged strip, the leaf falls again. Repeat with a charged acetate strip.

b) Charging by contact

Draw a charged polythene strip *firmly across the edge of the cap*. The leaf should rise and stay up when the strip is removed. If it does not, repeat the process but press harder. The electroscope has now become negatively charged by contact with the polythene strip, from which electrons have been transferred.

c) Insulators and conductors

Touch the cap of the charged electroscope with different things, such as a piece of paper, a wire, your finger, a comb, a cotton handkerchief, a piece of wood, a glass rod, a plastic pen, rubber tubing. Record your results.

When the leaf falls, charge is passing to or from the ground through you and the material touching the cap. If the fall is rapid the material is a **good conductor**; if the leaf falls slowly, the material is a poor conductor. If the leaf does not alter, the material is a **good insulator**.

● Electrons, insulators and conductors

In an insulator all electrons are bound firmly to their atoms; in a conductor some electrons can move freely from atom to atom. An insulator can be charged by rubbing because the charge produced cannot move from where the rubbing occurs, i.e. the electric charge is **static**. A conductor will become charged only if it is held with an insulating handle; otherwise electrons are transferred between the conductor and the ground via the person's body.

Good insulators include plastics such as polythene, cellulose acetate, Perspex and nylon. All metals and carbon are good conductors. In between are materials that are both poor conductors and (because they conduct to some extent) poor insulators. Examples are wood, paper, cotton, the human body and the Earth. Water conducts and if it were not present in materials like wood and on the surface of, for example, glass, these would be good insulators. Dry air insulates well.

● Electrostatic induction

This effect may be shown by bringing a negatively charged polythene strip near to an insulated metal sphere X which is touching a similar sphere Y (Figure 35.5a). Electrons in the spheres are repelled to the far side of Y.

If X and Y are separated, with the charged strip still in position, X is left with a positive charge (deficient of electrons) and Y with a negative charge (excess of electrons) (Figure 35.5b). The signs of the charges can be tested by removing the charged strip (Figure 35.5c), and taking X up to the cap of a positively charged electroscope. Electrons will be drawn towards X, making the leaf more positive so that it rises. If Y is taken towards the cap of a *negatively* charged electroscope the leaf again rises; can you explain why, in terms of electron motion?

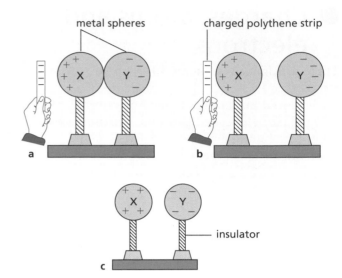

Figure 35.5 Electrostatic induction

● Attraction between uncharged and charged objects

The attraction of an uncharged object by a charged object near it is due to electrostatic induction.

In Figure 35.6a a small piece of aluminium foil is attracted to a negatively charged polythene rod held just above it. The charge on the rod pushes free electrons to the bottom of the foil (aluminium is a conductor), leaving the top of the foil short of electrons, i.e. with a net positive charge, and the bottom negatively charged. The top of the foil is nearer the rod than the bottom. Hence the force of attraction between the negative charge on the rod and the positive charge on the top of the foil is greater than the force of repulsion between the negative charge on the rod and the negative charge on the bottom of the foil. The foil is pulled to the rod.

A small scrap of paper, although an insulator, is also attracted by a charged rod. There are no free electrons in the paper but the charged rod pulls the electrons of the atoms in the paper slightly closer (by electrostatic induction) and so distorts the atoms. In the case of a negatively charged polythene rod, the paper behaves as if it had a positively charged top and a negative charge at the bottom.

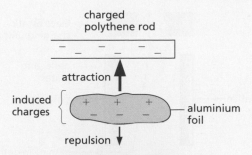

Figure 35.6a An uncharged object is attracted to a charged one.

Figure 35.6b A slow stream of water is bent by electrostatic attraction.

In Figure 35.6b a slow, uncharged stream of water is attracted by a charged polythene rod, due to the polar nature of water molecules (one end of a molecule is negatively charged while the other end is positively charged).

● Dangers of static electricity

a) Lightning

A tall building is protected by a lightning conductor consisting of a thick copper strip fixed on the outside of the building connecting metal spikes at the top to a metal plate in the ground (Figure 35.7).

Thunderclouds carry charges; a negatively charged cloud passing overhead repels electrons from the spikes to the Earth. The points of the spikes are left with a large positive charge (charge concentrates on sharp points) which removes electrons from nearby air molecules, so charging them positively and causing them to be repelled

from the spike. This effect, called **action at points**, results in an 'electric wind' of positive air molecules streaming upwards which can neutralise electrons discharging from the thundercloud in a lightning flash. If a flash occurs it is now less violent and the conductor gives it an easy path to ground.

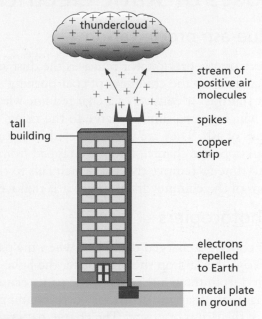

Figure 35.7 Lightning conductor

b) Refuelling

Sparks from static electricity can be dangerous when flammable vapour is present. For this reason, the tanks in an oil tanker may be cleaned in an atmosphere of nitrogen – otherwise oxygen in the air could promote a fire.

An aircraft in flight may become charged by 'rubbing' the air. Its tyres are made of conducting rubber which lets the charge pass harmlessly to ground on landing, otherwise an explosion could be 'sparked off' when the aircraft refuels. What precautions are taken at petrol pumps when a car is refuelled?

c) Operating theatres

Dust and germs are attracted by charged objects and so it is essential to ensure that equipment and medical personnel are well 'earthed' allowing electrons to flow to and from the ground, for example by conducting rubber.

▶▶

d) Computers

Computers require similar 'anti-static' conditions as they are vulnerable to electrostatic damage.

● Uses of static electricity

a) Flue-ash precipitation

An electrostatic precipitator removes the dust and ash that goes up the chimneys of coal-burning power stations. It consists of a charged fine wire mesh which gives a similar charge to the rising particles of ash. They are then attracted to plates with an opposite charge. These are tapped from time to time to remove the ash, which falls to the bottom of the chimney from where it is removed.

b) Photocopiers

These contain a charged drum and when the paper to be copied is laid on the glass plate, the light reflected from the white parts of the paper causes the charge to disappear from the corresponding parts of the drum opposite. The charge pattern remaining on the drum corresponds to the dark-coloured printing on the original. Special **toner** powder is then dusted over the drum and sticks to those parts which are still charged. When a sheet of paper passes over the drum, the particles of toner are attracted to it and fused into place by a short burst of heat.

c) Inkjet printers

In an inkjet printer tiny drops of ink are forced out of a fine nozzle, charged electrostatically and then passed between two oppositely charged plates; a negatively charged drop will be attracted towards the positive plate causing it to be deflected as shown in Figure 35.8. The amount of deflection and hence the position at which the ink strikes the page is determined by the charge on the drop and the p.d. between the plates; both of these are controlled by a computer. About 100 precisely located drops are needed to make up an individual letter but very fast printing speeds can be achieved.

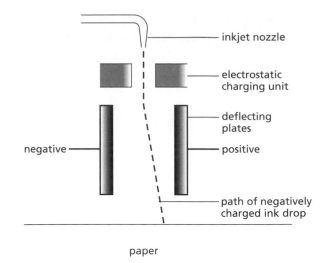

Figure 35.8 Inkjet printer

● van de Graaff generator

This produces a continuous supply of charge on a large metal dome when a rubber belt is driven by an electric motor or by hand, as shown in Figure 35.9a.

a) Demonstrations

In Figure 35.9a sparks jump between the dome and the discharging sphere. Electrons flow round a complete path (circuit) from the dome. Can you trace it? In part Figure 35.9b why does the 'hair' stand on end? In Figure 35.9c the 'windmill' revolves due to the reaction that arises from the 'electric wind' caused by the action at points effect, explained on p. 153 for the lightning conductor.

In Figure 35.9d the 'body' on the insulating stool first gets charged by touching the dome and then lights a neon lamp.

The dome can be discharged harmlessly by bringing your elbow close to it.

b) Action

Initially a positive charge is produced on the motor-driven Perspex roller because it is rubbing the belt. This induces a negative charge on the 'comb' of metal points P (Figure 35.9a). The charges are sprayed off by 'action at points' on to the outside of the belt and carried upwards. A positive charge is then induced in the comb of metal points, Q, and negative charge is repelled to the dome.

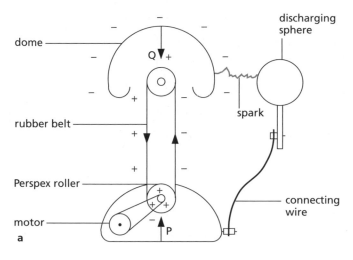

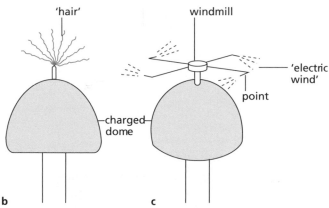

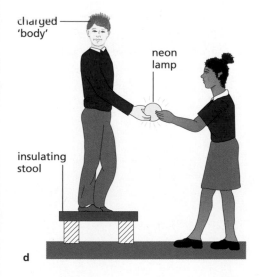

Figure 35.9

● Electric fields

When an electric charge is placed near to another electric charge it experiences a force. The electric force does not require contact between the two charges so we call it an 'action-at-a-distance force' – it acts through space. The region of space where an electric charge experiences a force due to other charges is called an **electric field**. If the electric force felt by a charge is the same everywhere in a region, the field is uniform; a uniform electric field is produced between two oppositely charged parallel metal plates (Figure 35.10). It can be represented by evenly spaced parallel lines drawn perpendicular to the metal surfaces. The direction of the field, denoted by arrows, is the direction of the force on a small positive charge placed in the field (negative charges experience a force in the opposite direction to the field).

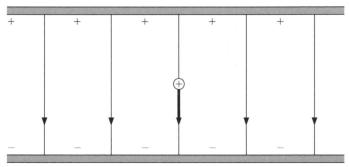

Figure 35.10 Uniform electric field

Moving charges are deflected by an electric field due to the electric force exerted on them; this occurs in the inkjet printer (Figure 35.8).

The electric field lines radiating from an isolated positively charged conducting sphere and a point charge are shown in Figures 35.11a, b; again the field lines emerge at right angles to the conducting surface.

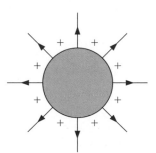

Figure 35.11a

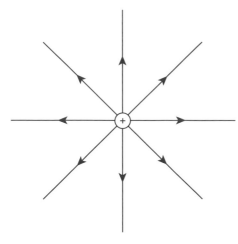

Figure 35.11b Radial electric field

Questions

1 Two identical conducting balls, suspended on nylon threads, come to rest with the threads making equal angles with the vertical, as shown in Figure 35.12.
   Which of these statements is true?
   This shows that:
   A the balls are equally and oppositely charged
   B the balls are oppositely charged but not necessarily equally charged
   C one ball is charged and the other is uncharged
   D the balls both carry the same type of charge
   E one is charged and the other may or may not be charged.

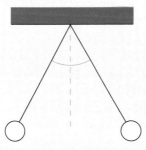

Figure 35.12

2 Explain in terms of electron movement what happens when a polythene rod becomes charged negatively by being rubbed with a cloth.

3 Which of statements A to E is true?
   In the process of electrostatic induction
   A a conductor is rubbed with an insulator
   B a charge is produced by friction
   C negative and positive charges are separated
   D a positive charge induces a positive charge
   E electrons are 'sprayed' into an object.

36 Electric current

- Effects of a current
- The ampere and the coulomb
- Circuit diagrams
- Series and parallel circuits
- Direct and alternating current
- Practical work: Measuring current

An **electric current** consists of moving electric charges. In Figure 36.1, when the van de Graaff machine is working, the table-tennis ball shuttles rapidly to and fro between the plates and the meter records a small current. As the ball touches each plate it becomes charged and is repelled to the other plate. In this way charge is carried across the gap. This also shows that 'static' charges, produced by friction, cause a deflection on a meter just as current electricity produced by a battery does.

In a metal, each atom has one or more loosely held electrons that are free to move. When a van de Graaff or a battery is connected across the ends of such a conductor, the free electrons drift slowly along it in the direction from the negative to the positive terminal of a battery. There is then a current of negative charge.

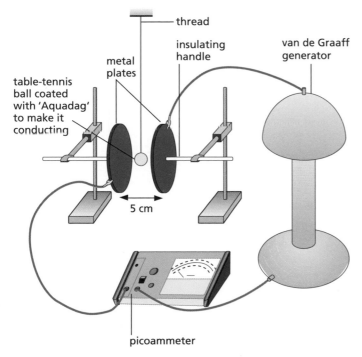

Figure 36.1 Demonstrating that an electric current consists of moving charges

● Effects of a current

An electric current has three effects that reveal its existence and which can be shown with the circuit of Figure 36.2.

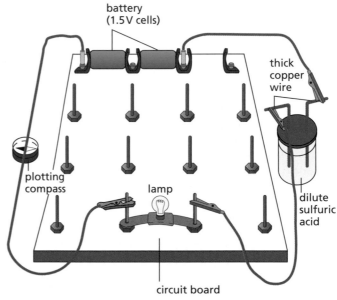

Figure 36.2 Investigating the effects of a current

a) Heating and lighting
The lamp lights because the small wire inside (the filament) is made white hot by the current.

b) Magnetic
The plotting compass is deflected when it is placed near the wire because a magnetic field is produced around any wire carrying a current.

c) Chemical
Bubbles of gas are given off at the wires in the acid because of the chemical action of the current.

● The ampere and the coulomb

The unit of current is the **ampere** (A) which is defined using the magnetic effect. One milliampere (mA) is one-thousandth of an ampere. Current is measured by an **ammeter**.

The unit of charge, the **coulomb** (C), is defined in terms of the ampere.

> One coulomb is the charge passing any point in a circuit when a steady current of 1 ampere flows for 1 second. That is, 1 C = 1 A s.

A charge of 3 C would pass each point in 1 s if the current were 3 A. In 2 s, 3 A × 2 s = 6 A s = 6 C would pass. In general, if a steady current I (amperes) flows for time t (seconds) the charge Q (coulombs) passing any point is given by

$$Q = I \times t$$

This is a useful expression connecting charge and current.

● Circuit diagrams

Current must have a complete path (a circuit) of conductors if it is to flow. Wires of copper are used to connect batteries, lamps, etc. in a circuit since copper is a good electrical conductor. If the wires are covered with insulation, such as plastic, the ends are bared for connecting up.

The signs or symbols used for various parts of an electric circuit are shown in Figure 36.3.

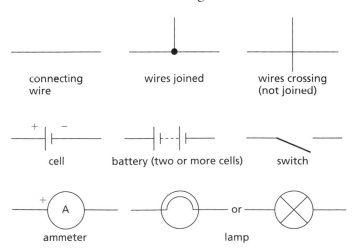

Figure 36.3 Circuit symbols

Before the electron was discovered scientists agreed to think of current as positive charges moving round a circuit in the direction from positive to negative of a battery. This agreement still stands. Arrows on circuit diagrams show the direction of what we call the **conventional current**, i.e. the direction in which **positive** charges would flow. Electrons flow in the opposite direction to the conventional current.

Practical work

Measuring current

(a) Connect the circuit of Figure 36.4a (on a circuit board if possible) ensuring that the + of the cell (the metal stud) goes to the + of the ammeter (marked red). Note the current.

(b) Connect the circuit of Figure 36.4b. The cells are **in series** (+ of one to – of the other), as are the lamps. Record the current. Measure the current at B, C and D by disconnecting the circuit at each point in turn and inserting the ammeter. What do you find?

(c) Connect the circuit of Figure 36.4c. The lamps are **in parallel**. Read the ammeter. Also measure the currents at P, Q and R. What is your conclusion?

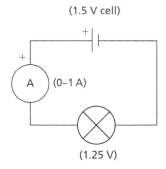

Figure 36.4a

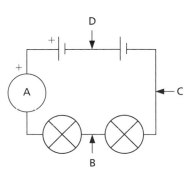

Figure 36.4b

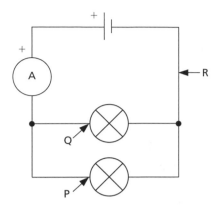

Figure 36.4c

● Series and parallel circuits

a) Series

In a **series circuit**, such as the one shown in Figure 36.4b, the different parts follow one after the other and there is just one path for the current to follow. You should have found in the previous experiment that the reading on the ammeter (e.g. 0.2 A) when in the position shown in the diagram is also obtained at B, C and D. That is, current is not used up as it goes round the circuit.

> The current is the same at all points in a series circuit.

b) Parallel

In a **parallel circuit**, as in Figure 36.4c, the lamps are side by side and there are alternative paths for the current. The current splits: some goes through one lamp and the rest through the other. The current from the source is larger than the current in each branch. For example, if the ammeter reading was 0.4 A in the position shown, then if the lamps are identical, the reading at P would be 0.2 A, and so would the reading at Q, giving a total of 0.4 A. Whether the current splits equally or not depends on the lamps (as we will see later); for example, it might divide so that 0.3 A goes one way and 0.1 A by the other branch.

> The sum of the currents in the branches of a parallel circuit equals the current entering or leaving the parallel section.

● Direct and alternating current

a) Difference

In a **direct current (d.c.)** the electrons flow in one direction only. Graphs for steady and varying d.c. are shown in Figure 36.5.

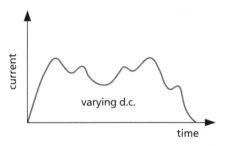

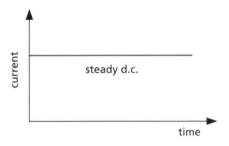

Figure 36.5 Direct current (d.c.)

In an **alternating current (a.c.)** the direction of flow reverses regularly, as shown in the graph in Figure 36.6. The circuit sign for a.c. is given in Figure 36.7.

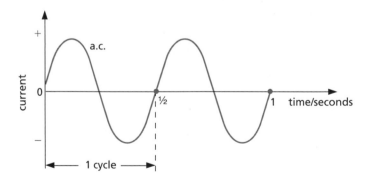

Figure 36.6 Alternating current (a.c.)

Figure 36.7 Symbol for alternating current

The pointer of an ammeter for measuring d.c. is deflected one way by the direct current. Alternating

159

current makes the pointer move to and fro about the zero if the changes are slow enough; otherwise no deflection can be seen.

Batteries give d.c.; generators can produce either d.c. or a.c.

b) Frequency of a.c.

The number of complete alternations or cycles in 1 second is the **frequency** of the alternating current. The unit of frequency is the **hertz** (Hz). The frequency of the a.c. in Figure 36.6 is 2 Hz, which means there are two cycles per second, or one cycle lasts $1/2 = 0.5$ s. The mains supply in the UK is a.c. of frequency 50 Hz; each cycle lasts 1/50th of a second. This regularity was used in the tickertape timer (Chapter 2) and is relied upon in mains-operated clocks.

Questions

1 If the current in a floodlamp is 5 A, what charge passes in
  a 1 s,
  b 10 s,
  c 5 minutes?
2 What is the current in a circuit if the charge passing each point is
  a 10 C in 2 s,
  b 20 C in 40 s,
  c 240 C in 2 minutes?
3 Study the circuits in Figure 36.8. The switch S is open (there is a break in the circuit at this point). In which circuit would lamps Q and R light but not lamp P?

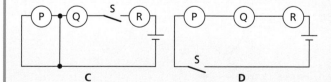

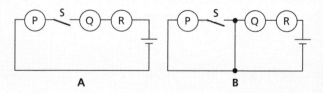

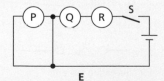

Figure 36.8

4 Using the circuit in Figure 36.9, which of the following statements is correct?
  A When S_1 and S_2 are closed, lamps A and B are lit.
  B With S_1 open and S_2 closed, A is lit and B is not lit.
  C With S_2 open and S_1 closed, A and B are lit.

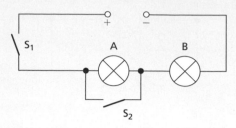

Figure 36.9

5 If the lamps are both the same in Figure 36.10 and if ammeter A_1 reads 0.50 A, what do ammeters A_2, A_3, A_4 and A_5 read?

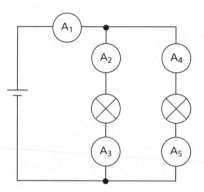

Figure 36.10

Checklist

After studying this chapter you should be able to

- describe a demonstration which shows that an electric current is a flow of charge,
- recall that an electric current in a metal is a flow of electrons from the negative to the positive terminal of the battery round a circuit,
- state the three effects of an electric current,
- state the unit of electric current and recall that current is measured by an ammeter,

- define the unit of charge in terms of the unit of current,

- recall the relation $Q = It$ and use it to solve problems,

- use circuit symbols for wires, cells, switches, ammeters and lamps,
- draw and connect simple series and parallel circuits, observing correct polarities for meters,
- recall that the current in a series circuit is the same everywhere in the circuit,
- state that for a parallel circuit, the current from the source is larger than the current in each branch,

- recall that the sum of the currents in the branches of a parallel circuit equals the current entering or leaving the parallel section,

- distinguish between electron flow and conventional current,

- distinguish between direct current and alternating current,
- recall that frequency of a.c. is the number of cycles per second.

(37) Potential difference

- Energy transfers and p.d.
- Model of a circuit
- The volt

- Cells, batteries and e.m.f.
- Voltages round a circuit
- Practical work: Measuring voltage

A battery transforms chemical energy to electrical energy. Because of the chemical action going on inside it, it builds up a surplus of electrons at one of its terminals (the negative) and creates a shortage at the other (the positive). It is then able to maintain a **flow of electrons**, i.e. an **electric current**, in any circuit connected across its terminals for as long as the chemical action lasts.

The battery is said to have a **potential difference** (**p.d.** for short) at its terminals. Potential difference is measured in **volts** (V) and the term **voltage** is sometimes used instead of p.d. The p.d. of a car battery is 12 V and the domestic mains supply in the UK is 230 V.

● Energy transfers and p.d.

In an electric circuit electrical energy is supplied from a source such as a battery and is transferred to other forms of energy by devices in the circuit. A lamp produces heat and light.

When each one of the circuits of Figure 37.1 is connected up, it will be found from the ammeter readings that the current is about the same (0.4 A) in each lamp. However, the mains lamp with a potential difference of 230 V applied to it gives much more light and heat than the car lamp with 12 V across it. In terms of energy, the mains lamp transfers a great deal more electrical energy in a second than the car lamp.

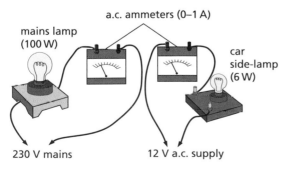

Figure 37.1 Investigating the effect of p.d. (potential difference) on energy transfer

Evidently the p.d. across a device affects the rate at which it transfers electrical energy. This gives us a way of defining the unit of potential difference: the volt.

● Model of a circuit

It may help you to understand the definition of the volt, i.e. what a volt is, if you *imagine* that the current in a circuit is formed by 'drops' of electricity, each having a charge of 1 coulomb and carrying equal-sized 'bundles' of electrical energy. In Figure 37.2, Mr Coulomb represents one such 'drop'. As a 'drop' moves around the circuit it gives up all its energy which is changed to other forms of energy. Note that **electrical energy, not charge or current, is 'used up'**.

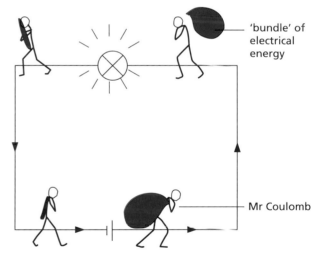

Figure 37.2 Model of a circuit

In our imaginary representation, Mr Coulomb travels round the circuit and unloads energy as he goes, most of it in the lamp. We think of him receiving a fresh 'bundle' every time he passes through the battery, which suggests he must be travelling very fast. In fact, as we found earlier (Chapter 36), the electrons drift along quite slowly. As soon as the circuit is complete, energy is delivered at once to the lamp, not by electrons directly from the battery but from electrons that were in the connecting wires. The model is helpful but is not an exact representation.

● The volt

The demonstrations of Figure 37.1 show that the greater the voltage at the terminals of a supply, the larger is the 'bundle' of electrical energy given to each coulomb and the greater is the rate at which light and heat are produced in a lamp.

> The p.d. between two points in a circuit is 1 volt if 1 joule of electrical energy is transferred to other forms of energy when 1 coulomb passes from one point to the other.

Figure 37.3 Compact batteries

That is, 1 volt = 1 joule per coulomb ($1\,V = 1\,J/C$). If 2 J are given up by each coulomb, the p.d. is 2 V. If 6 J are transferred when 2 C pass, the p.d. is $6\,J/2\,C = 3\,V$.

In general if E (joules) is the energy transferred (i.e. the work done) when charge Q (coulombs) passes between two points, the p.d. V (volts) between the points is given by

$$V = E/Q \quad \text{or} \quad E = Q \times V$$

If Q is in the form of a steady current I (amperes) flowing for time t (seconds) then $Q = I \times t$ (Chapter 36) and

$$E = I \times t \times V$$

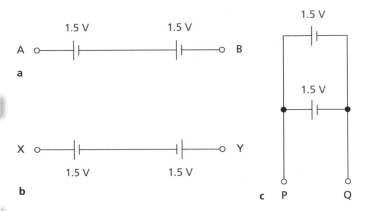

Figure 37.4

● Cells, batteries and e.m.f.

A 'battery' (Figure 37.3) consists of two or more **electric cells**. Greater voltages are obtained when cells are joined in series, i.e. + of one to − of next. In Figure 37.4a the two 1.5 V cells give a voltage of 3 V at the terminals A, B. Every coulomb in a circuit connected to this battery will have 3 J of electrical energy.

The cells in Figure 37.4b are in opposition and the voltage at X, Y is zero.

If two 1.5 V cells are connected in parallel, as in Figure 37.4c, the voltage at terminals P, Q is still 1.5 V but the arrangement behaves like a larger cell and will last longer.

The p.d. at the terminals of a battery decreases slightly when current is drawn from it. This effect is due to the internal resistance of the battery which transfers electrical energy to heat as current flows through it. The greater the current drawn, the larger the 'lost' voltage. When no current is drawn from a battery it is said to be an 'open circuit' and its terminal p.d. is a maximum. This maximum voltage is termed the **electromotive force** (**e.m.f.**) of the battery. Like potential difference, e.m.f. is measured in volts and can be written as

e.m.f. = 'lost' volts + terminal p.d.

In energy terms, the e.m.f. is defined as the number of joules of chemical energy transferred to electrical energy and heat when one coulomb of charge passes through the battery (or cell).

In Figure 37.2 the size of the energy bundle Mr Coulomb is carrying when he leaves the cell would be smaller if the internal resistance were larger.

Practical work

Measuring voltage

A **voltmeter** is an instrument for measuring voltage or p.d. It looks like an ammeter but has a scale marked in volts. Whereas an ammeter is inserted **in series** in a circuit to measure the current, a voltmeter is connected across that part of the circuit where the voltage is required, i.e. **in parallel**. (We will see later that a voltmeter should have a high resistance and an ammeter a low resistance.)

To prevent damage the + terminal (marked red) must be connected to the point nearest the + of the battery.

(a) Connect the circuit of Figure 37.5a. The voltmeter gives the voltage across the lamp. Read it.

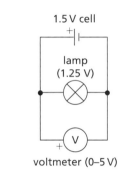

a

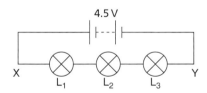

b

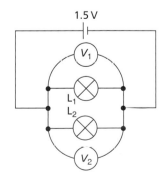

c

Figure 37.5

(b) Connect the circuit of Figure 37.5b. Measure:
   (i)   the voltage V between X and Y,
   (ii) the voltage V_1 across lamp L_1,
   (iii) the voltage V_2 across lamp L_2,
   (iv) the voltage V_3 across lamp L_3.
   How does the value of V compare with $V_1 + V_2 + V_3$?
(c) Connect the circuit of Figure 37.5c, so that two lamps L_1 and L_2 are in parallel across one 1.5 V cell. Measure the voltages, V_1 and V_2, across each lamp in turn. How do V_1 and V_2 compare?

● Voltages round a circuit

a) Series

In the previous experiment you should have found in the circuit of Figure 37.5b that
$$V = V_1 + V_2 + V_3$$
For example, if $V_1 = 1.4$ V, $V_2 = 1.5$ V and $V_3 = 1.6$ V, then V will be $(1.4 + 1.5 + 1.6)$ V $= 4.5$ V.

The voltage at the terminals of a battery equals the sum of the voltages across the devices in the external circuit from one battery terminal to the other.

b) Parallel

In the circuit of Figure 37.5c
$$V_1 = V_2$$

The voltages across devices in parallel in a circuit are equal.

Questions

1 The p.d. across the lamp in Figure 37.6 is 12 V. How many joules of electrical energy are changed into light and heat when
   a a charge of 1 C passes through it,
   b a charge of 5 C passes through it,
   c a current of 2 A flows in it for 10 s?

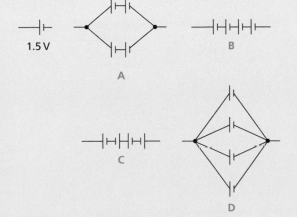

Figure 37.6

2 Three 2 V cells are connected in series and used as the supply for a circuit.
   a What is the p.d. at the terminals of the supply?
   b How many joules of electrical energy does 1 C gain on passing through
      (i) one cell,   **(ii)** all three cells?

3 Each of the cells shown in Figure 37.7 has a p.d. of 1.5 V. Which of the arrangements would produce a battery with a p.d. of 6 V?

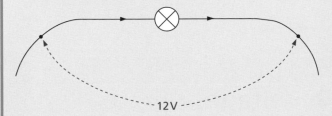

Figure 37.7

4 The lamps and the cells in all the circuits of Figure 37.8 are the same. If the lamp in **a** has its full, normal brightness, what can you say about the brightness of the lamps in **b**, **c**, **d**, **e** and **f**?

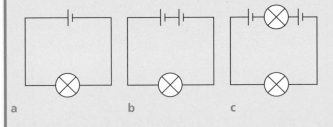

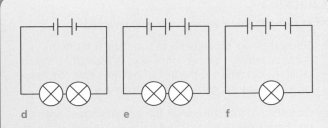

Figure 37.8

5 Three voltmeters V, V_1 and V_2 are connected as in Figure 37.9.
   a If V reads 18 V and V_1 reads 12 V, what does V_2 read?
   b If the ammeter A reads 0.5 A, how much electrical energy is changed to heat and light in lamp L_1 in one minute?
   c Copy Figure 37.9 and mark with a + the positive terminals of the ammeter and voltmeters for correct connection.

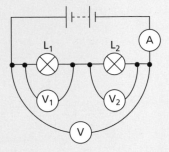

Figure 37.9

6 Three voltmeters are connected as in Figure 37.10.

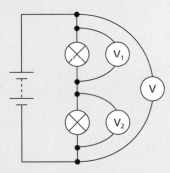

Figure 37.10

What are the voltmeter readings x, y and z in the table below (which were obtained with three different batteries)?

V/V	V_1/V	V_2/V
x	12	6
6	4	y
12	z	4

Checklist

After studying this chapter you should be able to

- describe simple experiments to show the transfer of electrical energy to other forms (e.g. in a lamp),
- recall the definition of the unit of p.d. and that p.d. (also called 'voltage') is measured by a voltmeter,

- demonstrate that the sum of the voltages across any number of components in series equals the voltage across all of those components,

- demonstrate that the voltages across any number of components in parallel are the same,

- work out the voltages of cells connected in series and parallel,

- explain the meaning of the term electromotive force (e.m.f.).

(38) Resistance

- The ohm
- Resistors
- I–V graphs: Ohm's law
- Resistors in series
- Resistors in parallel
- Resistor colour code
- Resistivity
- Potential divider
- Practical work: Measuring resistance

Electrons move more easily through some conductors than others when a p.d. is applied. The opposition of a conductor to current is called its **resistance**. A good conductor has a low resistance and a poor conductor has a high resistance. The resistance of a wire of a certain material

(i) increases as its length increases,
(ii) increases as its cross-sectional area decreases,
(iii) depends on the material.

A long thin wire has more resistance than a short thick one of the same material. Silver is the best conductor, but copper, the next best, is cheaper and is used for connecting wires and for domestic electric cables.

● The ohm

If the current in a conductor is I when the voltage across it is V, as shown in Figure 38.1a, its resistance R is defined by

$$R = \frac{V}{I}$$

This is a reasonable way to measure resistance since the smaller I is for a given V, the greater is R. If V is in volts and I in amperes, then R is in **ohms** (symbol Ω, the Greek letter omega). For example, if $I = 2\,A$ when $V = 12\,V$, then $R = 12\,V/2\,A$, that is, $R = 6\,\Omega$.

The ohm is the resistance of a conductor in which the current is 1 ampere when a voltage of 1 volt is applied across it.

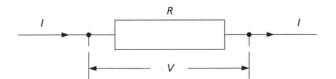

Figure 38.1a

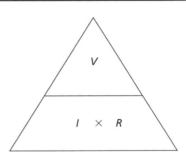

Figure 38.1b

Alternatively, if R and I are known, V can be found from

$$V = IR$$

Also, when V and R are known, I can be calculated from

$$I = \frac{V}{R}$$

The triangle in Figure 38.1b is an aid to remembering the three equations. It is used like the 'density triangle' in Chapter 5.

● Resistors

Conductors intended to have resistance are called **resistors** (Figure 38.2a) and are made either from wires of special alloys or from carbon. Those used in radio and television sets have values from a few ohms up to millions of ohms (Figure 38.2b).

Figure 38.2a Circuit symbol for a resistor

Figure 38.2b Resistor

Figure 38.2c Variable resistor (potentiometer)

Variable resistors are used in electronics (and are then called **potentiometers**) as volume and other controls (Figure 38.2c). Variable resistors that take larger currents, like the one shown in Figure 38.3, are useful in laboratory experiments. These consist of a coil of constantan wire (an alloy of 60% copper, 40% nickel) wound on a tube with a sliding contact on a metal bar above the tube.

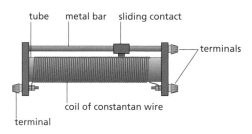

Figure 38.3 Large variable resistor

There are two ways of using such a variable resistor. It may be used as a **rheostat** for changing the current in a circuit; only one end connection and the sliding contact are then required. In Figure 38.4a moving the sliding contact to the left reduces the resistance and increases the current. This variable resistor can also act as a **potential divider** for changing the p.d. applied to a device; all three connections are then used. In Figure 38.4b any fraction from the total p.d. of the battery to zero can be 'tapped off' by moving the sliding contact down. Figure 38.5 shows the circuit diagram symbol for a variable resistor being used in rheostat mode.

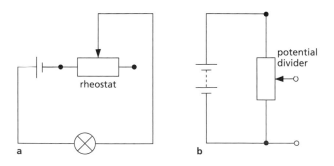

Figure 38.4 A variable resistor can be used as a rheostat or as a potential divider.

Figure 38.5 Circuit symbol for a variable resistor used as a rheostat

Practical work

Measuring resistance

The resistance R of a conductor can be found by measuring the current I in it when a p.d. V is applied across it and then using $R = V/I$. This is called the **ammeter–voltmeter** method.

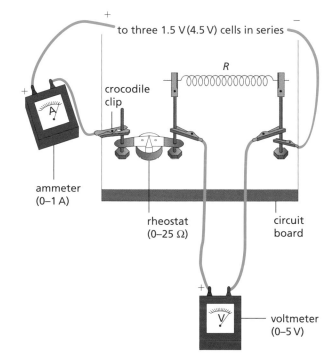

Figure 38.6

Set up the circuit of Figure 38.6 in which the unknown resistance R is 1 metre of SWG 34 constantan wire. Altering the rheostat changes both the p.d. V and the current I. Record in a table, with three columns, five values of I (e.g. 0.10, 0.15, 0.20, 0.25 and 0.3 A) and the corresponding values of V. Work out R for each pair of readings.

Repeat the experiment, but instead of the wire use **(i)** a lamp (e.g. 2.5 V, 0.3 A), **(ii)** a semiconductor diode (e.g. 1 N4001) connected first one way then the other way round, **(iii)** a thermistor (e.g. TH 7). (Semiconductor diodes and thermistors are considered in Chapter 41 in more detail.)

● *I–V* graphs: Ohm's law

The results of the previous experiment allow graphs of I against V to be plotted for different conductors.

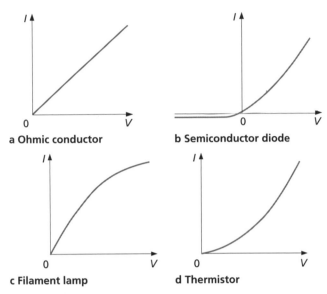

a Ohmic conductor

b Semiconductor diode

c Filament lamp

d Thermistor

Figure 38.7 *I–V* graphs

a) Metallic conductors

Metals and some alloys give *I–V* graphs that are a straight line through the origin, as in Figure 38.7a, provided that their temperature is constant. I is directly proportional to V, i.e. $I \propto V$. Doubling V doubles I, etc. Such conductors obey **Ohm's law**, stated as follows.

> The current in a metallic conductor is directly proportional to the p.d. across its ends if the temperature and other conditions are constant.

They are called **ohmic** or **linear conductors** and since $I \propto V$, it follows that $V/I =$ a constant (obtained from the slope of the *I–V* graph). The resistance of an ohmic conductor therefore does not change when the p.d. does.

b) Semiconductor diode

The typical *I–V* graph in Figure 38.7b shows that current passes when the p.d. is applied in one direction but is almost zero when it acts in the opposite direction. A diode has a small resistance when connected one way round but a very large resistance when the p.d. is reversed. It conducts in one direction only and is a **non-ohmic conductor.**

c) Filament lamp

A filament lamp is a non-ohmic conductor at high temperatures. For a filament lamp the *I–V* graph bends over as V and I increase (Figure 38.7c). That is, the resistance (V/I) increases as I increases and makes the filament hotter.

d) Variation of resistance with temperature

In general, an increase of temperature increases the resistance of metals, as for the filament lamp in Figure 38.7c, but it decreases the resistance of semiconductors. The resistance of semiconductor **thermistors** (see Chapter 41) decreases if their temperature rises, i.e. their *I–V* graph bends upwards, as in Figure 38.7d.

If a resistor and a thermistor are connected as a potential divider (Figure 38.8), the p.d. across the resistor increases as the temperature of the thermistor increases; the circuit can be used to monitor temperature, for example in a car radiator.

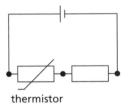

thermistor

Figure 38.8 Potential divider circuit for monitoring temperature

e) Variation of resistance with light intensity

The resistance of some semiconducting materials decreases when the intensity of light falling on them increases. This property is made use of in **light-dependent resistors** (LDRs) (see Chapter 41). The *I–V* graph for an LDR is similar to that shown in Figure 38.7d for a thermistor. Both thermistors and LDRs are non-ohmic conductors.

● Resistors in series

The resistors in Figure 38.9 are in series. The same current I flows through each and the total voltage V across all three is the sum of the separate voltages across them, i.e.

$$V = V_1 + V_2 + V_3$$

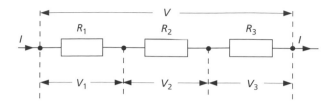

Figure 38.9 Resistors in series

But $V_1 = IR_1$, $V_2 = IR_2$ and $V_3 = IR_3$. Also, if R is the combined resistance, $V = IR$, and so

$$IR = IR_1 + IR_2 + IR_3$$

Dividing both sides by I,

$$R = R_1 + R_2 + R_3$$

● Resistors in parallel

The resistors in Figure 38.10 are in parallel. The **voltage V between the ends of each is the same** and the total current I equals the sum of the currents in the separate branches, i.e.

$$I = I_1 + I_2 + I_3$$

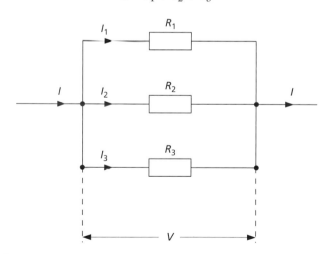

Figure 38.10 Resistors in parallel

But $I_1 = V/R_1$, $I_2 = V/R_2$ and $I_3 = V/R_3$. Also, if R is the combined resistance, $I = V/R$,

$$\frac{V}{R} = \frac{V}{R_1} + \frac{V}{R_2} + \frac{V}{R_3}$$

Dividing both sides by V,

$$\frac{1}{R} = \frac{1}{R_1} + \frac{1}{R_2} + \frac{1}{R_3}$$

For the simpler case of *two* resistors in parallel

$$\frac{1}{R} = \frac{1}{R_1} + \frac{1}{R_2} = \frac{R_2}{R_1 R_2} + \frac{R_1}{R_1 R_2}$$

$$\therefore \qquad \frac{1}{R} = \frac{R_2 + R_1}{R_1 R_2}$$

Inverting both sides,

$$R = \frac{R_1 R_2}{R_1 + R_2} = \frac{\text{product of resistances}}{\text{sum of resistances}}$$

The combined resistance of two resistors in parallel is less than the value of either resistor alone. Check this is true in the following Worked example. Lamps are connected in parallel rather than in series in a lighting circuit. Can you suggest why? (See p.180 for the advantages.)

● Worked example

A p.d. of 24 V from a battery is applied to the network of resistors in Figure 38.11a.

a What is the combined resistance of the 6 Ω and 12 Ω resistors in parallel?

b What is the current in the 8 Ω resistor?

c What is the voltage across the parallel network?

d What is the current in the 6 Ω resistor?

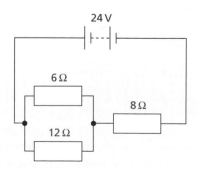

Figure 38.11a

a Let R_1 = resistance of $6\,\Omega$ and $12\,\Omega$ in parallel. Then

$$\frac{1}{R_1} = \frac{1}{6} + \frac{1}{12} = \frac{2}{12} + \frac{1}{12} = \frac{3}{12}$$

$$\therefore \quad R_1 = \frac{12}{3} = 4\,\Omega$$

b Let R = total resistance of circuit = $4 + 8$, that is, $R = 12\,\Omega$. The equivalent circuit is shown in Figure 38.11b, and if I is the current in it then, since $V = 24\,V$

$$I = \frac{V}{R} = \frac{24\,V}{12\,\Omega} = 2\,A$$

$\therefore$ current in $8\,\Omega$ resistor = $2\,A$

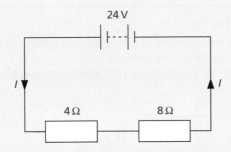

Figure 38.11b

c Let V_1 = voltage across parallel network in Figure 38.11a. Then

$$V_1 = I \times R_1 = 2\,A \times 4\,\Omega = 8\,V$$

d Let I_1 = current in $6\,\Omega$ resistor, then since $V_1 = 8\,V$

$$I_1 = \frac{V_1}{6\,\Omega} = \frac{8\,V}{6\,\Omega} = \frac{4}{3}\,A$$

● Resistor colour code

Resistors have colour coded bands as shown in Figure 38.12. In the orientation shown the first two bands on the left give digits 2 and 7; the third band gives the number of noughts (3) and the fourth band gives the resistor's 'tolerance' (or accuracy, here $\pm 10\%$). So the resistor has a value of $27\,000\,\Omega\,(\pm 10\%)$.

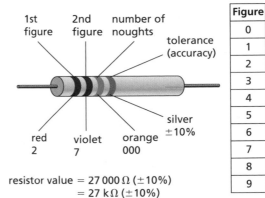

Figure	Colour
0	black
1	brown
2	red
3	orange
4	yellow
5	green
6	blue
7	violet
8	grey
9	white

Tolerance	
$\pm 5\%$	gold
$\pm 10\%$	silver
$\pm 20\%$	no band

Figure 38.12 Colour code for resistors

● Resistivity

Experiments show that the resistance R of a wire of a given material is

(i) directly proportional to its length l, i.e. $R \propto l$,
(ii) inversely proportional to its cross-sectional area A, i.e. $R \propto 1/A$ (doubling A halves R).

Combining these two statements, we get

$$R \propto \frac{1}{A} \quad \text{or} \quad R = \frac{\rho l}{A}$$

where ρ is a constant, called the **resistivity** of the material. If we put $l = 1\,m$ and $A = 1\,m^2$, then $\rho = R$.

> The resistivity of a material is numerically equal to the resistance of a 1 m length of the material with cross-sectional area 1 m².

The unit of ρ is the **ohm-metre** ($\Omega\,m$), as can be seen by rearranging the equation to give $\rho = AR/l$ and inserting units for A, R and l. Knowing ρ for a material, the resistance of any sample of it can be calculated. The resistivities of metals increase at higher temperatures; for most other materials they decrease.

● Worked example

Calculate the resistance of a copper wire 1.0 km long and 0.50 mm diameter if the resistivity of copper is $1.7 \times 10^{-8}\,\Omega\,\mathrm{m}$.

Converting all units to metres, we get

length $l = 1.0\,\mathrm{km} = 1000\,\mathrm{m} = 10^3\,\mathrm{m}$
diameter $d = 0.50\,\mathrm{mm} = 0.50 \times 10^{-3}\,\mathrm{m}$

If r is the radius of the wire, the cross-sectional area $A = \pi r^2 = \pi (d/2)^2 = (\pi/4)d^2$, so

$$A = \frac{\pi}{4}\left(0.50 \times 10^{-3}\right)^2\,\mathrm{m}^2 \approx 0.20 \times 10^{-6}\,\mathrm{m}^2$$

Then

$$R = \frac{\rho l}{A} = \frac{\left(1.7 \times 10^{-8}\,\Omega\,\mathrm{m}\right) \times \left(10^3\,\mathrm{m}\right)}{0.20 \times 10^{-6}\,\mathrm{m}^2} = 85\,\Omega$$

● Potential divider

In the circuit shown in Figure 38.13, two resistors R_1 and R_2 are in series with a supply of voltage V. The current in the circuit is

$$I = \frac{\text{supply voltage}}{\text{total resistance}} = \frac{V}{(R_1 + R_2)}$$

So the voltage across R_1 is

$$V_1 = I \times R_1 = \frac{V \times R_1}{(R_1 + R_2)} = V \times \frac{R_1}{(R_1 + R_2)}$$

and the voltage across R_2 is

$$V_2 = I \times R_2 = \frac{V \times R_2}{(R_1 + R_2)} = V \times \frac{R_2}{(R_1 + R_2)}$$

Also the ratio of the voltages across the two resistors is

$$\frac{V_1}{V_2} = \frac{R_1}{R_2}$$

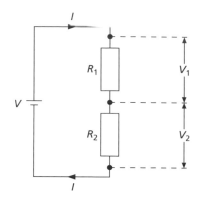

Figure 38.13 Potential divider circuit

Returning to Figure 38.8 (p. 169), can you now explain why the voltage across the resistor increases when the resistance of the thermistor decreases?

Questions

1 What is the resistance of a lamp when a voltage of 12 V across it causes a current of 4 A?
2 Calculate the p.d. across a 10 Ω resistor carrying a current of 2 A.
3 The p.d. across a 3 Ω resistor is 6 V. What is the current flowing (in ampere)?

  A $\frac{1}{2}$ B 1 C 2 D 6 E 8

4 The resistors R_1, R_2, R_3 and R_4 in Figure 38.14 are all equal in value. What would you expect the voltmeters A, B and C to read, assuming that the connecting wires in the circuit have negligible resistance?

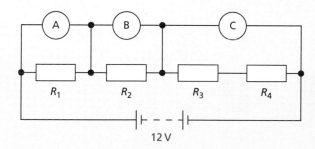

Figure 38.14

5 Calculate the effective resistance between A and B in Figure 38.15.

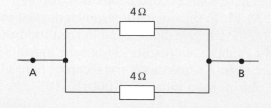

Figure 38.15

6 What is the effective resistance in Figure 38.16 between
   a A and B,
   b C and D?

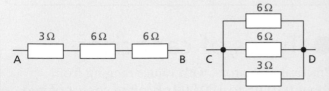
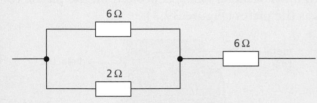

Figure 38.16

7 Figure 38.17 shows three resistors. Their combined resistance in ohms is

A $1\frac{5}{7}$ B 14   C $1\frac{1}{5}$ D $7\frac{1}{2}$ E $6\frac{2}{3}$

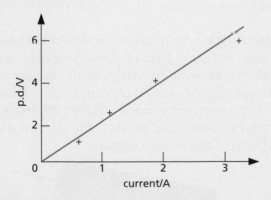

Figure 38.17

8 a The graph in Figure 38.18 illustrates how the p.d. across the ends of a conductor is related to the current in it.
     (i) What law may be deduced from the graph?
     (ii) What is the resistance of the conductor?
   b Draw diagrams to show how six 2 V lamps could be lit to normal brightness when using a
     (i)   2 V supply,
     (ii) 6 V supply,
     (iii) 12 V supply.

Figure 38.18

9 When a 4 Ω resistor is connected across the terminals of a 12 V battery, the number of coulombs passing through the resistor per second is

A 0.3   B 3   C 4   D 12   E 48

Checklist

After studying this chapter you should be able to

- define resistance and state the factors on which it depends,
- recall the unit of resistance,
- solve simple problems using $R = V/I$,
- describe experiments using the ammeter–voltmeter method to measure resistance, and study the relationship between current and p.d. for (a) metallic conductors, (b) semiconductor diodes, (c) filament lamps, (d) thermistors, (e) LDRs,
- plot I–V graphs from the results of such experiments and draw appropriate conclusions from them,
- use the formulae for resistors in series,
- recall that the combined resistance of two resistors in parallel is less than that of either resistor alone,

- calculate the effective resistance of two resistors in parallel,

- relate the resistance of a wire to its length and diameter,
- calculate voltages in a potential divider circuit.

(39) Capacitors

- Capacitance
- Types of capacitor
- Charging and discharging a capacitor
- Effect of capacitors in d.c. and a.c. circuits

A **capacitor** stores electric charge and is useful in many electronic circuits. In its simplest form it consists of two parallel metal plates separated by an insulator, called the **dielectric** (Figure 39.1a). Figure 39.1b shows the circuit symbol for a capacitor.

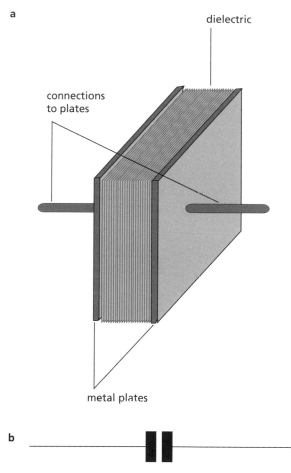

Figure 39.1a A parallel-plate capacitor; b symbol for a capacitor

● Capacitance

The more charge a capacitor can store, the greater is its **capacitance** (C). The capacitance is large when the plates have a large area and are close together. It is measured in **farads** (F) but smaller units such as the microfarad (µF) are more convenient.

$$1\,\mu F = 1 \text{ millionth of a farad} = 10^{-6}\,F$$

● Types of capacitor

Practical capacitors, with values ranging from about 0.01 µF to 100 000 µF, often consist of two long strips of metal foil separated by long strips of dielectric, rolled up like a 'Swiss roll', as in Figure 39.2. The arrangement allows plates of large area to be close together in a small volume. Plastics (e.g. polyesters) are commonly used as the dielectric, with films of metal being deposited on the plastic to act as the plates (Figure 39.3).

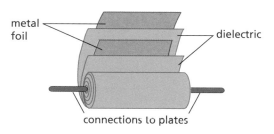

Figure 39.2 Construction of a practical capacitor

Figure 39.3 Polyester capacitor

The **electrolytic** type of capacitor shown in Figure 39.4a has a very thin layer of aluminium oxide as the dielectric between two strips of aluminium foil, giving large capacitances. It is polarised, i.e. it has positive and negative terminals (Figure 39.4b), and these *must* be connected to the + and – terminals, respectively, of the voltage supply.

Figure 39.4a Electrolytic capacitor; b symbol showing polarity

174

● Charging and discharging a capacitor

a) Charging

A capacitor can be charged by connecting a battery across it. In Figure 39.5a, the + terminal of the battery attracts electrons (since they have a negative charge) from plate X and the – terminal of the battery repels electrons to plate Y. A positive charge builds up on plate X (since it loses electrons) and an equal negative charge builds up on Y (since it gains electrons).

During the charging, there is a *brief* flow of electrons round the circuit from X to Y (but not through the dielectric). A momentary current would be detected by a sensitive ammeter. The voltage builds up between X and Y and opposes the battery voltage. Charging stops when these two voltages are equal; the electron flow, i.e. the charging current, is then zero. The variation of *current* with time (for both charging or discharging a capacitor) has a similar shape to the curve shown in Figure 39.7b.

During the charging process, electrical energy is transferred from the battery to the capacitor, which then stores the energy.

b) Discharging

When a conductor is connected across a charged capacitor, as in Figure 39.5b, there is a brief flow of electrons from the negatively charged plate to the positively charged one, i.e. from Y to X. The charge stored by the capacitor falls to zero, as does the voltage across it. The capacitor has transferred its stored energy to the conductor. The 'delay' time taken for a capacitor to fully charge or discharge through a resistor is made use of in many electronic circuits.

c) Demonstration

The circuit in Figure 39.6 has a two-way switch S. When S is in position 1 the capacitor C charges up, and discharges when S is in position 2. The larger the values of R and C the longer it takes for the capacitor to charge or discharge; with the values shown in Figure 39.6, the capacitor will take 2 to 3 minutes to fully charge or discharge. The direction of the deflection of the centre-zero milliammeter reverses for each process. The corresponding changes of capacitor charge (measured by the voltage across it) with time are shown by the graphs in Figures 39.7a and b. These can be plotted directly if the voltmeter is replaced by a datalogger and computer.

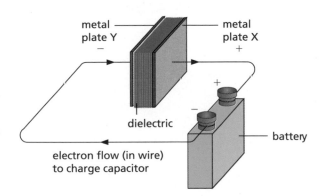

Figure 39.5a Charging a capacitor

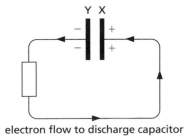

Figure 39.5b Discharging a capacitor

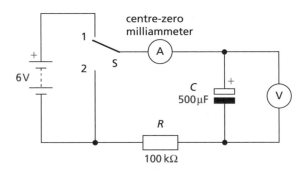

Figure 39.6 Demonstration circuit for charging and discharging a capacitor

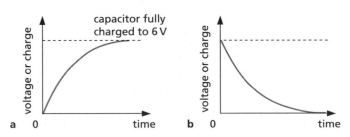

Figure 39.7 Graphs: **a** charging; **b** discharging

● Effect of capacitors in d.c. and a.c. circuits

a) Direct current circuit

In Figure 39.8a the supply is d.c. but the lamp does not light, that is, a capacitor blocks direct current.

b) Alternating current circuit

In Figure 39.8b the supply is a.c. and the lamp lights, suggesting that a capacitor passes alternating current. In fact, no current actually passes *through* the capacitor since its plates are separated by an insulator. But as the a.c. reverses direction, the capacitor charges and discharges, causing electrons to flow to and fro rapidly in the wires joining the plates. Thus, effectively, a.c. flows round the circuit, lighting the lamp.

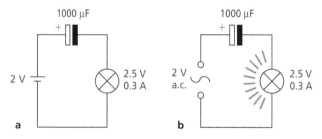

Figure 39.8 A capacitor blocks direct current and allows a flow of alternating current.

Checklist

After studying this chapter you should be able to

- state what a capacitor does,
- state the unit of capacitance,
- describe in terms of electron motion how a capacitor can be charged and discharged, and sketch graphs of the capacitor voltage with time for charging and discharging through a resistor,
- recall that a capacitor blocks d.c. but passes a.c. and explain why.

Questions

1 a Describe the basic construction of a capacitor.
  b What does a capacitor do?
  c State two ways of increasing the capacitance of a capacitor.
  d Name a unit of capacitance.
2 a When a capacitor is being charged, is the value of the charging current maximum or zero
    (i) at the start, or
    (ii) at the end of charging?
  b When a capacitor is discharging, is the value of the current in the circuit maximum or zero
    (i) at the start, or
    (ii) at the end of charging?
3 How does a capacitor behave in a circuit with
  a a d.c. supply,
  b an a.c. supply?

40 Electric power

- Power in electric circuits
- Electric lighting
- Electric heating
- Joulemeter

- House circuits
- Paying for electricity
- Dangers of electricity
- Practical work: Measuring electric power.

● Power in electric circuits

In many circuits it is important to know the rate at which electrical energy is transferred into other forms of energy. Earlier (Chapter 13) we said that **energy transfers were measured by the work done** and power was defined by the equation

$$\text{power} = \frac{\text{work done}}{\text{time taken}} = \frac{\text{energy transfer}}{\text{time taken}}$$

In symbols

$$P = \frac{E}{t} \qquad (1)$$

where if E is in joules (J) and t in seconds (s) then P is in J/s or watts (W).

From the definition of p.d. (Chapter 37) we saw that if E is the electrical energy transferred when there is a steady current I (in amperes) for time t (in seconds) in a device (e.g. a lamp) with a p.d. V (in volts) across it, as in Figure 40.1, then

$$E = ItV \qquad (2)$$

Substituting for E in (1) we get

$$P = \frac{E}{t} = \frac{ItV}{t}$$

or

$$P = IV$$

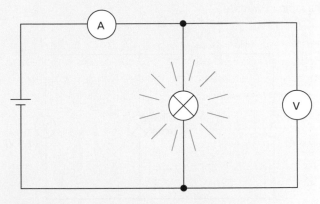

Figure 40.1

Therefore to calculate the power P of an electrical appliance we multiply the current I in it by the p.d. V across it. For example if a lamp on a 240 V supply has a current of 0.25 A in it, its power is 240 V × 0.25 A = 60 W. The lamp is transferring 60 J of electrical energy into heat and light each second. Larger units of power are the **kilowatt** (kW) and the **megawatt** (MW) where

$$1\,\text{kW} = 1000\,\text{W} \quad \text{and} \quad 1\,\text{MW} = 1\,000\,000\,\text{W}$$

In units

$$\text{watts} = \text{amperes} \times \text{volts} \qquad (3)$$

It follows from (3) that since

$$\text{volts} = \frac{\text{watts}}{\text{amperes}} \qquad (4)$$

the volt can be defined as a **watt per ampere** and p.d. calculated from (4).

If all the energy is transferred to heat in a resistor of resistance R, then $V = IR$ and the rate of production of heat is given by

$$P = V \times I = IR \times I = I^2 R$$

That is, if the current is doubled, four times as much heat is produced per second. Also, $P = V^2/R$.

Measuring electric power

a) Lamp

Connect the circuit of Figure 40.2. Note the ammeter and voltmeter readings and work out the electric power supplied to the lamp in watts.

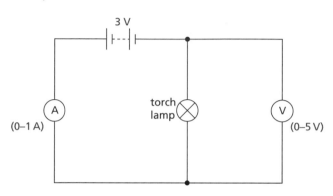

Figure 40.2

b) Motor

Replace the lamp in Figure 40.2 by a small electric motor. Attach a known mass m (in kg) to the axle of the motor with a length of thin string and find the time t (in s) required to raise the mass through a known height h (in m) at a steady speed. Then the power output P_o (in W) of the motor is given by

$$P_o = \frac{\text{work done in raising mass}}{\text{time taken}} = \frac{mgh}{t}$$

If the ammeter and voltmeter readings I and V are noted while the mass is being raised, the power input P_i (in W) can be found from

$$P_i = IV$$

The efficiency of the motor is given by

$$\text{efficiency} = \frac{P_o}{P_i} \times 100\%$$

Also investigate the effect of a greater mass on: (i) the speed, (ii) the power output and (iii) the efficiency of the motor at its rated p.d.

● Electric lighting

a) Filament lamps

The **filament** is a small coil of tungsten wire (Figure 40.3) which becomes white hot when there is a current in it. The higher the temperature of the filament, the greater is the proportion of electrical energy transferred to light and for this reason it is made of tungsten, a metal with a high melting point (3400 °C).

Most lamps are gas-filled and contain nitrogen and argon, not air. This reduces evaporation of the tungsten which would otherwise condense on the bulb and blacken it. The coil is coiled compactly so that it is cooled less by convection currents in the gas.

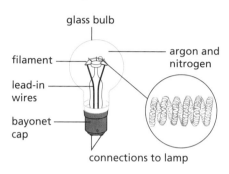

Figure 40.3 A filament lamp

b) Fluorescent strips

A filament lamp transfers only 10% of the electrical energy supplied to light; the other 90% becomes heat. Fluorescent strip lamps (Figure 40.4a) are five times as efficient and may last 3000 hours compared with the 1000-hour life of filament lamps. They cost more to install but running costs are less.

When a fluorescent strip lamp is switched on, the mercury vapour emits invisible ultraviolet radiation which makes the powder on the inside of the tube fluoresce (glow), i.e. visible light is emitted. Different powders give different colours.

c) Compact fluorescent lamps

These energy-saving fluorescent lamps (Figure 40.4b) are available to fit straight into normal light sockets, either bayonet or screw-in. They last up to eight times longer (typically 8000 hours) and use about five times less energy than filament lamps for the same light output. For example, a 20 W compact fluorescent is equivalent to a 100 W filament lamp.

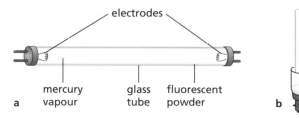

Figure 40.4 Fluorescent lamps

● Electric heating

a) Heating elements

In domestic appliances such as electric fires, cookers, kettles and irons the 'elements' (Figure 40.5) are made from Nichrome wire. This is an alloy of nickel and chromium which does not oxidise (and so become brittle) when the current makes it red hot.

The elements in **radiant** electric fires are at red heat (about 900 °C) and the radiation they emit is directed into the room by polished reflectors. In **convector** types the element is below red heat (about 450 °C) and is designed to warm air which is drawn through the heater by natural or forced convection. In **storage** heaters the elements heat fire-clay bricks during the night using 'off-peak' electricity. On the following day these cool down, giving off the stored heat to warm the room.

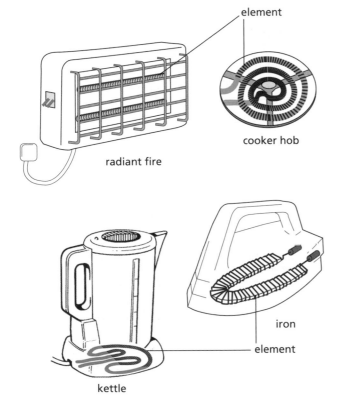

Figure 40.5 Heating elements

b) Three-heat switch

This is sometimes used to control heating appliances. It has three settings and uses two identical elements. On 'high', the elements are in parallel across the supply voltage (Figure 40.6a); on 'medium', there is only current in one (Figure 40.6b); on 'low', they are in series (Figure 40.6c).

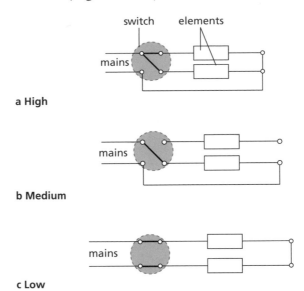

Figure 40.6 Three-heat switch

c) Fuses

A **fuse** protects a circuit. It is a short length of wire of material with a low melting point, often 'tinned copper', which melts and breaks the circuit when the current in it exceeds a certain value. Two reasons for excessive currents are 'short circuits' due to worn insulation on connecting wires and overloaded circuits. Without a fuse the wiring would become hot in these cases and could cause a fire. **A fuse should ensure that the current-carrying capacity of the wiring is not exceeded.** In general the thicker a cable is, the more current it can carry, but each size has a limit.

Two types of fuse are shown in Figure 40.7a. **Always switch off before replacing a fuse**, and always replace with one of the same value as recommended by the manufacturer of the appliance.

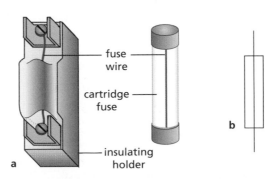

Figure 40.7a Two types of fuse; **b** the circuit symbol for a fuse

● Joulemeter

Instead of using an ammeter and a voltmeter to measure the electrical energy transferred by an appliance, a **joulemeter** can be used to measure it directly in joules. The circuit connections are shown in Figure 40.8. A household electricity meter (Figure 40.12) is a joulemeter.

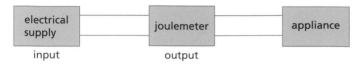

Figure 40.8 Connections to a joulemeter

● House circuits

Electricity usually comes to our homes by an underground cable containing two wires, the **live** (L) and the **neutral** (N). The neutral is earthed at the local sub-station and so there is no p.d. between it and earth. The supply is a.c. (Chapter 36) and the live wire is alternately positive and negative. Study the typical house circuits shown in Figure 40.9.

a) Circuits in parallel

Every circuit is connected in parallel with the supply, i.e. across the live and neutral, and receives the full mains p.d. of 230 V (in the UK). The advantages of having appliances connected in parallel, rather than in series, can be seen by studying the lighting circuit in Figure 40.9.

(i) The p.d. across each lamp is fixed (at the mains p.d.), so the lamp shines with the same brightness irrespective of how many other lamps are switched on.

(ii) Each lamp can be turned on and off independently; if one lamp fails, the others can still be operated.

b) Switches and fuses

These are always in the live wire. If they were in the neutral, light switches and power sockets would be 'live' when switches were 'off' or fuses 'blown'. A fatal shock could then be obtained by, for example, touching the element of an electric fire when it was switched off.

c) Staircase circuit

The light is controlled from two places by the two two-way switches.

d) Ring main circuit

The live and neutral wires each run in two complete rings round the house and the power sockets, each rated at 13 A, are tapped off from them. Thinner wires can be used since the current to each socket flows by two paths, i.e. from both directions in the ring. The ring has a 30 A fuse and if it has, say, ten sockets, then all can be used so long as the total current does not exceed 30 A, otherwise the wires overheat. A house may have several ring circuits, each serving a different area.

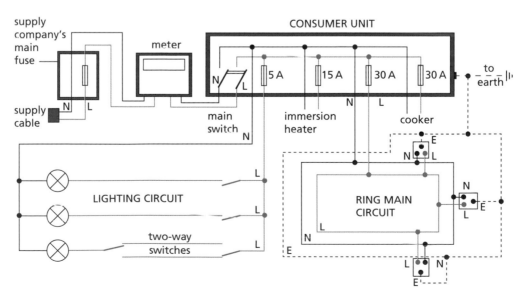

Figure 40.9 Electric circuits in a house

e) Fused plug

Only one type of plug is used in a UK ring main circuit. It is wired as in Figure 40.10a. Note the colours of the wire coverings: L – brown, N – blue, E – green and yellow. It has its own cartridge fuse, 3 A (red) for appliances with powers up to 720 W, or 13 A (brown) for those between 720 W and 3 kW.

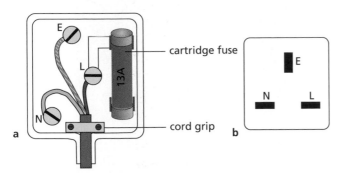

Figure 40.10 a Wiring of a plug; **b** socket

Typical power ratings for various appliances are shown in Table 40.1, p. 182. Calculation of a current in a device allows the correct size of fuse to be chosen.

In some countries the fuse is placed in the appliance rather than in the plug.

f) Safety in electrical circuits

Earthing

A ring main has a third wire which goes to the top sockets on all power points (Figure 40.9) and is earthed by being connected either to a **metal** water pipe entering the house or to an earth connection on the supply cable. This third wire is a safety precaution to prevent electric shock should an appliance develop a fault.

The earth pin on a three-pin plug is connected to the metal case of the appliance which is thus joined to earth by a path of almost zero resistance. If then, for example, the element of an electric fire breaks or sags and touches the case, a large current flows to earth and 'blows' the fuse. Otherwise the case would become 'live' and anyone touching it would receive a shock which might be fatal, especially if they were 'earthed' by, say, standing in a damp environment, such as on a wet concrete floor.

Circuit breakers

Figure 40.11 Circuit breakers

Circuit breakers (Figure 40.11) are now used instead of fuses in consumer units. They contain an electromagnet (Chapter 45) which, when the current exceeds the rated value of the circuit breaker, becomes strong enough to separate a pair of contacts and breaks the circuit. They operate much faster than fuses and have the advantage that they can be reset by pressing a button.

The **residual current circuit breaker** (RCCB), also called a **residual current device** (RCD), is an adapted circuit breaker which is used when the resistance of the earth path between the consumer and the substation is not small enough for a fault-current to blow the fuse or trip the circuit breaker. It works by detecting any difference between the currents in the live and neutral wires; when these become unequal due to an earth fault (i.e. some of the current returns to the substation via the case of the appliance and earth) it breaks the circuit before there is any danger. They have high sensitivity and a quick response.

An RCD should be plugged into a socket supplying power to a portable appliance such as an electric lawnmower or hedge trimmer. In these cases the risk of electrocution is greater because the user is generally making a good earth connection through the feet.

Double insulation

Appliances such as vacuum cleaners, hairdryers and food mixers are usually **double insulated**.

Connection to the supply is by a two-core insulated cable, with no earth wire, and the appliance is enclosed in an insulating plastic case. Any metal attachments that the user might touch are fitted into this case so that they do not make a direct connection with the internal electrical parts, such as a motor. There is then no risk of a shock should a fault develop.

● Paying for electricity

Electricity supply companies charge for the **electrical energy** they supply. A joule is a very small amount of energy and a larger unit, the **kilowatt-hour** (kWh), is used.

> A kilowatt-hour is the electrical energy used by a 1 kW appliance in 1 hour.

$$1 \text{ kWh} = 1000 \text{ J/s} \times 3600 \text{ s}$$
$$= 3\,600\,000 \text{ J} = 3.6 \text{ MJ}$$

A 3 kW electric fire working for 2 hours uses 6 kWh of electrical energy – usually called 6 'units'. Electricity meters, which are joulemeters, are marked in kWh: the latest have digital readouts like the one in Figure 40.12. At present a 'unit' costs about 8p in the UK.

Typical powers of some appliances are given in Table 40.1.

Table 40.1 Power of some appliances

DVD player	20 W	iron	1 kW
laptop computer	50 W	fire	1, 2, 3 kW
light bulbs	60, 100 W	kettle	2 kW
television	100 W	immersion heater	3 kW
fridge	150 W	cooker	6.4 kW

Note that the current required by a 6.4 kW cooker is given by

$$I = \frac{P}{V} = \frac{6400 \text{ W}}{230 \text{ V}} = 28 \text{ A}$$

This is too large a current to draw from the ring main and so a separate circuit must be used.

Figure 40.12 Electricity meter with digital display

● Dangers of electricity

a) Electric shock

Electric shock occurs if current flows from an electric circuit through a person's body to earth. This can happen if there is **damaged insulation** or **faulty wiring**. The typical resistance of dry skin is about $10\,000\,\Omega$, so if a person touches a wire carrying electricity at 240 V, an estimate of the current flowing through them to earth would be $I = V/R = 240/10\,000 = 0.024 \text{ A} = 24 \text{ mA}$. For wet skin, the resistance is lowered to about $1000\,\Omega$ (since water is a good conductor of electricity) so the current would increase to around 240 mA.

It is the **size of the current** (not the voltage) and the **length of time** for which it acts which determine the strength of an electric shock. The path the current takes influences the effect of the shock; some parts of the body are more vulnerable than others. A current of 100 mA through the heart is likely to be fatal.

Damp conditions increase the severity of an electric shock because water lowers the resistance of the path to earth; wearing shoes with insulating rubber soles or standing on a dry insulating floor increases the resistance between a person and earth and will reduce the severity of an electric shock.

To avoid the risk of getting an electric shock:

- Switch off the electrical supply to an appliance before starting repairs.
- Use plugs that have an earth pin and a cord grip; a rubber or plastic case is preferred.
- Do not allow appliances or cables to come into contact with water. For example holding a hairdryer with wet hands in a bathroom can be dangerous. Keep electrical appliances well away from baths and swimming pools!
- Do not have long cables trailing across a room, under a carpet that is walked over regularly or in other situations where the insulation can become damaged. Take particular care when using electrical cutting devices (such as hedge cutters) not to cut the supply cable.

In case of an electric shock, take the following action:

1 **Switch off the supply** if the shocked person is still touching the equipment.
2 **Send for qualified medical assistance.**
3 **If breathing or heartbeat has stopped, commence CPR** (cardiopulmonary resuscitation) by applying chest compressions at the rate of about 100 a minute until there are signs of chest movement or medical assistance arrives.

b) Fire risks

If flammable material is placed too close to a hot appliance such as an electric heater, it may catch fire. Similarly if the electrical wiring in the walls of a house becomes overheated, a fire may start. Wires become hot when they carry electrical currents – the larger the current carried, the hotter a particular wire will become, since the rate of production of heat equals I^2R (see p. 177).

To reduce the risk of fire through **overheated cables**, the maximum current in a circuit should be limited by taking these precautions:

- Use plugs that have the correct fuse.
- Do not attach too many appliances to a circuit.
- Don't overload circuits by using too many adapters.
- Appliances such as heaters use large amounts of power (and hence current), so do not connect them

to a lighting circuit designed for low current use. (Thick wires have a lower resistance than thin wires so are used in circuits expected to carry high currents.)

Damaged insulation or faulty wiring which leads to a large current flowing to earth through flammable material can also start a fire.

The factors leading to fire or electric shock can be summarised as follows:

damaged insulation	→ electric shock and fire risk
overheated cables	→ fire risk
damp conditions	→ increased severity of electric shocks

Questions

1 How much electrical energy in joules does a 100 watt lamp transfer in
   a 1 second,
   b 5 seconds,
   c 1 minute?
2 a What is the power of a lamp rated at 12 V 2 A?
   b How many joules of electrical energy are transferred per second by a 6 V 0.5 A lamp?
3 The largest number of 100 W lamps connected in parallel which can safely be run from a 230 V supply with a 5 A fuse is
   A 2   B 5   C 11   D 12   E 0
4 What is the maximum power in kilowatts of the appliance(s) that can be connected safely to a 13 A 230 V mains socket?
5 The circuits of Figures 40.13a and b show 'short circuits' between the live (L) and neutral (N) wires. In both, the fuse has blown but whereas circuit a is now safe, b is still dangerous even though the lamp is out which suggests the circuit is safe. Explain.

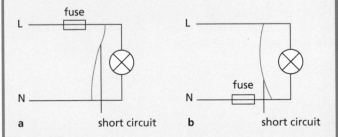

Figure 40.13

6 What steps should be taken before replacing a blown fuse in a plug?
7 What size fuse (3 A or 13 A) should be used in a plug connected to
   a a 150 W television,
   b a 900 W iron,
   c a 2 kW kettle,
   if the supply is 230 V?   ▶▶

8 What is the cost of heating a tank of water with a 3000 W immersion heater for 80 minutes if electricity costs 10p per kWh?

9 a Below is a list of wattages of various appliances. State which is most likely to be the correct one for each of the appliances named.

60 W   250 W   850 W   2 kW   3.5 kW

   (i)   kettle
   (ii) table lamp
   (iii) iron

 b What will be the current in a 920 W appliance if the supply voltage is 230 V?

Checklist

After studying this chapter you should be able to

- recall the relations $E = ItV$ and $P = IV$ and use them to solve simple problems on energy transfers,

- describe experiments to measure electric power,

- describe electric lamps, heating elements and fuses,
- recall that a joulemeter measures electrical energy,
- describe with the aid of diagrams a house wiring system and explain the functions and positions of switches, fuses, circuit breakers and earth,
- state the advantages of connecting lamps in parallel in a lighting circuit,
- wire a mains plug and recall the international insulation colour code,
- perform calculations of the cost of electrical energy in joules and kilowatt-hours,
- recall the hazards of damaged insulation, damp conditions and overheating of cables and the associated risks.

- Electronic systems
- Input transducers
- Output transducers
- Semiconductor diode

- Transistor
- Transistor as a switch
- Practical work: Transistor switching circuits: light-operated, temperature-operated.

The use of electronics in our homes, factories, offices, schools, banks, shops and hospitals is growing all the time. The development of semiconductor devices such as transistors and integrated circuits ('chips') has given us, among other things, automatic banking machines, laptop computers, programmable control devices, robots, computer games, digital cameras (Figure 41.1a) and heart pacemakers (Figure 41.1b).

Figure 41.1a Digital camera

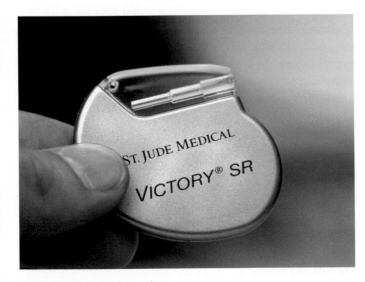

Figure 41.1b Heart pacemaker

● Electronic systems

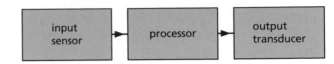

Figure 41.2 Electronic system

Any electronic system can be considered to consist of the three parts shown in the block diagram of Figure 41.2, i.e.

(i) an **input sensor** or **input transducer**,
(ii) a **processor** and
(iii) an **output transducer**.

A 'transducer' is a device for converting a non-electrical input into an electrical signal or vice versa.

The **input sensor** detects changes in the environment and converts them from their present form of energy into electrical energy. Input sensors or transducers include LDRs (light-dependent resistors), thermistors, microphones and switches that respond, for instance, to pressure changes.

The **processor** decides on what action to take on the electrical signal it receives from the input sensor. It may involve an operation such as counting, amplifying, timing or storing.

The **output transducer** converts the electrical energy supplied by the processor into another form. Output transducers include lamps, LEDs (light-emitting diodes), loudspeakers, motors, heaters, relays and cathode ray tubes.

In a radio, the input sensor is the aerial that sends an electrical signal to processors in the radio. These processors, among other things, amplify the signal so that it can enable the output transducer, in this case a loudspeaker, to produce sound.

● Input transducers

a) Light-dependent resistor (LDR)

The action of an LDR depends on the fact that the resistance of the semiconductor cadmium sulfide decreases as the intensity of the light falling on it increases.

An LDR and a circuit showing its action are shown in Figures 41.3a and b. Note the circuit symbol for an LDR, sometimes seen without a circle. When light from a lamp falls on the 'window' of the LDR, its resistance decreases and the increased current lights the lamp.

LDRs are used in photographic exposure meters and in series with a resistor to provide an input signal for a transistor (Figure 41.16, p. 190) or other switching circuit.

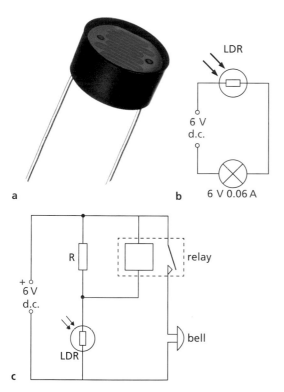

Figure 41.3 a LDR; **b** LDR demonstration circuit; **c** light-operated intruder alarm

Figure 41.3c shows how an LDR can be used to switch a 'relay' (Chapter 45.) The LDR forms part of a potential divider across the 6 V supply. When light falls on the LDR, the resistance of the LDR, and hence the voltage across it, decreases. There is a corresponding increase in the voltage across resistor R and the relay; when the voltage across the relay coil reaches a high enough p.d. (its operating p.d.) it acts as a switch and the normally open contacts close, allowing current to flow to the bell, which rings. If the light is removed,

the p.d. across resistor R and the relay drops below the operating p.d. of the relay so that the relay contacts open again; power to the bell is cut and it stops ringing.

b) Thermistor

A thermistor contains semiconducting metallic oxides whose resistance decreases markedly when the temperature rises. The temperature may rise either because the thermistor is directly heated or because a current is in it.

Figure 41.4a shows one type of thermistor. Figure 41.4b shows the symbol for a thermistor in a circuit to demonstrate how the thermistor works. When the thermistor is heated with a match, the lamp lights.

A thermistor in series with a meter marked in °C can measure temperatures (Chapter 38). Used in series with a resistor it can provide an input signal to a transistor (Figure 41.18, p. 191) or other switching circuit.

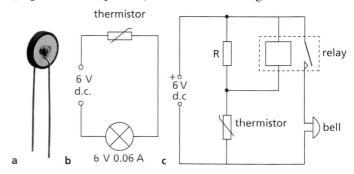

Figure 41.4 a Thermistor; **b** thermistor demonstration circuit; **c** high-temperature alarm

Figure 41.4c shows how a thermistor can be used to switch a relay. The thermistor forms part of a potential divider across the d.c. source. When the temperature rises, the resistance of the thermistor falls, and so does the p.d. across it. The voltage across resistor R and the relay increases. When the voltage across the relay reaches its operating p.d. the normally open contacts close, so that the circuit to the bell is completed and it rings. If a variable resistor is used in the circuit, the temperature at which the alarm sounds can be varied.

● Output transducers

a) Relays

A switching circuit cannot supply much power to an appliance so a relay is often included; this allows the small current provided by the switching circuit to control the larger current needed to operate a buzzer as in a temperature-operated switch (Figure 41.18, p. 191) or other device. Relays controlled by a switching circuit can also be used to switch on the mains supply for electrical appliances in the home. In Figure 41.5 if the output of the switching circuit is 'high' (5 V), a small current flows to the relay which closes the mains switch; the relay also isolates the low voltage circuit from the high voltage mains supply.

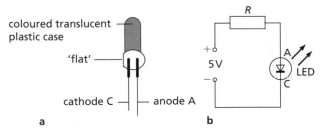

Figure 41.5 Use of a relay to switch mains supply

b) Light-emitting diode (LED)

An LED, shown in Figure 41.6a, is a diode made from the semiconductor gallium arsenide phosphide. When forward biased (with the cathode C connected to the negative terminal of the voltage supply, as shown in Figure 41.6b), the current in it makes it emit red, yellow or green light. No light is emitted on reverse bias (when the anode A is connected to the negative terminal of the voltage supply). If the reverse bias voltage exceeds 5 V, it may cause damage.

In use an LED must have a suitable resistor R in series with it (e.g. 300 Ω on a 5 V supply) to limit the current (typically 10 mA). Figure 41.6b shows the symbol for an LED (again the use of the circle is optional) in a demonstration circuit.

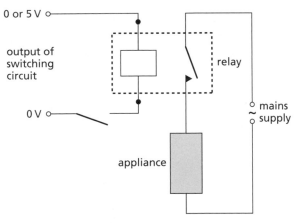

Figure 41.6 LED and demonstration circuit

LEDs are used as indicator lamps on computers, radios and other electronic equipment. Many clocks, calculators, video recorders and measuring instruments have seven-segment red or green numerical displays (Figure 41.7a). Each segment is an LED and, depending on which have a voltage across them, the display lights up the numbers 0 to 9, as in Figure 41.7b.

LEDs are small, reliable and have a long life; their operating speed is high and their current requirements are very low.

Diode lasers operate in a similar way to LEDs but emit coherent laser light; they are used in optical fibre communications as transmitters.

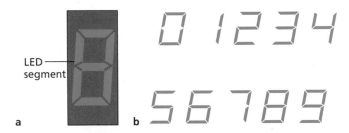

Figure 41.7 LED numerical display

● Semiconductor diode

A diode is a device that lets current pass in one direction only. One is shown in Figure 41.8 with its symbol. (You will also come across the symbol without its outer circle.) The wire nearest the band is the **cathode** and the one at the other end is the **anode**.

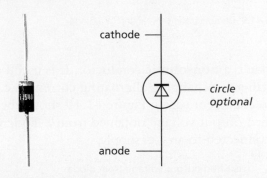

Figure 41.8 A diode and its symbol

The typical *I–V* graph is shown in Figure 38.7b (p. 169). The diode conducts when the anode goes to the + terminal of the voltage supply and the cathode to the – terminal (Figure 41.9a). It is then **forward-biased**; its resistance is small and conventional current passes in the direction of the arrow on its symbol. If the connections are the other way round, it does not conduct; its resistance is large and it is **reverse-biased** (Figure 41.9b). ▶▶

The lamp in the circuit shows when the diode is conducting, as the lamp lights up. It also acts as a resistor to limit the current when the diode is forward-biased. Otherwise the diode might overheat and be damaged.

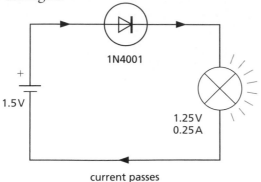

a

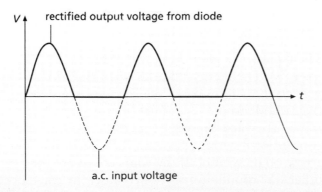

b

Figure 41.9 Demonstrating the action of a diode

A diode is a **non-ohmic** conductor. It is useful as a **rectifier** for changing alternating current (a.c.) to direct current (d.c.). Figure 41.10 shows the rectified output voltage obtained from a diode when it is connected to an a.c. supply.

Figure 41.10 Rectification by a diode

● Transistor

Transistors are the small semiconductor devices which have revolutionised electronics. They are made both as separate components in their cases, like those in Figure 41.11a, and also as parts of **integrated circuits** (ICs) in which millions may be 'etched' on a 'chip' of silicon (Figure 41.11b).

Transistors have three connections called the **base** (B), the **collector** (C) and the **emitter** (E). In the transistor symbol shown in Figure 41.12, the arrow indicates the direction in which conventional current flows in it when C and B are connected to a battery + terminal, and E to a battery − terminal. Again, the outer circle of the symbol is not always included.

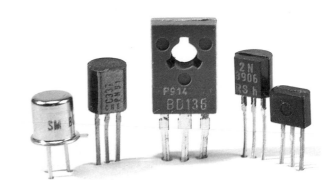

Figure 41.11a Transistor components

Figure 41.11b Integrated circuits which may each contain millions of transistors

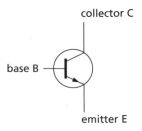

Figure 41.12 Symbol for a transistor

There are two current paths through a transistor. One is the **base–emitter path** and the other is the **collector–emitter** (via base) **path**. The transistor's usefulness arises from the fact that it can link circuits connected to each path so that the current in one controls that in the other, just like a relay.

Its action can be shown using the circuit of Figure 41.13. When **S is open**, the base current I_B is zero and neither L_1 nor L_2 lights up, showing that the collector current I_C is also zero even though the battery is correctly connected across the C–E path.

When **S is closed**, B is connected through R to the battery + terminal and L_2 lights up but not L_1. This shows there is now collector current (which is in L_2) and that it is much greater than the base current (which is in L_1 but is too small to light it).

Therefore, **in a transistor the base current I_B switches on and controls the much greater collector current I_C.**

Resistor R has to be in the circuit to limit the base current which would otherwise create so large a collector current as to destroy the transistor by overheating.

● Transistor as a switch

a) Advantages

Transistors have many advantages over other electrically operated switches such as relays. They are small, cheap, reliable, have no moving parts, their life is almost indefinite (in well-designed circuits) and they can switch on and off millions of times a second.

b) 'On' and 'off' states

A transistor is considered to be 'off' when the collector current is zero or very small. It is 'on' when the collector current is much larger. The resistance of the collector–emitter path is large when the transistor is 'off' (as it is for an ordinary mechanical switch) and small (ideally it should be zero) when it is 'on'.

To switch a transistor 'on' requires the base voltage (and therefore the base current) to exceed a certain minimum value (about +0.6 V base voltage).

c) Basic switching circuits

Two are shown in Figures 41.14a, b. The 'on' state is shown by the lamp in the collector circuit becoming fully lit.

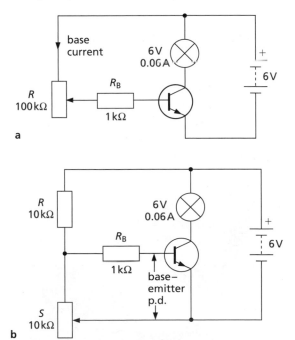

a

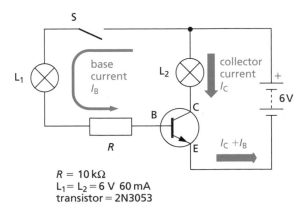

$R = 10\,k\Omega$
$L_1 = L_2 = 6\,V\ 60\,mA$
transistor $= 2N3053$

Figure 41.13 Demonstration circuit

b

Figure 41.14 Transistor switching circuits ▶▶

Rheostat control is used in the circuit in Figure 41.14a. 'Switch-on' occurs by reducing R until the base current is large enough to make the collector current light the lamp. (The base resistor R_B is *essential* in case R is made zero and results in +6V from the battery being applied directly to the base. This would produce very large base and collector currents and destroy the transistor by overheating.)

Potential divider control is used in the circuit in Figure 41.14b. Here 'switch-on' is obtained by adjusting the variable resistance S until the p.d. across S (which is the base–emitter p.d. and depends on the value of S compared with that of R) exceeds +0.6V or so.

Note In a potential divider the p.d.s across the resistors are in the ratio of their resistances (see Chapter 38). For example, in the circuit shown in Figure 41.14b, the p.d. across R and S in series is 6V. If $R = 10\,\text{k}\Omega$ and S is set to $5\,\text{k}\Omega$, then the p.d. across R, that is V_R, is 4V and the p.d. across S, that is V_S, is 2V. So $V_R/V_S = 4\text{V}/2\text{V} = 2/1$.

In general

$$\frac{V_R}{V_S} = \frac{R}{S} \quad \text{and} \quad V_S = (V_R + V_S) \times \frac{S}{(R + S)}$$

Also see question 2 on p. 191.

Practical work

Transistor switching circuits

The components can be mounted on a circuit board, for example an 'S-DeC' as in Figure 41.15a. The diagrams in Figure 41.15b show how to lengthen transistor leads and also how to make connections (without soldering) to parts that have 'tags', for example, variable resistors.

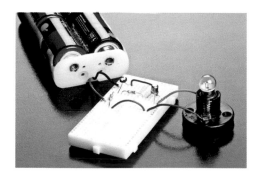

Figure 41.15a Partly built transistor switching circuit

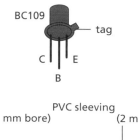

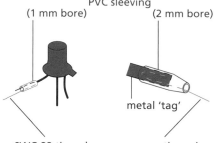

SWG 22 tinned copper connecting wire held in contact by sleeving

Figure 41.15b Lengthening transistor leads and making connections to tags

In many control circuits, devices such as LDRs and thermistors are used in potential divider arrangements to detect small changes of light intensity and temperature, respectively. These changes then enable a transistor to act as a simple processor by controlling the current to an output transducer, such as a lamp or a buzzer.

a) Light-operated switch
In the circuit of Figure 41.16 the LDR is part of a potential divider. The lamp comes on when the LDR is shielded: more of the battery p.d. acts across the increased resistance of the LDR (i.e. more than 0.6V) and less across R. In the dark, the base–emitter p.d. increases as does the base current and so also the collector current.

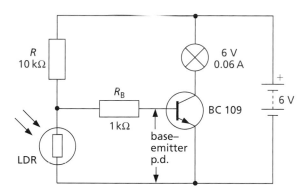

Figure 41.16 Light-operated switch

If the LDR and R are interchanged the lamp goes off in the dark and the circuit could act as a light-operated intruder alarm.

If a variable resistor is used for R, the light level at which switching occurs can be changed.

b) Temperature-operated switch
In the low-temperature-operated switch of Figure 41.17, a thermistor and resistor form a potential divider across the

6V supply. When the temperature of the thermistor falls, its resistance increases and so does the p.d. across it, i.e. the base–emitter p.d. rises. When it reaches 0.6V, the transistor switches on and the collector current becomes large enough to operate the lamp. The circuit could act as a frost-warning device.

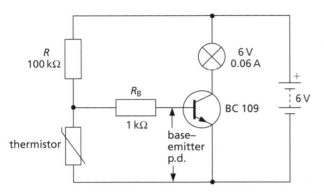

Figure 41.17 Low-temperature-operated switch

If the thermistor and resistor are interchanged, the circuit can be used as a high-temperature alarm (Figure 41.18).

When the temperature of the thermistor rises, its resistance decreases and a larger share of the 6V supply acts across R, i.e. the base–emitter p.d. increases. When it exceeds 0.6V or so the transistor switches on and collector current (too small to ring the buzzer directly) goes through the relay coil. The relay contacts close, enabling the buzzer to obtain, directly from the 6V supply, the larger current it needs.

The diode D protects the transistor from damage: when the collector current falls to zero at switch off this induces a large p.d. in the relay coil (see Chapter 43). The diode is forward-biased by the induced p.d. (which tries to maintain the current in the relay coil) and, because of its low forward resistance (e.g. 1Ω), offers an easy path for the current produced. To the 6V supply the diode is reverse-biased and its high resistance does not short-circuit the relay coil when the transistor is on.

If R is variable the temperature at which switching occurs can be changed.

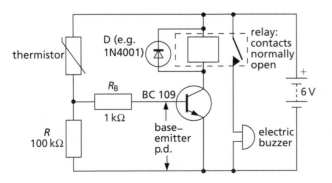

Figure 41.18 High-temperature-operated switch

Questions

1 Figure 41.19a shows a lamp, a semiconductor diode and a cell connected in series. The lamp lights when the diode is connected in this direction. Say what happens to each of the lamps in b, c and d. Give reasons for your answers.

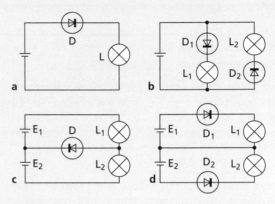

Figure 41.19

2 What are the readings V_1 and V_2 on the high-resistance voltmeters in the potential divider circuit of Figure 41.20 if
  a $R_1 = R_2 = 10\,\text{k}\Omega$,
  b $R_1 = 10\,\text{k}\Omega$, $R_2 = 50\,\text{k}\Omega$,
  c $R_1 = 20\,\text{k}\Omega$, $R_2 = 10\,\text{k}\Omega$?

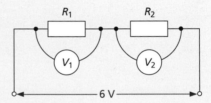

Figure 41.20

3 A simple moisture-warning circuit is shown in Figure 41.21, in which the moisture detector consists of two closely spaced copper rods.

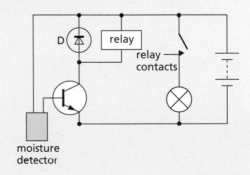

Figure 41.21

▶▶

a Describe how the circuit works when the detector gets wet.
b Warning lamps are often placed in the collector circuit of a transistor. Why is a relay used here?
c What is the function of D?

Checklist

After studying this chapter you should be able to

• recall the functions of the input sensor, processor and output transducer in an electronic system and give some examples,
• describe the action of an LDR and a thermistor and show an understanding of their use as input transducers,
• understand the use of a relay in a switching circuit,

• explain what is meant by a diode being forward biased and reverse biased and recall that a diode can produce rectified a.c.

42 Digital electronics

- Analogue and digital electronics
- Logic gates
- Logic gate control systems

- Problems to solve
- Electronics and society

● Analogue and digital electronics

There are two main types of electronic circuits, devices or systems – analogue and digital.

In **analogue circuits**, voltages (and currents) can have any value within a certain range over which they can be varied smoothly and continuously, as shown in Figure 42.1a. They include amplifier-type circuits.

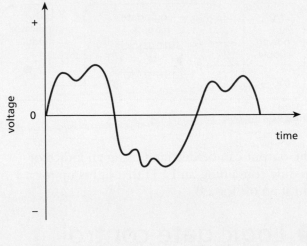

a

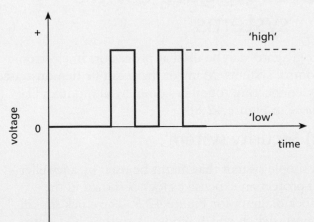

b

Figure 42.1

In **digital circuits**, voltages have only one of two values, either 'high' (e.g. 5 V) or 'low' (e.g. near 0 V), as shown in Figure 42.1b. They include switching-type circuits such as those we have considered in Chapter 41.

A **variable resistor** is an analogue device which, in a circuit with a lamp, allows the lamp to have a wide range of light levels. A **switch** is a digital device which allows a lamp to be either 'on' or 'off'.

Analogue meters display their readings by the deflection of a pointer over a continuous scale (see Figure 47.4a, p. 220). **Digital meters** display their readings as digits, i.e. numbers, which change by one digit at a time (see Figure 47.4b, p. 220).

● Logic gates

Logic gates are switching circuits used in computers and other electronic systems. They 'open' and give a 'high' output voltage, i.e. a signal (e.g. 5 V), depending on the combination of voltages at their inputs, of which there is usually more than one.

There are five basic types, all made from transistors in integrated circuit form. The behaviour of each is described by a **truth table** showing what the output is for all possible inputs. 'High' (e.g. 5 V) and 'low' (e.g. near 0 V) outputs and inputs are represented by 1 and 0, respectively, and are referred to as **logic levels** 1 and 0.

a) NOT gate or inverter

This is the simplest gate, with one input and one output. It produces a 'high' output if the input is 'low', i.e. the output is then NOT high, and vice versa. Whatever the input, the gate inverts it. The symbol and truth table are given in Figure 42.2.

►►

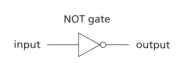

Input	Output
0	1
1	0

Figure 42.2 NOT gate symbol and truth table

b) OR, NOR, AND, NAND gates

All these have two or more inputs and one output. The truth tables and symbols for 2-input gates are shown in Figure 42.3. Try to remember the following.

OR: output is 1 if input A **OR** input B **OR** both are 1

NOR: output is 1 if neither input A **NOR** input B is 1

AND: output is 1 if input A **AND** input B are 1

NAND: output is 1 if input A **AND** input B are **NOT** both 1

OR gate   NOR gate

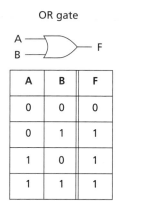

A	B	F
0	0	0
0	1	1
1	0	1
1	1	1

A	B	F
0	0	1
0	1	0
1	0	0
1	1	0

AND gate   NAND gate

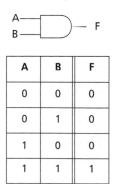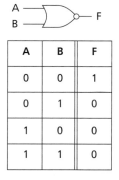

A	B	F
0	0	0
0	1	0
1	0	0
1	1	1

A	B	F
0	0	1
0	1	1
1	0	1
1	1	0

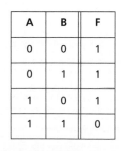

Figure 42.3 Symbols and truth tables for 2-input gates

Note from the truth tables that the outputs of the NOR and NAND gates are the inverted outputs of the OR and AND gates, respectively. They have a small circle at the output end of their symbols to show this inversion.

c) Testing logic gates

The truth tables for the various gates can be conveniently checked by having the logic gate integrated circuit (IC) mounted on a small board with sockets for the power supply, inputs A and B and output F (Figure 42.4). A 'high' input (i.e. logic level 1) is obtained by connecting the input socket to the positive of the power supply, e.g. +5 V and a 'low' input (i.e. logic level 0) by connecting to 0 V.

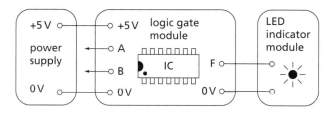

Figure 42.4 Modules for testing logic gates

The output can be detected using an indicator module containing an LED that lights up for a 1 and stays off for a 0.

● Logic gate control systems

Logic gates can be used as processors in electronic control systems. Many of these can be demonstrated by connecting together commercial modules like those in Figure 42.8b

a) Security system

A simple system that might be used by a jeweller to protect an expensive clock is shown in the block diagram for Figure 42.5. The clock sits on a push switch which sends a 1 to the NOT gate, unless the clock is lifted when a 0 is sent. In that case the output from the NOT gate is a 1 which rings the bell.

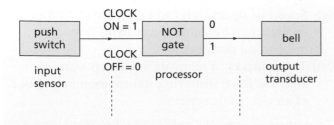

Figure 42.5 Simple alarm system

b) Safety system for a machine operator

A safety system could prevent a machine (e.g. an electric motor) from being switched on before another switch had been operated, for example, by a protective safety guard being in the correct position. In Figure 42.6, when switches A *and* B are on, they supply a 1 to each input of the AND gate which can then start the motor.

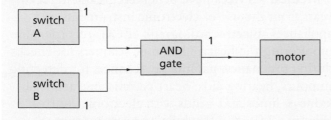

Figure 42.6 Safety system for controlling a motor

c) Heater control system

The heater control has to switch on the heating system when it is

(i) **cold**, i.e. the temperature is below a certain value and the output from the temperature sensor is 0, and

(ii) **daylight**, i.e. the light sensor output is 1.

With these outputs from the sensors applied to the processor in Figure 42.7, the AND gate has two 1 inputs. The output from the AND gate is then 1 and will turn on the heater control. Any other combination of sensor outputs produces a 0 output from the AND gate, as you can check.

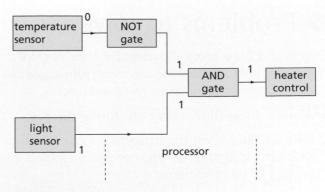

Figure 42.7 Heater control system

d) Street lights

A system is required that allows the street lights either to be turned on manually by a switch at any time, or automatically by a light sensor when it is dark. The arrangement in Figure 42.8a achieves this since the OR gate gives a 1 output when either or both of its inputs are 1.

The system can be demonstrated using the module shown in Figure 42.8b.

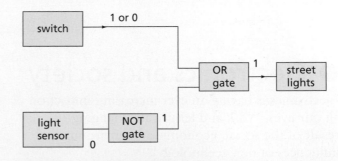

Figure 42.8a Control system with manual override

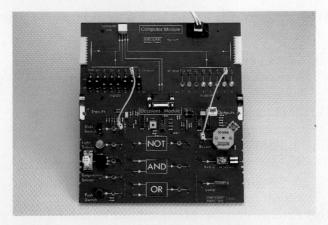

Figure 42.8b Module for demonstrating street lights

▶▶

● Problems to solve

Design and draw block diagrams for logic control systems to indicate how the following jobs could be done. If possible build them using modules.

1 Allow a doorbell to work only during the **day**.

2 Give warning when the temperature of a domestic hot water system is too **high** or when a switch is pressed to **test** the alarm.

3 Switch on a bathroom heater when it is **cold** and **light**.

4 Sound an alarm when it is **cold** or a switch is **pressed**.

5 Give warning if the temperature of a room **falls** during the **day** and also allow a **test** switch to check the alarm works.

6 Give warning of **frosty** conditions at **night** to a gardener who is sometimes very tired after a hard day and may want to switch off the alarm.

● Electronics and society

Electronics is having an ever-increasing impact on all our lives. Work and leisure are changing as a result of the social, economic and environmental influences of new technology.

a) Reasons for the impact

Why is electronics having such a great impact? Some of the reasons are listed below.

(i) **Mass production** of large quantities of semiconductor devices (e.g. ICs) allows them to be made very cheaply.

(ii) **Miniaturisation** of components means that even complex systems can be compact.

(iii) **Reliability** of electronic components is a feature of well-designed circuits. There are no moving parts to wear out and systems can be robust.

(iv) **Energy consumption** and use of natural resources is often much less than for their non-electronic counterparts. For example, the transistor uses less power than a relay.

(v) **Speed of operation** can be millions of times greater than for other alternatives (e.g. mechanical devices).

(vi) **Transducers** of many different types are available for transferring information in and out of an electronic system.

To sum up, electronic systems tend to be cheaper, smaller, more reliable, less wasteful, much faster and can respond to a wider range of signals than other systems.

b) Some areas of impact

At home devices such as washing machines, burglar alarms, telephones, cookers and sewing machines contain electronic components. Central heating systems and garage doors may have automatic electronic control. For home entertainment, DVD players, interactive digital televisions or computers with internet connections and electronic games are finding their way into more and more homes.

Medical services have benefited greatly in recent years from the use of electronic instruments and appliances. Electrocardiograph (ECG) recorders for monitoring the heart, ultrasonic scanners for checks during pregnancy, gamma ray scanners for detecting tumours, hearing aids, heart pacemakers, artificial kidneys, limbs and hands with electronic control (Figure 42.9), and 'keyhole' surgery are some examples.

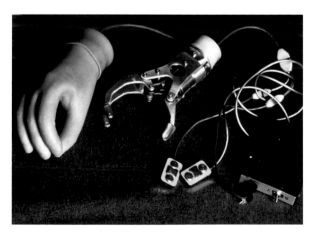

Figure 42.9 Electronically controlled artificial hands

In industry microprocessor-controlled equipment is taking over. Robots are widely used for car assembly work, and to do dull, routine, dirty jobs such as welding and paint spraying. In many cases production lines and even whole factories, such as sugar refineries and oil refineries, are almost entirely automated. Computer-aided design (CAD) of products is increasing (Figure 42.10), even in the clothing industry. Three-dimensional printers programmed by CAD files can now produce solid objects in a variety of materials for use as prototypes or components in industries ranging from aerospace to entertainment.

In offices, banks and shops computers are used for word processing, data control and communications via **email**: text, numbers and pictures are transmitted by electronic means, often by high-speed digital links. Cash dispensers and other automated services at banks are a great convenience for their customers. Bar codes (like the one on the back cover of this book) on packaged products are used by shops for stock control in conjunction with a bar code reader (which uses a laser) and a data recorder connected to a computer. A similar system is operated by libraries to record the issue and return of books. Libraries provide electronic databases and internet facilities for research.

Figure 42.10 Computer-aided design of clothing

Communications have been transformed. Satellites enable events on one side of the world to be seen and heard on the other side, as they happen. Digital telephone and communication links, smart phones, tablets, social media and cloud computing are the order of the day.

Leisure activities have been affected by electronic developments. For some people, leisure means participating in or attending sporting activities and here the electronic scoreboard is likely to be in evidence. For the golf enthusiast, electronic machines claim to analyse 'swings' and reduce handicaps. For others, leisure means listening to music, whose production, recording and listening facilities have been transformed by the digital revolution. Electronically synthesised music has become the norm for popular recordings. The lighting and sound effects in stage shows are programmed by computer. For the cinemagoer, special effects in film production have been vastly improved by computer-generated animated images (Figure 42.11). The availability of home computers and games consoles in recent years has enabled a huge market in computer games and home-learning resources to develop.

Figure 42.11 Computer animation brings the tiger into the scene

c) Consequences of the impact

Most of the social and economic consequences of electronics are beneficial but a few cause problems.

An **improved quality of life** has resulted from the greater convenience and reliability of electronic systems, with increased life expectancy and leisure time, and fewer dull, repetitive jobs.

Better communication has made the world a smaller place. The speed with which news can be reported to our homes by radio, television and the internet enables the public to be better informed.

Databases have been developed. These are memories which can store huge amounts of information for rapid transmission from one place to another. For example, the police can obtain in seconds, by radio, details of a car they are following. Databases raise questions, however, about invasion of privacy and security.

▶▶

Employment is affected by the demand for new equipment – new industry and jobs are created to make and maintain it – but when electronic systems replace mechanical ones, redundancy and/or retraining needs arise. Conditions of employment and long-term job prospects can also be affected for many people, especially certain manual and clerical workers. One industrial robot may replace four factory workers.

The public attitude to the electronics revolution is not always positive. Modern electronics is a 'hidden' technology with parts that are enclosed in a tiny package (or 'black box') and do not move. It is also a 'throwaway' technology in which the whole lot is discarded and replaced – by an expert – if a part fails, and rapid advances in design technology cause equipment to quickly become obsolete. For these reasons it may be regarded as mysterious and unfriendly – people feel they do not understand what makes it tick.

d) The future

The only certain prediction about the future is that new technologies will be developed and these, like present ones, will continue to have a considerable influence on our lives.

Today the development of 'intelligent' computers is being pursued with great vigour, and voice recognition techniques are already in use. Optical systems, which are more efficient than electronic ones, are being increasingly developed for data transmission, storage and processing of information.

2 What do the symbols **A** to **E** represent in Figure 42.12?

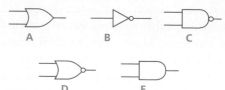

Figure 42.12

3 Design and draw the block diagrams for logic control systems to:
  a wake you at the crack of dawn and which you can also switch off,
  b protect the contents of a drawer which you can still open without setting off the alarm.

Checklist

After studying this chapter you should be able to

- explain and use the terms analogue and digital,
- state that logic gates are switching circuits containing transistors and other components,
- describe the action of NOT, OR, NOR, AND and NAND logic gates and recall their truth tables,
- design and draw block diagrams of logic control systems for given requirements.

Questions

1 The combined truth tables for four logic gates **A, B, C, D** are given below. State what kind of gate each one is.

Inputs		Outputs			
		A	B	C	D
0	0	0	0	1	1
0	1	0	1	1	0
1	0	0	1	1	0
1	1	1	1	0	0

- Electromagnetic induction
- Faraday's law
- Lenz's law
- Simple a.c. generator (alternator)

- Simple d.c. generator (dynamo)
- Practical generators
- Applications of electromagnetic induction

The effect of producing electricity from magnetism was discovered in 1831 by Faraday and is called **electromagnetic induction**. It led to the construction of generators for producing electrical energy in power stations.

● Electromagnetic induction

Two ways of investigating the effect follow.

a) Straight wire and U-shaped magnet

First the wire is held at rest between the poles of the magnet. It is then moved in each of the six directions shown in Figure 43.1 and the meter observed. Only *when it is moving upwards* (direction 1) or *downwards* (direction 2) is there a deflection on the meter, indicating an induced current in the wire. The deflection is in opposite directions in these two cases and only lasts while the wire is in motion.

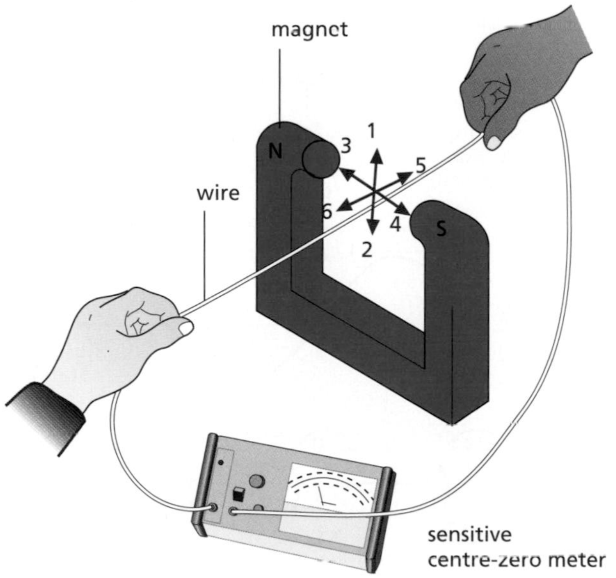

b) Bar magnet and coil

The magnet is pushed into the coil, one pole first (Figure 43.2), then held still inside it. It is then withdrawn. The meter shows that current is induced in the coil in one direction as the magnet is *moved in* and in the opposite direction as it is *moved out*. There is no deflection when the magnet is at rest. The results are the same if the coil is moved instead of the magnet, i.e. only *relative motion* is needed.

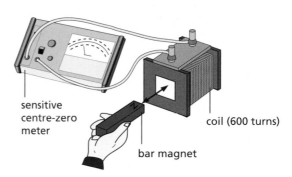

sensitive centre-zero meter

coil (600 turns)

bar magnet

Figure 43.2 A current is induced in the coil when the magnet is moved in or out.

● Faraday's law

To 'explain' electromagnetic induction Faraday suggested that a voltage is induced in a conductor whenever it 'cuts' magnetic field lines, i.e. moves *across* them, but not when it moves along them or is at rest. If the conductor forms part of a complete circuit, an induced current is also produced.

Faraday found, and it can be shown with apparatus like that in Figure 43.2, that the induced p.d. or voltage increases with increases of

(i) the **speed of motion** of the magnet or coil,
(ii) the **number of turns** on the coil,
(iii) the **strength of the magnet**.

These facts led him to state a law:

> The size of the induced p.d. is directly proportional to the rate at which the conductor cuts magnetic field lines.

magnet

N
3 1
5
wire
6
4
2
S

sensitive centre-zero meter

Figure 43.1 A current is induced in the wire when it is moved up or down between the magnet poles.

● Lenz's law

The direction of the induced current can be found by a law stated by the Russian scientist, Lenz.

> The direction of the induced current is such as to oppose the change causing it.

In Figure 43.3a the magnet approaches the coil, north pole first. According to Lenz's law the induced current should flow in a direction that makes the coil behave like a magnet with its top a north pole. The downward motion of the magnet will then be opposed since like poles repel.

When the magnet is withdrawn, the top of the coil should become a south pole (Figure 43.3b) and attract the north pole of the magnet, so hindering its removal. The induced current is thus in the opposite direction to that when the magnet approaches.

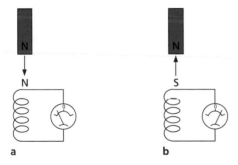

Figure 43.3 The induced current opposes the motion of the magnet.

Lenz's law is an example of the principle of conservation of energy. If the currents caused opposite poles from those that they do make, electrical energy would be created from nothing. As it is, mechanical energy is provided, by whoever moves the magnet, to overcome the forces that arise.

For a straight wire moving at right angles to a magnetic field a more useful form of Lenz's law is **Fleming's right-hand rule** (the 'dynamo rule') (Figure 43.4).

> Hold the thumb and first two fingers of the right hand at right angles to each other with the **F**irst finger pointing in the direction of the **F**ield and the thu**M**b in the direction of **M**otion of the wire, then the se**C**ond finger points in the direction of the induced **C**urrent.

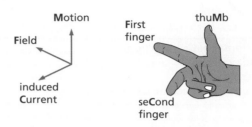

Figure 43.4 Fleming's right-hand (dynamo) rule

● Simple a.c. generator (alternator)

The simplest alternating current (a.c.) generator consists of a rectangular coil between the poles of a C-shaped magnet (Figure 43.5a). The ends of the coil are joined to two **slip rings** on the axle and against which **carbon brushes** press.

When the coil is rotated it cuts the field lines and a voltage is induced in it. Figure 43.5b shows how the voltage varies over one complete rotation.

As the coil moves through the vertical position with **ab** uppermost, **ab** and **cd** are moving along the lines (**bc** and **da** do so always) and no cutting occurs. The induced voltage is zero.

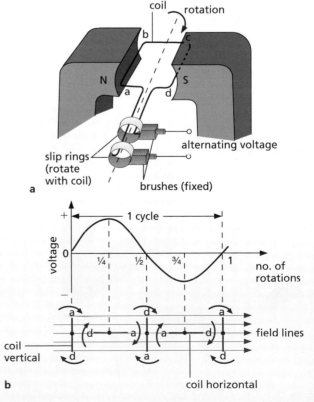

Figure 43.5 A simple a.c. generator and its output

During the first quarter rotation the p.d. increases to a maximum when the coil is horizontal. Sides **ab** and **dc** are then cutting the lines at the greatest rate.

In the second quarter rotation the p.d. decreases again and is zero when the coil is vertical with **dc** uppermost. After this, the direction of the p.d. reverses because, during the next half rotation, the motion of **ab** is directed upwards and **dc** downwards.

An alternating voltage is generated which acts first in one direction and then the other; it causes alternating current (a.c.) to flow in a circuit connected to the brushes. The **frequency** of an a.c. is the number of complete cycles it makes each second and is measured in **hertz** (Hz), i.e. 1 cycle per second = 1 Hz. If the coil rotates twice per second, the a.c. has frequency 2 Hz. The mains supply is a.c. of frequency 50 Hz.

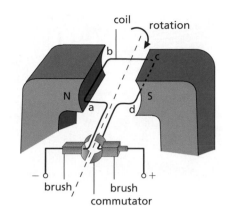

a

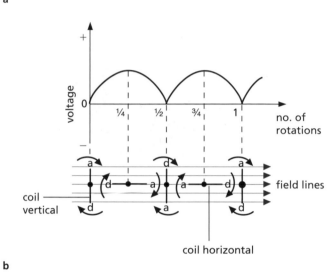

b

Figure 43.6 A simple d.c. generator and its output

● Simple d.c. generator (dynamo)

An a.c. generator becomes a direct current (d.c.) one if the slip rings are replaced by a **commutator** (like that in a d.c. motor, see p. 216), as shown in Figure 43.6a.

The brushes are arranged so that as the coil goes through the vertical, changeover of contact occurs from one half of the split ring of the commutator to the other. But it is when the coil goes through the vertical position that the voltage induced in the coil reverses, so one brush is always positive and the other negative.

The voltage at the brushes is shown in Figure 43.6b; although varying in value, it never changes direction and would produce a direct current (d.c.) in an external circuit.

In construction the simple d.c. dynamo is the same as the simple d.c. motor and one can be used as the other. When an electric motor is working it also acts as a dynamo and creates a voltage which opposes the applied voltage. The current in the coil is therefore much less once the motor is running.

● Practical generators

In actual generators several coils are wound in evenly spaced slots in a soft iron cylinder and electromagnets usually replace permanent magnets.

a) Power stations

In power station alternators the electromagnets rotate (the rotor, Figure 43.7a) while the coils and their iron core are at rest (the stator, Figure 43.7b). The large p.d.s and currents (e.g. 25 kV at several thousand amps) induced in the stator are led away through stationary cables, otherwise they would quickly destroy the slip rings by sparking. Instead the relatively small d.c. required by the rotor is fed via the slip rings from a small dynamo (the exciter) which is driven by the same turbine as the rotor.

a Rotor (electromagnets)

b Stator (induction coils)

Figure 43.7 The rotor and stator of a power station alternator

In a thermal power station (Chapter 15), the turbine is rotated by high-pressure steam obtained by heating water in a coal- or oil-fired boiler or in a nuclear reactor (or by hot gas in a gas-fired power station). A block diagram of a thermal power station is shown in Figure 43.8. The energy transfer diagram was given in Figure 15.7, p. 63.

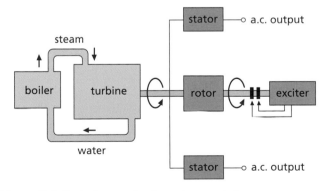

Figure 43.8 Block diagram of a thermal power station

b) Cars

Most cars are now fitted with alternators because they give a greater output than dynamos at low engine speeds.

c) Bicycles

The rotor of a bicycle generator is a permanent magnet and the voltage is induced in the coil, which is at rest (Figure 43.9).

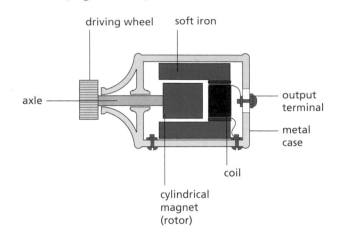

Figure 43.9 Bicycle generator

● Applications of electromagnetic induction

a) Moving-coil microphone

The moving-coil loudspeaker shown in Figure 46.7 (p. 218) can be operated in reverse mode as a microphone. When sound is incident on the paper cone it vibrates, causing the attached coil to move in and out between the poles of the magnet. A varying electric current, representative of the sound, is then induced in the coil by electromagnetic induction.

b) Magnetic recording

Magnetic tapes or disks are used to record information in sound systems and computers. In the recording head shown in Figure 43.10, the tape becomes magnetised when it passes over the gap in the pole piece of the electromagnet and retains a magnetic record of the electrical signal applied to the coil from a microphone or computer. In playback mode, the varying magnetisation on the moving tape or disk induces a corresponding electrical signal in the coil as a result of electromagnetic induction.

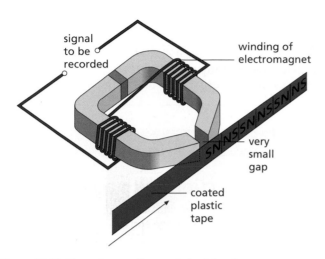

signal
to be
recorded
winding of
electromagnet
SNNSSNNSSNNS
very
small
gap
coated
plastic
tape

Figure 43.10 Magnetic recording or playback head

Questions

1 A simple generator is shown in Figure 43.11.
   a What are A and B called and what is their purpose?
   b What changes can be made to increase the p.d. generated?

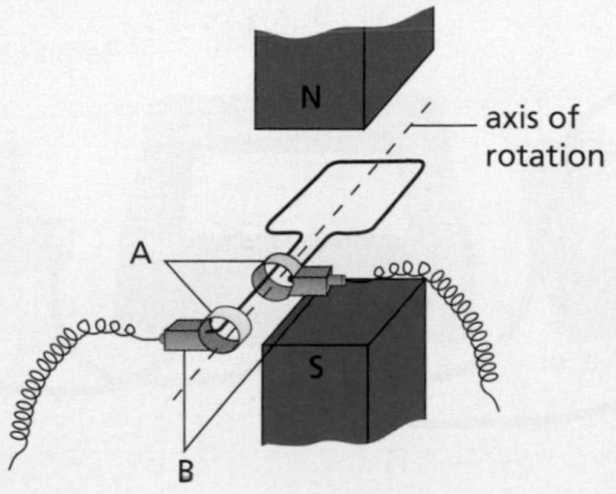

N
axis of
rotation
A
S
B

Figure 43.11

2 Describe the deflections observed on the sensitive, centre-zero galvanometer G (Figure 43.12) when the copper rod XY is connected to its terminals and is made to vibrate up and down (as shown by the arrows), between the poles of a U-shaped magnet, at right angles to the magnetic field. Explain what is happening.

G
Y
N
S
X

Figure 43.12

Checklist

After studying this chapter you should be able to

- describe experiments to show electromagnetic induction,
- recall Faraday's explanation of electromagnetic induction,
- predict the direction of the induced current using Lenz's law or Fleming's right-hand rule,
- draw a diagram of a simple a.c. generator and sketch a graph of its output.

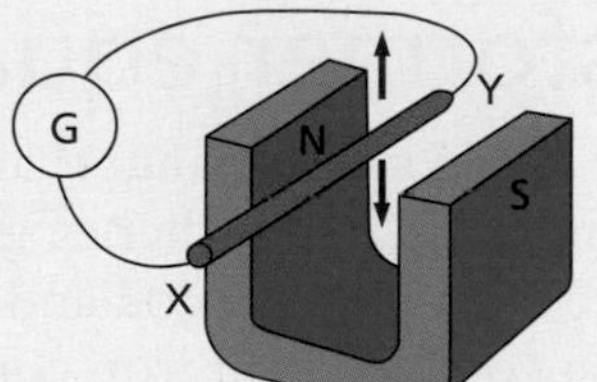

⟨44⟩ Transformers

- Mutual induction
- Transformer equation
- Energy losses in a transformer

- Transmission of electrical power
- Applications of eddy currents
- Practical work: Mutual induction with a.c.

● Mutual induction

When the current in a coil is switched on or off or changed, a voltage is induced in a neighbouring coil. The effect, called **mutual induction**, is an example of electromagnetic induction and can be shown with the arrangement of Figure 44.1. Coil A is the **primary** and coil B the **secondary**.

Switching on the current in the primary sets up a magnetic field and as its field lines 'grow' outwards from the primary they 'cut' the secondary. A p.d. is induced in the secondary until the current in the primary reaches its steady value. When the current is switched off in the primary, the magnetic field dies away and we can imagine the field lines cutting the secondary as they collapse, again inducing a p.d. in it. Changing the primary current by *quickly* altering the rheostat has the same effect.

The induced p.d. is increased by having a soft iron rod in the coils or, better still, by using coils wound on a complete iron ring. More field lines then cut the secondary due to the magnetisation of the iron.

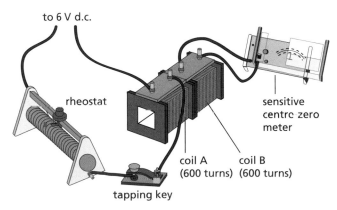

Figure 44.1 A changing current in a primary coil (A) induces a current in a secondary coil (B).

Practical work

Mutual induction with a.c.

An alternating current is changing all the time and if it flows in a primary coil, an alternating voltage and current are induced in a secondary coil.

Connect the circuit of Figure 44.2. The 1 V high current power unit supplies a.c. to the primary and the lamp detects the secondary current.

Find the effect on the brightness of the lamp of

(i)   pulling the C-cores apart slightly,
(ii) increasing the secondary turns to 15,
(iii) decreasing the secondary turns to 5.

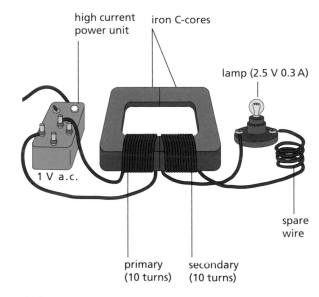

Figure 44.2

● Transformer equation

A **transformer** transforms (changes) an *alternating* voltage from one value to another of greater or smaller value. It has a primary coil and a secondary coil wound on a complete soft iron core, either one on top of the other (Figure 44.3a) or on separate limbs of the core (Figure 44.3b).

204

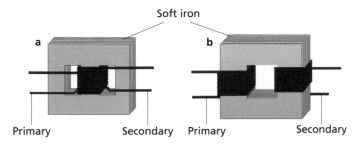

Figure 44.3 Primary and secondary coils of a transformer

An alternating voltage applied to the primary induces an alternating voltage in the secondary. The value of the secondary voltage can be shown, for a transformer in which all the field lines cut the secondary, to be given by

$$\frac{\text{secondary voltage}}{\text{primary voltage}} = \frac{\text{secondary turns}}{\text{primary turns}}$$

In symbols

$$\frac{V_s}{V_p} = \frac{N_s}{N_p}$$

A **step-up transformer** has more turns on the secondary than the primary and V_s is greater than V_p (Figure 44.4a). For example, if the secondary has twice as many turns as the primary, V_s is about twice V_p. In a **step-down transformer** there are fewer turns on the secondary than the primary and V_s is less than V_p (Figure 44.4b).

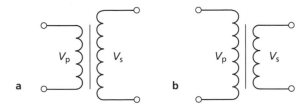

Figure 44.4 Symbols for a transformer: **a** step-up ($V_s > V_p$); **b** step-down ($V_p > V_s$)

● Energy losses in a transformer

If the p.d. is stepped up in a transformer, the current is stepped down in proportion. This must be so if we assume that all the electrical energy given to the primary appears in the secondary, i.e. that energy is conserved and the transformer is 100% efficient or 'ideal' (many approach this efficiency). Then

power in primary = power in secondary

$$V_p \times I_p = V_s \times I_s$$

where I_p and I_s are the primary and secondary currents, respectively.

$$\therefore \quad \frac{I_s}{I_p} = \frac{V_p}{V_s}$$

So, for the ideal transformer, if the p.d. is doubled the current is halved. In practice, it is more than halved, because of small energy losses in the transformer arising from the following three causes.

a) Resistance of windings

The windings of copper wire have some resistance and heat is produced by the current in them. Large transformers like those in Figure 44.5 have to be oil-cooled to prevent overheating.

Figure 44.5 Step-up transformers at a power station

▶▶

b) Eddy currents

The iron core is in the changing magnetic field of the primary and currents, called **eddy currents**, are induced in it which cause heating. These are reduced by using a **laminated** core made of sheets, insulated from one another to have a high resistance.

c) Leakage of field lines

All the field lines produced by the primary may not cut the secondary, especially if the core has an air gap or is badly designed.

● Worked example

A transformer steps down the mains supply from 230 V to 10 V to operate an answering machine.

a What is the turns ratio of the transformer windings?
b How many turns are on the primary if the secondary has 100 turns?
c What is the current in the primary if the transformer is 100% efficient and the current in the answering machine is 2 A?

a Primary voltage, $V_p = 230\,V$
Secondary voltage, $V_s = 10\,V$

$$\text{Turns ratio} = \frac{N_s}{N_p} = \frac{V_s}{V_p} = \frac{10\,V}{230\,V} = \frac{1}{23}$$

b Secondary turns, $N_s = 100$
From **a**,

$$\frac{N_s}{N_p} = \frac{1}{23}$$

$$\therefore \qquad N_p = 23 \times N_s = 23 \times 100$$
$$= 2300 \text{ turns}$$

c Efficiency = 100%

$\therefore$ power in primary = power in secondary
$$V_p \times I_p = V_s \times I_s$$

$$\therefore \quad I_p = \frac{V_s \times I_s}{V_p} = \frac{10\,V \times 2\,A}{230\,V} = \frac{2}{23}\,A = 0.09\,A$$

Note In this ideal transformer the current is stepped up in the same ratio as the voltage is stepped down.

● Transmission of electrical power

a) Grid system

The **National Grid** is a network of cables throughout Britain, mostly supported on pylons, that connects over 100 power stations to consumers. In the largest modern stations, electricity is generated at 25 000 V (25 kilovolts = 25 kV) and stepped up at once in a transformer to 275 or 400 kV to be sent over long distances on the Supergrid. Later, the p.d. is reduced by substation transformers for distribution to local users (Figure 44.6).

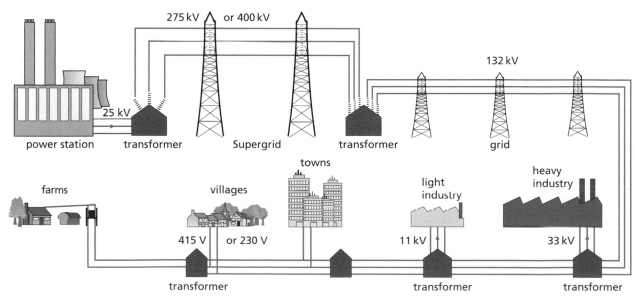

Figure 44.6 The National Grid transmission system in Britain

At the National Control Centre, engineers direct the flow and re-route it when breakdown occurs. This makes the supply more reliable and cuts costs by enabling smaller, less efficient stations to be shut down at off-peak periods.

b) Use of high alternating p.d.s

The efficiency with which transformers step alternating p.d.s up and down accounts for the use of a.c. rather than d.c. in power transmission. High voltages are used in the transmission of electric power to reduce the amount of energy 'lost' as heat.

Power cables have resistance, and so electrical energy is transferred to heat during the transmission of electricity from the power station to the user. The power 'lost' as heat in cables of resistance R is I^2R, so I should be kept low to reduce energy loss. Since power = IV, if 400 000 W of electrical power has to be sent through cables, it might be done, for example, either as 1 A at 400 000 V or as 1000 A at 400 V. Less energy will be transferred to heat if the power is transmitted at the lower current and higher voltage, i.e. 1 A at 400 000 V. High p.d.s require good insulation but are readily produced by a.c. generators.

● Applications of eddy currents

Eddy currents are the currents induced in a piece of metal when it cuts magnetic field lines. They can be quite large due to the low resistance of the metal. They have their uses as well as their disadvantages.

a) Car speedometer

The action depends on the eddy currents induced in a thick aluminium disc (Figure 44.7), when a permanent magnet, near it but *not touching it*, is rotated by a cable driven from the gearbox of the car. The eddy currents in the disc make it rotate in an attempt to reduce the relative motion between it and the magnet (see Chapter 43). The extent to which the disc can turn, however, is controlled by a spring. The faster the magnet rotates the more the disc turns before it is stopped by the spring. A pointer fixed to the disc moves over a scale marked in mph (or km/h) and gives the speed of the car.

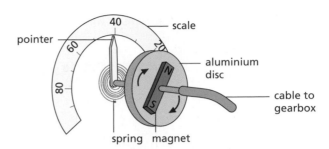

Figure 44.7 Car speedometer

b) Metal detector

The metal detector shown in Figure 44.8 consists of a large primary coil (A), through which an a.c. current is passed, and a smaller secondary coil (B). When the detector is swept over a buried metal object (such as a nail, coin or pipe) the fluctuating magnetic field lines associated with the alternating current in coil A 'cut' the hidden metal and induce eddy currents in it. The changing magnetic field lines associated with these eddy currents cut the secondary coil B in turn and induce a current which can be used to operate an alarm. The coils are set at right angles to each other so that their magnetic fields do not interact.

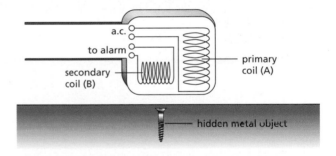

Figure 44.8 Metal detector

Questions

1 Two coils of wire, A and B, are placed near one another (Figure 44.9). Coil A is connected to a switch and battery. Coil B is connected to a centre-reading moving-coil galvanometer, G.
  a If the switch connected to coil A were closed for a few seconds and then opened, the galvanometer connected to coil B would be affected. Explain and describe, step by step, what would actually happen.
  b What changes would you expect if a bundle of soft iron wires was placed through the centre of the coils? Give a reason for your answer.

▶▶

c What would happen if more turns of wire were wound on the coil B?

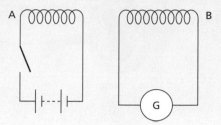

Figure 44.9

2 The main function of a step-down transformer is to
  A decrease current
  B decrease voltage
  C change a.c. to d.c.
  D change d.c. to a.c.
  E decrease the resistance of a circuit.
3 a Calculate the number of turns on the secondary of a step-down transformer which would enable a 12 V lamp to be used with a 230 V a.c. mains power, if there are 460 turns on the primary.
  b What current will flow in the secondary when the primary current is 0.10 A? Assume there are no energy losses.
4 A transformer has 1000 turns on the primary coil. The voltage applied to the primary coil is 230 V a.c. How many turns are on the secondary coil if the output voltage is 46 V a.c.?
  A 20   B 200   C 2000   D 4000   E 8000

Checklist

After studying this chapter you should be able to

- explain the principle of the transformer,
- recall the transformer equation $V_s/V_p = N_s/N_p$ and use it to solve problems,
- recall that for an ideal transformer $V_p \times I_p = V_s \times I_s$ and use the relation to solve problems,
- recall the causes of energy losses in practical transformers,
- explain why high voltage a.c. is used for transmitting electrical power.

45 Electromagnets

- Oersted's discovery
- Field due to a straight wire
- Field due to a circular coil
- Field due to a solenoid
- Magnetisation and demagnetisation

- Electromagnets
- Electric bell
- Relay, reed switch and circuit breaker
- Telephone
- Practical work: Simple electromagnet

● Oersted's discovery

In 1819 Oersted accidentally discovered the magnetic effect of an electric current. His experiment can be repeated by holding a wire over and parallel to a compass needle that is pointing N and S (Figure 45.1). The needle moves when the current is switched on. Reversing the current causes the needle to move in the opposite direction.

Evidently around a wire carrying a current there is a magnetic field. As with the field due to a permanent magnet, we represent the field due to a current by **field lines** or **lines of force**. Arrows on the lines show the direction of the field, i.e. the direction in which a N pole points.

Different field patterns are given by differently shaped conductors.

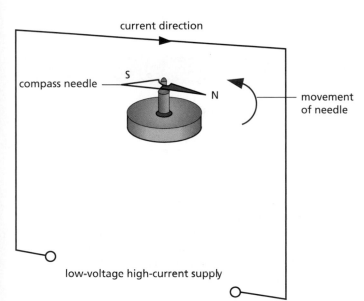

Figure 45.1 An electric current produces a magnetic effect.

● Field due to a straight wire

If a straight vertical wire passes through the centre of a piece of card held horizontally and there is a current in the wire (Figure 45.2), iron filings sprinkled on the card settle in concentric circles when the card is gently tapped.

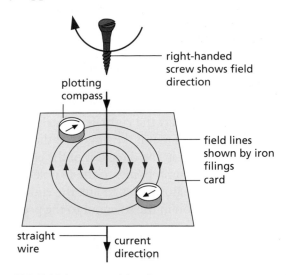

Figure 45.2 Field due to a straight wire

Plotting compasses placed on the card settle along the field lines and show the direction of the field at different points. When the current direction is reversed, the compasses point in the opposite direction showing that the direction of the field reverses when the current reverses.

If the current direction is known, the direction of the field can be predicted by the **right-hand screw rule**:

If a right-handed screw moves forwards in the direction of the current (conventional), the direction of rotation of the screw gives the direction of the field.

● Field due to a circular coil

The field pattern is shown in Figure 45.3. At the centre of the coil the field lines are straight and at right angles to the plane of the coil. The right-hand screw rule again gives the direction of the field at any point.

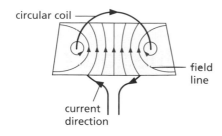

Figure 45.3 Field due to a circular coil

● Field due to a solenoid

A **solenoid** is a long cylindrical coil. It produces a field similar to that of a bar magnet; in Figure 45.4a, end A behaves like a N pole and end B like a S pole. The polarity can be found as before by applying the right-hand screw rule to a short length of one turn of the solenoid. Alternatively the **right-hand grip rule** can be used. This states that if the fingers of the right hand grip the solenoid in the direction of the current (conventional), the thumb points to the N pole (Figure 45.4b). Figure 45.4c shows how to link the end-on view of the current direction in the solenoid to the polarity.

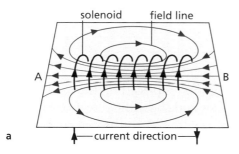

Figure 45.4a Field due to a solenoid

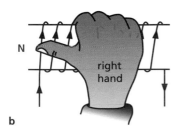

Figure 45.4b The right right-hand grip rule

(i) View from A   (ii) View from B

c

Figure 45.4c End-on views

Inside the solenoid in Figure 45.4a, the field lines are closer together than they are outside the solenoid. This indicates that the magnetic field is stronger inside a solenoid than outside it.

The field inside a solenoid can be made very strong if it has a large number of turns or a large current. Permanent magnets can be made by allowing molten ferromagnetic metal to solidify in such fields.

● Magnetisation and demagnetisation

A ferromagnetic material can be magnetised by placing it inside a solenoid and gradually increasing the current. This increases the magnetic field strength in the solenoid (the density of the field lines increases), and the material becomes magnetised. Reversing the direction of current flow reverses the direction of the magnetic field and reverses the polarity of the magnetisation. A magnet can be demagnetised by placing it inside a solenoid through which the current is repeatedly reversed and reduced.

Practical work

Simple electromagnet

An **electromagnet** is a coil of wire wound on a soft iron core. A 5 cm iron nail and 3 m of PVC-covered copper wire (SWG 26) are needed.

(a) Leave about 25 cm at one end of the wire (for connecting to the circuit) and then wind about 50 cm as a single layer on the nail. **Keep the turns close together and always wind in the same direction.** Connect the circuit of Figure 45.5, setting the rheostat at its maximum resistance.
Find the number of paper clips the electromagnet can support when the current is varied between 0.2 A and 2.0 A. Record the results in a table. How does the 'strength' of the electromagnet depend on the current?

(b) Add another two layers of wire to the nail, winding in the *same direction* as the first layer. Repeat the experiment. What can you say about the 'strength' of an electromagnet and the number of turns of wire?

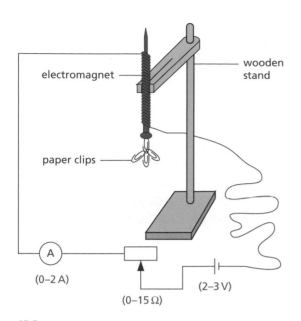

Figure 45.5

(c) Place the electromagnet on the bench and under a sheet of paper. Sprinkle iron filings on the paper, tap it gently and observe the field pattern. How does it compare with that given by a bar magnet?

(d) Use the right-hand screw (or grip) rule to predict which end of the electromagnet is a N pole. Check with a plotting compass.

● Electromagnets

The magnetism of an electromagnet is *temporary* and can be switched on and off, unlike that of a permanent magnet. It has a core of soft iron which is magnetised only when there is current in the surrounding coil.

The strength of an electromagnet increases if
(i)   the **current** in the coil increases,
(ii) the **number of turns** on the coil increases,
(iii) the poles are moved **closer together**.

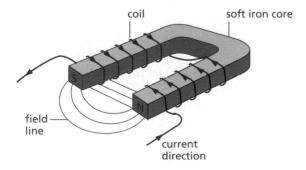

Figure 45.6 C-core or horseshoe electromagnet

In C-core (or horseshoe) electromagnets condition **(iii)** is achieved (Figure 45.6). Note that the coil on each limb of the core is wound in *opposite* directions.

As well as being used in cranes to lift iron objects, scrap iron, etc. (Figure 45.7), electromagnets are an essential part of many electrical devices.

Figure 45.7 Electromagnet being used to lift scrap metal

● Electric bell

When the circuit in Figure 45.8 is completed, by someone pressing the bell push, current flows in the coils of the electromagnet which becomes magnetised and attracts the soft iron bar (the armature).

The hammer hits the gong but the circuit is now broken at the point C of the contact screw.

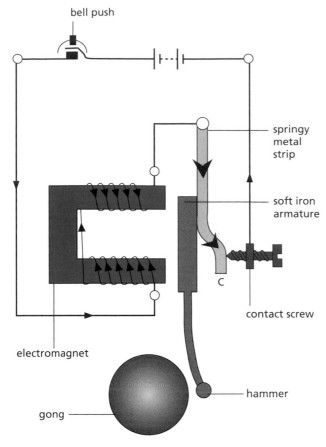

Figure 45.8 Electric bell

The electromagnet loses its magnetism (becomes demagnetised) and no longer attracts the armature. The springy metal strip is then able to pull the armature back, remaking contact at C and so completing the circuit again. This cycle is repeated so long as the bell push is depressed, and continuous ringing occurs.

● Relay, reed switch and circuit breaker

a) Relay

A **relay** is a switch based on the principle of an electromagnet. It is useful if we want one circuit to control another, especially if the current and power are larger in the second circuit (see question 3, p. 214). Figure 45.9 shows a typical relay. When a current is in the coil from the circuit connected to AB, the soft iron core is magnetised and attracts the

L-shaped iron armature. This rocks on its pivot and closes the contacts at C in the circuit connected to DE. The relay is then 'energised' or 'on'.

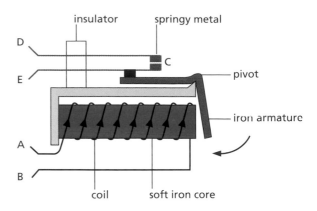

Figure 45.9 Relay

The current needed to operate a relay is called the **pull-on** current and the **drop-off** current is the smaller current in the coil when the relay just stops working. If the coil resistance, R, of a relay is $185\,\Omega$ and its operating p.d. V is $12\,V$, then the pull-on current $I = V/R = 12/185 = 0.065\,A = 65\,mA$. The symbols for relays with normally open and normally closed contacts are given in Figure 45.10.

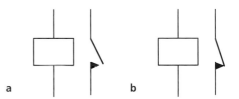

Figure 45.10 Symbols for a relay: **a** open; **b** closed

Some examples of the use of relays in circuits appear in Chapter 41.

b) Reed switch

One such switch is shown in Figure 45.11a. When current flows in the coil, the magnetic field produced magnetises the strips (called **reeds**) of magnetic material. The ends become opposite poles and one reed is attracted to the other, so completing the circuit connected to AB. The reeds separate when the current in the coil is switched off. This type of reed switch is sometimes called a **reed relay**.

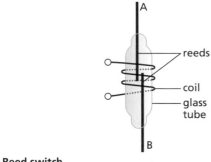

a Reed switch

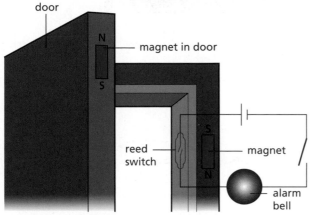

b Burglar alarm activated by a reed switch

Figure 45.11

Reed switches are also operated by permanent magnets. Figure 45.11b shows the use of a normally open reed switch as a burglar alarm. How does it work?

c) Circuit breaker

A **circuit breaker** (p. 181) acts in a similar way to a normally closed relay; when the current in the electromagnet exceeds a critical value, the contact points are separated and the circuit is broken. In the design shown in Figure 40.11, when the iron bolt is attracted far enough towards the electromagnet, the plunger is released and the push switch opens, breaking contact to the rest of the circuit.

● Telephone

A telephone contains a microphone at the speaking end and a receiver at the listening end.

a) Carbon microphone

When someone speaks into a carbon microphone (Figure 45.12), sound waves cause the diaphragm to move backwards and forwards. This varies the pressure on the carbon granules between the movable carbon dome which is attached to the diaphragm and the fixed carbon cup at the back. When the pressure increases, the granules are squeezed closer together and their electrical resistance decreases. A decrease of pressure has the opposite effect. The current passing through the microphone varies in a similar way to the sound wave variations.

b) Receiver

The coils are wound in opposite directions on the two S poles of a magnet (Figure 45.13). If the current goes round one in a clockwise direction, it goes round the other anticlockwise, so making one S pole stronger and the other weaker. This causes the iron armature to rock on its pivot towards the stronger S pole. When the current reverses, the armature rocks the other way due to the S pole which was the stronger before becoming the weaker. These armature movements are passed on to the diaphragm, making it vibrate and produce sound of the same frequency as the alternating current in the coil (received from the microphone).

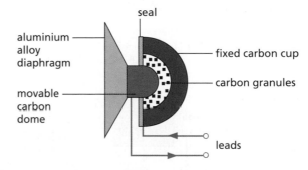

Figure 45.12 Carbon microphone

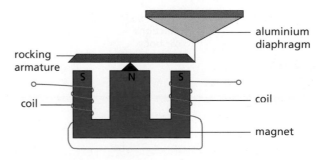

Figure 45.13 Telephone receiver

Questions

1 The vertical wire in Figure 45.14 is at right angles to the card. In what direction will a plotting compass at A point when
a there is no current in the wire,
b the current direction is upwards?

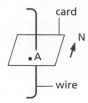

Figure 45.14

2 Figure 45.15 shows a solenoid wound on a core of soft iron. Will the end A be a N pole or S pole when the current is in the direction shown?

Figure 45.15

3 Part of the electrical system of a car is shown in Figure 45.16.
a Why are connections made to the car body?
b There are *two* circuits in parallel with the battery. What are they?
c Why is wire A thicker than wire B?
d Why is a relay used?

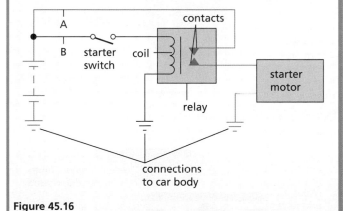

Figure 45.16

Checklist

After studying this chapter you should be able to

• describe and draw sketches of the magnetic fields round current-carrying, straight and circular conductors and solenoids,
• recall the right-hand screw and right-hand grip rules for relating current direction and magnetic field direction,
• describe the effect on the magnetic field of changing the magnitude and direction of the current in a solenoid,
• identify regions of different magnetic field strength around a solenoid,
• make a simple electromagnet,
• describe uses of electromagnets,
• explain the action of an electric bell, a relay, a reed switch and a circuit breaker.

46 Electric motors

- ● The motor effect
- ● Fleming's left-hand rule
- ● Simple d.c. electric motor

- ● Practical motors
- ● Moving-coil loudspeaker
- ● Practical work: A model motor

Electric motors form the heart of a whole host of electrical devices ranging from domestic appliances such as vacuum cleaners and washing machines to electric trains and lifts. In a car the windscreen wipers are usually driven by one and the engine is started by another.

● The motor effect

A wire carrying a current in a magnetic field experiences a force. If the wire can move, it does so.

a) Demonstration

In Figure 46.1 the flexible wire is loosely supported in the strong magnetic field of a C-shaped magnet (permanent or electromagnet). When the switch is closed, current flows in the wire which jumps upwards as shown. If either the direction of the current or the direction of the field is reversed, the wire moves downwards. **The force increases if the strength of the field increases and if the current increases.**

b) Explanation

Figure 46.2a is a side view of the magnetic field lines due to the wire and the magnet. Those due to the wire are circles and we will assume their direction is as shown. The dotted lines represent the field lines of the magnet and their direction is towards the right.

The resultant field obtained by combining both fields is shown in Figure 46.2b. There are more lines below than above the wire since both fields act in the same direction below but they are in opposition above. If we *suppose* the lines are like stretched elastic, those below will try to straighten out and in so doing will exert an upward force on the wire.

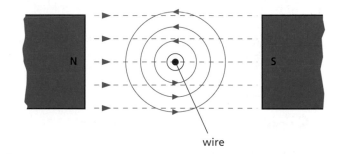

a

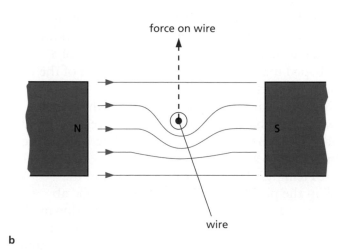

b

Figure 46.2

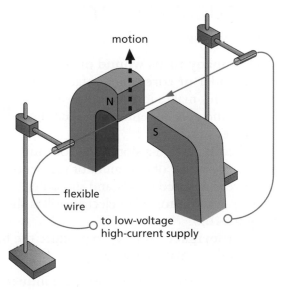

Figure 46.1 A wire carrying a current in a magnetic field experiences a force.

215

Fleming's left-hand rule

The direction of the force or thrust on the wire can be found by this rule which is also called the **motor rule** (Figure 46.3).

Hold the thumb and first two fingers of the left hand at right angles to each other with the **F**irst finger pointing in the direction of the **F**ield and the se**C**ond finger in the direction of the **C**urrent, then the **Th**umb points in the direction of the **Th**rust.

If the wire is not at right angles to the field, the force is smaller and is zero if the wire is parallel to the field.

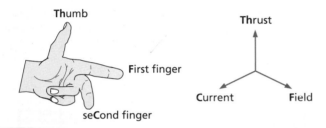

Figure 46.3 Fleming's left-hand (motor) rule

Simple d.c. electric motor

A simple motor to work from direct current (d.c.) consists of a rectangular coil of wire mounted on an axle which can rotate between the poles of a C-shaped magnet (Figure 46.4). Each end of the coil is connected to half of a split ring of copper, called a **commutator**, which rotates with the coil. Two carbon blocks, the **brushes**, are pressed lightly against the commutator by springs. The brushes are connected to an electrical supply.

If Fleming's left-hand rule is applied to the coil in the position shown, we find that side **ab** experiences an upward force and side **cd** a downward force. (No forces act on **ad** and **bc** since they are parallel to the field.) These two forces form a **couple** which rotates the coil in a clockwise direction until it is vertical.

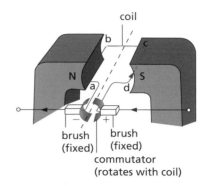

Figure 46.4 Simple d.c. motor

The brushes are then in line with the gaps in the commutator and the current stops. However, because of its inertia, the coil overshoots the vertical and the commutator halves change contact from one brush to the other. This reverses the current through the coil and so also the directions of the forces on its sides. Side **ab** is on the right now, acted on by a downward force, while **cd** is on the left with an upward force. The coil thus carries on rotating clockwise.

The more turns there are on the coil, or the larger the current through it, the greater is the couple on the coil and the faster it turns. The coil will also turn faster if the strength of the magnetic field is increased.

Practical motors

Practical motors have:

(a) **a coil of many turns wound on a soft iron cylinder or core** which rotates with the coil. This makes it more powerful. The coil and core together are called the **armature**.
(b) **several coils** each in a slot in the core and each having a pair of commutator segments. This gives increased power and smoother running. The motor of an electric drill is shown in Figure 46.5.
(c) **an electromagnet** (usually) to produce the field in which the armature rotates.

Most electric motors used in industry are **induction motors**. They work off a.c. (alternating current) on a different principle from the d.c. motor.

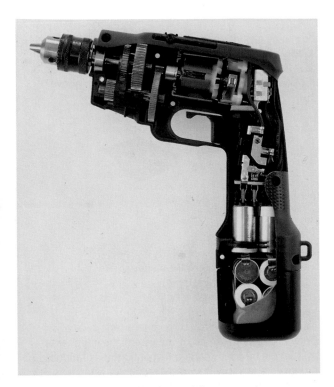

Figure 46.5 Motor inside an electric drill

Practical work

A model motor

The motor shown in Figure 46.6 is made from a kit.

1 Wrap Sellotape round one end of the metal tube which passes through the wooden block.
2 Cut two rings off a piece of narrow rubber tubing; slip them on to the Sellotaped end of the metal tube.
3 Remove the insulation from one end of a 1.5-metre length of SWG 26 PVC-covered copper wire and fix it under both rubber rings so that it is held tight against the Sellotape. This forms one end of the coil.
4 Wind 10 turns of the wire in the slot in the wooden block and finish off the second end of the coil by removing the PVC and fixing this too under the rings but on the opposite side of the tube from the first end. The bare ends act as the **commutator**.
5 Push the axle through the metal tube of the wooden base so that the block spins freely.
6 Arrange two 0.5-metre lengths of wire to act as **brushes** and leads to the supply, as shown. Adjust the brushes so that they are vertical and each touches one bare end of the coil when the plane of the coil is horizontal. **The motor will not work if this is not so**.
7 Slide the base into the magnet with *opposite poles facing*. Connect to a 3 V battery (or other low-voltage d.c. supply) and a slight push of the coil should set it spinning at high speed.

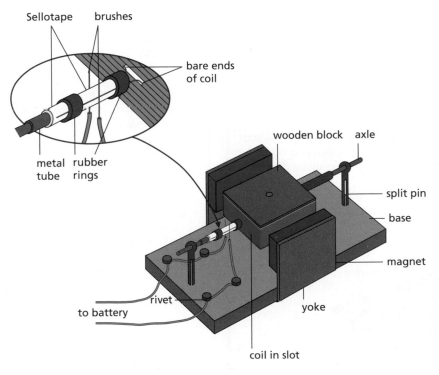

Figure 46.6 A model motor

● Moving-coil loudspeaker

Varying currents from a radio, disc player, etc. pass through a short cylindrical coil whose turns are at right angles to the magnetic field of a magnet with a central pole and a surrounding ring pole (Figure 46.7a).

A force acts on the coil which, according to Fleming's left-hand rule, makes it move in and out. A paper cone attached to the coil moves with it and sets up sound waves in the surrounding air (Figure 46.7b.

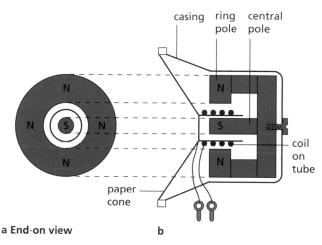

a End-on view   b

Figure 46.7 Moving-coil loudspeaker

Questions

1 The current direction in a wire running between the N and S poles of a magnet lying horizontally is shown in Figure 46.8. The force on the wire due to the magnet is directed
   A from N to S
   B from S to N
   C opposite to the current direction
   D in the direction of the current
   E vertically upwards.

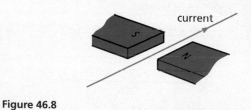

Figure 46.8

2 In the simple electric motor of Figure 46.9, the coil rotates anticlockwise as seen by the eye from the position X when current flows in the coil. Is the current flowing clockwise or anticlockwise around the coil when viewed from above?

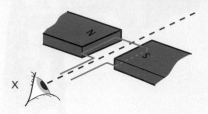

Figure 46.9

3 An electric motor is a device which transfers
   A mechanical energy to electrical energy
   B heat energy to electrical energy
   C electrical energy to heat only
   D heat to mechanical energy
   E electrical energy to mechanical energy and heat.
4 a Draw a labelled diagram of the essential components of a simple motor. Explain how continuous rotation is produced and show how the direction of rotation is related to the direction of the current.
   b State what would happen to the direction of rotation of the motor you have described if
      (i)   the current was reversed,
      (ii) the magnetic field was reversed,
      (iii) both current and field were reversed simultaneously.

Checklist

After studying this chapter you should be able to

• describe a demonstration to show that a force acts on a current-carrying conductor in a magnetic field, and recall that it increases with the strength of the field and the size of the current,

• draw the resultant field pattern for a current-carrying conductor which is at right angles to a uniform magnetic field,

• explain why a rectangular current-carrying coil experiences a couple in a uniform magnetic field,

• draw a diagram of a simple d.c. electric motor and explain how it works,

• describe a practical d.c. motor.

Electric meters

- Moving-coil galvanometer
- Ammeters and shunts
- Voltmeters and multipliers
- Multimeters
- Reading a voltmeter

Moving-coil galvanometer

A **galvanometer** detects small currents or small p.d.s, often of the order of milliamperes (mA) or millivolts (mV).

In the moving-coil **pointer-type** meter, a coil is pivoted between the poles of a permanent magnet (Figure 47.1a). Current enters and leaves the coil by hair springs above and below it. When there is a current, a couple acts on the coil (as in an electric motor), causing it to rotate until stopped by the springs. The greater the current, the greater the deflection which is shown by a pointer attached to the coil.

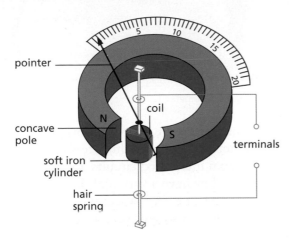

a

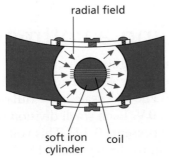

b View from above

Figure 47.1 Moving-coil pointer-type galvanometer

The soft iron cylinder at the centre of the coil is fixed and along with the concave poles of the magnet it produces a **radial** field (Figure 47.1b), i.e. the

field lines are directed to and from the centre of the cylinder. The scale on the meter is then even or linear, i.e. all divisions are the same size.

The sensitivity of a galvanometer is increased by having

(i) more turns on the coil,
(ii) a stronger magnet,
(iii) weaker hair springs or a wire suspension,
(iv) as a pointer, a long beam of light reflected from a mirror on the coil.

The last two are used in **light-beam** meters which have a full-scale deflection of a few microamperes (μA). ($1\,\mu$A$=10^{-6}$A)

Ammeters and shunts

An **ammeter** is a galvanometer that has a known low resistance, called a **shunt**, in parallel with it to take most of the current (Figure 47.2). An ammeter is placed in series in a circuit and must have a *low resistance* otherwise it changes the current to be measured.

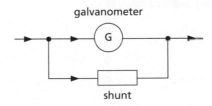

Figure 47.2 An ammeter

Voltmeters and multipliers

A **voltmeter** is a galvanometer having a known high resistance, called a **multiplier**, in series with it (Figure 47.3). A voltmeter is placed in *parallel* with the part of the circuit across which the p.d. is to be measured and must have a *high resistance* – otherwise the total resistance of the whole circuit is reduced so changing the current and the p.d. measured.

219

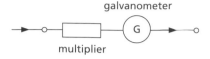

Figure 47.3 A voltmeter

● Multimeters

A **multimeter** can have analogue or digital displays (see Figures 47.4a and 47.4b) and can be used to measure a.c. or d.c. currents or voltages and also resistance. The required function is first selected, say a.c. current, and then a suitable range chosen. For example if a current of a few milliamps is expected, the 10 mA range might be selected and the value of the current (in mA) read from the display; if the reading is off-scale, the sensitivity should be reduced by changing to the higher, perhaps 100 mA, range.

For the measurement of resistance, the resistance function is chosen and the appropriate range selected. The terminals are first short-circuited to check the zero of resistance, then the unknown resistance is disconnected from any circuit and reconnected across the terminals of the meter in place of the short circuit.

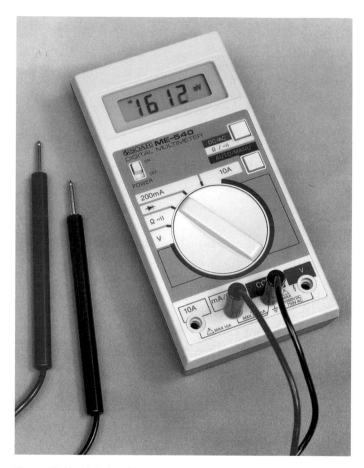

Figure 47.4b Digital multimeter

Analogue multimeters are adapted moving-coil galvanometers. Digital multimeters are constructed from integrated circuits. On the voltage setting they have a very high input resistance (10 MΩ), i.e. they affect most circuits very little and so give very accurate readings.

Figure 47.4a Analogue multimeter

● Reading a voltmeter

The face of an analogue voltmeter is represented in Figure 47.5. The voltmeter has two scales. The 0–5 scale has a **full-scale deflection** of 5.0 V. Each small division on the 0–5 scale represents 0.1 V. This voltmeter scale can be read to the nearest 0.1 V. However the human eye is very good at judging a half division, so we are able to estimate the voltmeter reading to the nearest 0.05 V with considerable precision.

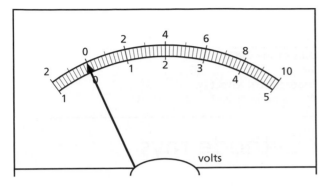

Figure 47.5 An analogue voltmeter scale

Every measuring instrument has a calibrated scale. When you write an account of an experiment (see p. x, *Scientific enquiry*) you should include details about each scale that you use.

Questions

1 What does a galvanometer do?
2 Why should the resistance of
  a an ammeter be very small,
  b a voltmeter be very large?
3 The scales of a voltmeter are shown in Figure 47.6.

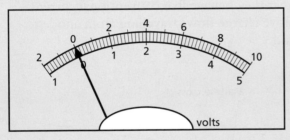

Figure 47.6

  a What are the two ranges available when using the voltmeter?
  b What do the small divisions between the numbers 3 and 4 represent?
  c Which scale would you use to measure a voltage of 4.6 V?
  d When the voltmeter reads 4.0 V where should you position your eye to make the reading?
  e When making the reading for 4.0 V an observer's eye is over the 0 V mark. Explain why the value obtained by this observer is higher than 4.0 V.

48 Electrons

- Thermionic emission
- Cathode rays
- Deflection of an electron beam
- Cathode ray oscilloscope (CRO)

- Uses of the CRO
- X-rays
- Photoelectric effect
- Waves or particles?

The discovery of the **electron** was a landmark in physics and led to great technological advances.

● Thermionic emission

The evacuated bulb in Figure 48.1 contains a small coil of wire, the **filament**, and a metal plate called the **anode** because it is connected to the positive of the 400 V d.c. power supply. The negative of the supply is joined to the filament which is also called the **cathode**. The filament is heated by current from a 6 V supply (a.c. or d.c.).

With the circuit as shown, the meter deflects, indicating current flow in the circuit containing the gap between anode and cathode. The current stops if *either* the 400 V supply is reversed to make the anode negative *or* the filament is not heated.

This demonstration shows that negative charges, in the form of electrons, escape from the filament when it is hot because they have enough energy to get free from the metal surface. The process is known as **thermionic emission** and the bulb as a thermionic diode (since it has two electrodes). There is a certain minimum **threshold energy** (depending on the metal) which the electrons must have to escape. Also, the higher the temperature of the metal, the greater the number of electrons emitted. The electrons are attracted to the anode if it is positive and are able to reach it because there is a vacuum in the bulb.

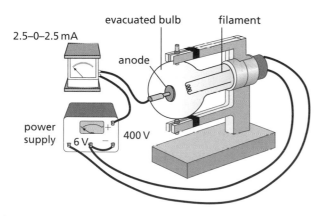

Figure 48.1 Demonstrating thermionic emission

● Cathode rays

Beams of electrons moving at high speed are called **cathode rays**. Their properties can be studied using the 'Maltese cross tube' (Figure 48.2).

Electrons emitted by the hot cathode are accelerated towards the anode but most pass through the hole in it and travel on along the tube. Those that miss the cross cause the screen to fluoresce with green or blue light and cast a shadow of the cross on it. The cathode rays evidently travel in straight lines.

If the N pole of a magnet is brought up to the neck of the tube, the rays (and the fluorescent shadow) can be shown to move upwards. The rays are clearly deflected by a magnetic field and, using Fleming's left-hand rule (Chapter 46), we see that they behave like conventional current (positive charge flow) travelling from anode to cathode.

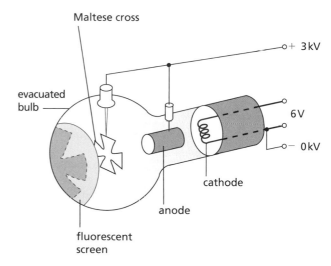

Figure 48.2 Maltese cross tube

There is also an optical shadow of the cross, due to light emitted by the cathode. This is unaffected by the magnet.

● Deflection of an electron beam

a) By a magnetic field

In Figure 48.3 the evenly spaced crosses represent a uniform magnetic field (i.e. one of the same strength throughout the area shown) acting into and perpendicular to the paper. An electron beam entering the field at right angles to the field experiences a force due to the motor effect (Chapter 46) whose direction is given by Fleming's left-hand rule. This indicates that the force acts at right angles to the direction of the beam and makes it follow a *circular* path as shown (the beam being treated as conventional current in the opposite direction).

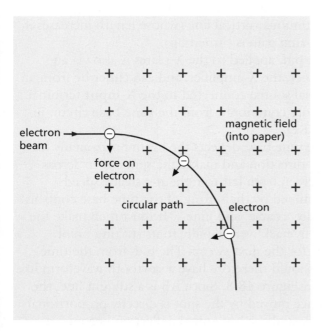

Figure 48.3 Path of an electron beam at right angles to a magnetic field

b) By an electric field

An electric field is a region where an electric charge experiences a force due to other charges (see p. 155). In Figure 48.4 the two metal plates behave like a capacitor that has been charged by connection to a voltage supply. If the charge is evenly spread over the plates, a uniform electric field is created between them and is represented by parallel, equally spaced lines; the arrows indicate the direction in which a positive charge would move.

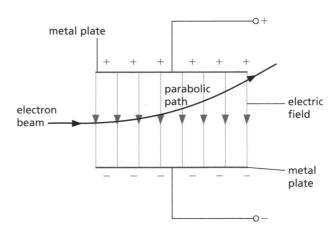

Figure 48.4 Path of an electron beam incident perpendicular to an electric field

If a beam of electrons enters the field in a direction perpendicular to the field, the negatively charged beam is attracted towards the positively charged plate and follows a **parabolic** path, as shown. In fact its behaviour is not unlike that of a projectile (Chapter 4) in which the horizontal and vertical motions can be treated separately.

c) Demonstration

The deflection tube in Figure 48.5 can be used to show the deflection of an electron beam in electric and magnetic fields. Electrons from a hot cathode strike a fluorescent screen S set at an angle. A p.d. applied across two horizontal metal plates Y_1Y_2 creates a *vertical* electric field which deflects the rays upwards if Y_1 is positive (as shown) and downwards if it is negative.

When there is current in the two coils X_1X_2 (in series) outside the tube, a *horizontal* magnetic field is produced across the tube. It can be used instead of a magnet to deflect the rays, or to cancel the deflection due to an electric field.

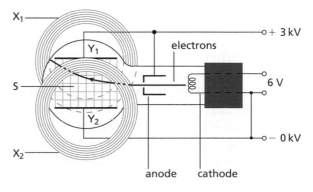

Figure 48.5 Deflection tube

● Cathode ray oscilloscope (CRO)

Historically the CRO is one of the most important scientific instruments ever developed. It contains a cathode ray tube that has three main parts (Figure 48.6).

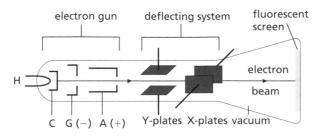

Figure 48.6 Main parts of a CRO

a) Electron gun

This consists of a **heater** H, a **cathode** C, another electrode called the **grid** G and two or three **anodes** A. G is at a negative voltage with respect to C and controls the number of electrons passing through its central hole from C to A; it is the **brilliance** or **brightness** control. The anodes are at high positive voltages relative to C; they accelerate the electrons along the highly evacuated tube and also **focus** them into a narrow beam.

b) Fluorescent screen

A bright spot of light is produced on the screen where the beam hits it.

c) Deflecting system

Beyond A are two pairs of deflecting plates to which p.d.s can be applied. The **Y-plates** are horizontal but create a vertical electric field which deflects the beam vertically. The **X-plates** are vertical and deflect the beam horizontally.

The p.d. to create the electric field between the Y-plates is applied to the **Y-input** terminals (often marked 'high' and 'low') on the front of the CRO. The input is usually amplified by an amount that depends on the setting of the **Y-amp gain** control, before it is applied to the Y-plates. It can then be made large enough to give a suitable vertical deflection of the beam.

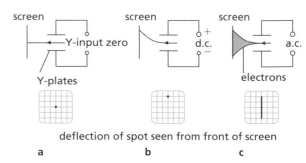

Figure 48.7 Deflection of the electron beam

In Figure 48.7a the p.d. between the Y-plates is zero, as is the deflection. In part b of the figure, the d.c. input p.d. makes the upper plate positive and it attracts the beam of negatively charged electrons upwards. In part c the 50 Hz a.c. input makes the beam move up and down so rapidly that it produces a continuous vertical line (whose length increases if the Y-amp gain is turned up).

The p.d. applied to the X-plates is also via an amplifier, the X-amplifier, and can either be from an external source connected to the **X-input** terminal or, more commonly, from the **time base** circuit in the CRO.

The time base deflects the beam horizontally in the X-direction and makes the spot sweep across the screen from left to right at a steady speed determined by the setting of the time base controls (usually 'coarse' and 'fine'). It must then make the spot 'fly back' very rapidly to its starting point, ready for the next sweep. The p.d. from the time base should therefore have a sawtooth waveform like that in Figure 48.8. Since AB is a straight line, the distance moved by the spot is directly proportional to time and the horizontal deflection becomes a measure of time, i.e. a time axis or base.

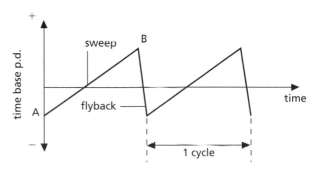

Figure 48.8 Time base waveform

In Figures 48.9a, b and c, the time base is on, applied to the X-plates. For the trace in part a, the Y-input p.d. is zero, for the trace in part b the Y-input is d.c. which makes the upper Y-plate positive. In both cases the spot traces out a horizontal line which appears to be continuous if the flyback is fast enough. For the trace in part c the Y-input is a.c., that is, the Y-plates are alternately positive and negative and the spot moves accordingly.

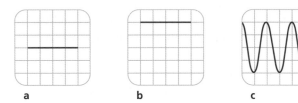

a   b   c

Figure 48.9 Deflection of the spot with time base on

● Uses of the CRO

A small CRO is shown in Figure 48.10.

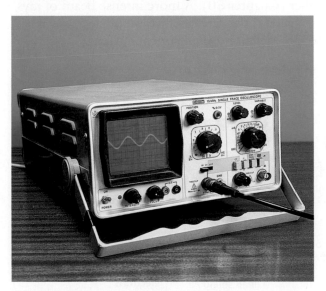

Figure 48.10 Single-beam CRO

a) Practical points

The **brilliance** or **intensity** control, which is sometimes the **on/off** switch as well, should be as low as possible when there is just a spot on the screen. Otherwise screen 'burn' occurs which damages the fluorescent material. If possible it is best to defocus the spot when not in use, or draw it into a line by running the time base.

When preparing the CRO for use, set the **brilliance**, **focus**, **X-shift** and **Y-shift** controls (which allow the spot to be moved 'manually' over the screen in the X and Y directions, respectively) to their mid-positions. The **time base** and **Y-amp gain** controls can then be adjusted to suit the input.

When the **a.c./d.c. selector** switch is in the 'd.c.' (or 'direct') position, both d.c. and a.c. can pass to the Y-input. In the 'a.c.' (or 'via C') position, a capacitor blocks d.c. in the input but allows a.c. to pass.

b) Measuring p.d.s

A CRO can be used as a d.c./a.c. voltmeter if the p.d. to be measured is connected across the Y-input terminals; **the deflection of the spot is proportional to the p.d.**

For example, if the Y-amp gain control is on, say, 1 V/div, a deflection of one vertical division on the screen graticule (like graph paper with squares for measuring deflections) would be given by a 1 V d.c. input. A line one division long (time base off) would be produced by an a.c. input of 1 V peak-to-peak, i.e. peak p.d. = 0.5 V.

Increasingly the CRO is being replaced by a data-logger and computer with software which simulates the display on a CRO screen by plotting the p.d. against time.

c) Displaying waveforms

In this widely used role, the time base is on and the CRO acts as a 'graph-plotter' to show the waveform, i.e. the variation with time, of the p.d. applied to its Y-input. The displays in Figures 48.11a and b are of alternating p.d.s with sine waveforms. For the trace in part a, the time base frequency *equals* that of the input and one complete wave is obtained. For the trace in part b, it is *half* that of the input and two waves are formed. If the traces are obtained with the Y-amp gain control on, say, 0.5 V/div, the peak-to-peak voltage of the a.c. = 3 divs × 0.5 V/div, that is, 1.5 V, and the peak p.d. = 0.75 V.

Sound waveforms can be displayed if a microphone is connected to the Y-input terminals (see Chapter 33).

▶▶

a

b

Figure 48.11 Alternating p.d. waveforms on the CRO

d) Measuring time intervals and frequency

These can be measured if the CRO has a calibrated time base. For example, when the time base is set on 10 ms/div, the spot takes 10 milliseconds to move one division horizontally across the screen graticule. If this is the time base setting for the waveform in Figure 48.11b then, since one complete wave occupies two horizontal divisions, we can say

time for one complete wave $= 2\,\text{divs} \times 10\,\text{ms/div}$

$$= 20\,\text{ms}$$

$$= \frac{20}{1000} = \frac{1}{50}\,\text{s}$$

$\therefore$ number of complete waves per second $= 50$

$\therefore$ frequency of a.c. applied to Y-input $= 50\,\text{Hz}$

● X-rays

X-rays are produced when high-speed electrons are stopped by matter.

a) Production

In an X-ray tube, Figure 48.12, electrons from a hot filament are accelerated across a vacuum to the anode by a large p.d. (up to 100 kV). The anode is a copper block with a 'target' of a high-melting-point metal such as tungsten on which the electrons are focused by the electric field between the anode and the concave cathode. The tube has a lead shield with a small exit for the X-rays.

The work done (see p. 163) in transferring a charge Q through a p.d. V is

$$E = Q \times V$$

This will equal the k.e. of the electrons reaching the anode if $Q =$ charge on an electron $(= 1.6 \times 10^{-19}\,\text{C})$ and V is the accelerating p.d. Less than 1% of the k.e. of the electrons becomes X-ray energy; the rest heats the anode which has to be cooled.

High p.d.s give short wavelength, very penetrating (**hard**) X-rays. Less penetrating (**soft**) rays, of longer wavelength, are obtained with lower p.d.s. The absorption of X-rays by matter is greatest by materials of high density having a large number of outer electrons in their atoms, i.e. of high atomic number (Chapter 50). A more intense beam of rays is produced if the rate of emission of electrons is raised by increasing the filament current.

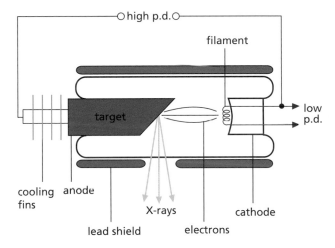

Figure 48.12 X-ray tube

b) Properties and nature

X-rays:

(i) readily penetrate matter – up to 1 mm of lead,

(ii) are not deflected by electric or magnetic fields,

(iii) ionise a gas, making it a conductor, e.g. a charged electroscope discharges when X-rays pass through the surrounding air,

(iv) affect a photographic film,

(v) cause fluorescence,
(vi) give interference and diffraction effects.

These facts (and others) suggest that X-rays are electromagnetic waves of very short wavelength.

c) Uses

These were considered earlier (Chapter 32).

● Photoelectric effect

Electrons are emitted by certain metals when electromagnetic radiation of small enough wavelength falls on them. The effect is called **photoelectric emission**. It happens, for example, when zinc is exposed to ultraviolet.

The photoelectric effect only occurs for a given metal if the frequency of the incident electromagnetic radiation exceeds a certain **threshold frequency**. We can explain this by assuming that

(i) all electromagnetic radiation is emitted and absorbed as packets of energy, called **photons**, and
(ii) the energy of a photon is directly proportional to its frequency.

Ultraviolet (UV) photons would therefore have more energy than light photons since UV has a higher frequency than light. The behaviour of zinc (and most other substances) in not giving photoelectric emission with light but with UV would therefore be explained: a photon of light has less than the minimum energy required to cause the zinc to emit an electron.

The absorption of a photon by an atom results in the electron gaining energy and the photon disappearing. If the photon has more than the minimum amount of energy required to enable an electron to escape, the excess appears as k.e. of the emitted electron.

> energy of photon = energy needed for electron to escape
> + k.e. of electron

The **photoelectric effect** is the process by which X-ray photons are absorbed by matter; in effect it causes **ionisation** (Chapter 49) since electrons are ejected and positive ions remain. Photons not absorbed by the metal pass through with unchanged energy.

● Waves or particles?

The wave theory of electromagnetic radiation can account for properties such as interference, diffraction and polarisation which the photon theory cannot. On the other hand, the wave theory does not explain the photoelectric effect and the photon theory does.

It would seem that electromagnetic radiation has a dual nature and has to be regarded as waves on some occasions and as 'particles' (photons) on others.

Questions

1 a In Figure 48.13a, to which terminals on the power supply must plates A and B be connected to deflect the cathode rays downwards?
  b In Figure 48.13b, in which direction will the cathode rays be deflected?

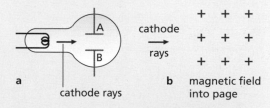

a   cathode rays
b   magnetic field into page

Figure 48.13

2 An electron, charge e and mass m, is accelerated in a cathode ray tube by a p.d. of 1000 V. Calculate
  a the kinetic energy gained by the electron,
  b the speed it acquires.
  ($e = 1.6 \times 10^{-19}$ C, $m = 9.1 \times 10^{-31}$ kg)

Checklist

After studying this chapter you should be able to

- explain the term cathode rays,
- describe experiments to show that cathode rays are deflected by magnetic and electric fields.

Section 5 Atomic physics

Chapters

49 Radioactivity

50 Atomic structure

- Ionising effect of radiation
- Geiger–Müller (GM) tube
- Alpha, beta and gamma radiation
- Particle tracks

- Radioactive decay
- Uses of radioactivity
- Dangers and safety

The discovery of radioactivity in 1896 by the French scientist Becquerel was accidental. He found that uranium compounds emitted radiation that: **(i)** affected a photographic plate even when it was wrapped in black paper, and **(ii)** ionised a gas. Soon afterwards Marie Curie discovered the radioactive element radium. We now know that radioactivity arises from unstable nuclei (Chapter 50) which may occur naturally or be produced in reactors. Radioactive materials are widely used in industry, medicine and research.

We are all exposed to natural **background radiation** caused partly by radioactive materials in rocks, the air and our bodies, and partly by cosmic rays from outer space (see p. 235).

● Ionising effect of radiation

A charged electroscope discharges when a lighted match or a radium source (**held in forceps**) is brought near the cap (Figures 49.1a and b).

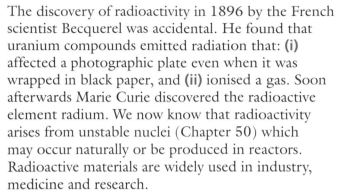

Figure 49.1

In the first case the flame knocks electrons out of surrounding air molecules leaving them as positively charged **ions**, i.e. air molecules which have lost one or more electrons (Figure 49.2); in the second case radiation causes the same effect, called **ionisation**. The positive ions are attracted to the cap if it is negatively charged; if it is positively charged the electrons are attracted. As a result in either case the charge on the electroscope is neutralised, i.e. it loses its charge.

Figure 49.2 Ionisation

● Geiger–Müller (GM) tube

The ionising effect is used to detect radiation.

When radiation enters a **GM tube** (Figure 49.3), either through a thin end-window made of mica, or, if the radiation is very penetrating, through the wall, it creates argon ions and electrons. These are accelerated towards the electrodes and cause more ionisation by colliding with other argon atoms.

On reaching the electrodes, the ions produce a current pulse which is amplified and fed either to a **scaler** or a **ratemeter**. A scaler counts the pulses and shows the total received in a certain time. A ratemeter gives the counts per second (or minute), or **count-rate**, directly. It usually has a loudspeaker which gives a 'click' for each pulse.

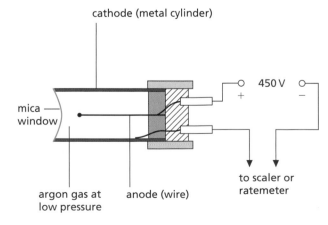

Figure 49.3 Geiger–Müller (GM) tube

Alpha, beta and gamma radiation

Experiments to study the penetrating power, ionising ability and behaviour of radiation in magnetic and electric fields show that a radioactive substance emits one or more of three types of radiation – called **alpha** (α), **beta** (β– or β+) and **gamma** (γ) rays.

Penetrating power can be investigated as in Figure 49.4 by observing the effect on the count-rate of placing one of the following in turn between the GM tube and the lead sheet:

(i) a sheet of thick paper (the radium source, lead and tube must be close together for this part),
(ii) a sheet of aluminium 2 mm thick,
(iii) a further sheet of lead 2 cm thick.

Radium (Ra-226) emits α-particles, β-particles and γ-rays. Other sources can be tried, such as americium, strontium and cobalt.

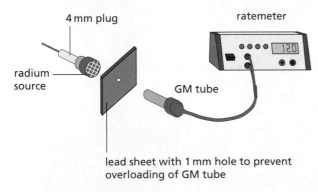

Figure 49.4 Investigating the penetrating power of radiation

a) Alpha particles

These are stopped by a thick sheet of paper and have a range in air of only a few centimetres since they cause intense ionisation in a gas due to frequent collisions with gas molecules. They are deflected by electric and strong magnetic fields in a direction and by an amount which suggests they are helium atoms minus two electrons, i.e. helium ions with a double positive charge. From a particular substance, they are all emitted with the same speed (about 1/20th of that of light).

Americium (Am-241) is a pure α source.

b) Beta particles

These are stopped by a few millimetres of aluminium and some have a range in air of several metres. Their ionising power is much less than that of α-particles. As well as being deflected by electric fields, they are more easily deflected by magnetic fields. Measurements show that β–-particles are streams of **high-energy electrons**, like cathode rays, emitted with a range of speeds up to that of light. Strontium (Sr-90) emits β–-particles only.

The magnetic deflection of β–-particles can be shown as in Figure 49.5. With the GM tube at A and without the magnet, the count-rate is noted. Inserting the magnet reduces the count-rate but it increases again when the GM tube is moved sideways to B.

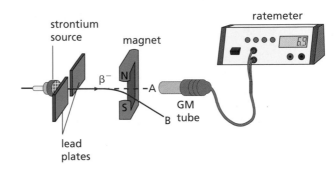

Figure 49.5 Demonstrating magnetic deflection of β–-particles

c) Gamma rays

These are the most penetrating and are stopped only by many centimetres of lead. They ionise a gas even less than β-particles and are not deflected by electric and magnetic fields. They give interference and diffraction effects and are electromagnetic radiation travelling at the speed of light. Their wavelengths are those of very short X-rays, from which they differ only because they arise in atomic nuclei whereas X-rays come from energy changes in the electrons outside the nucleus.

Cobalt (Co-60) emits γ-rays and β–-particles but can be covered with aluminium to provide pure γ-rays.

Comparing alpha, beta and gamma radiation

In a collision, α-particles, with their relatively large mass and charge, have more of a chance of knocking an electron from an atom and causing ionisation than the lighter β-particles; γ-rays, which have no charge, are even less likely than β-particles to produce ionisation.

A GM tube detects β-particles and γ-rays and energetic α-particles; a charged electroscope detects only α-particles. All three types of radiation cause fluorescence.

The behaviour of the three kinds of radiation in a magnetic field is summarised in Figure 49.6a. The deflections (not to scale) are found from Fleming's left-hand rule, taking negative charge moving to the right as equivalent to positive (conventional) current to the left.

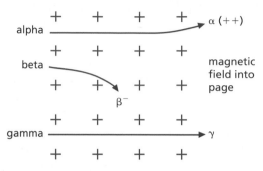

Figure 49.6a Deflection of α-, β- and γ-radiation in a magnetic field

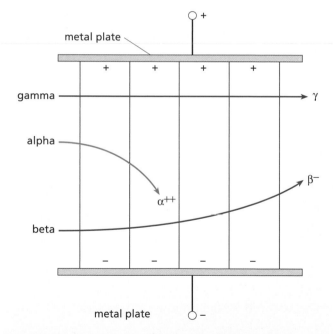

Figure 49.6b Deflection of α-, β- and γ-radiation in a uniform electric field

Figure 49.6b shows the behaviour of α-particles, β⁻-radiation and γ-rays in a uniform electric field: α-particles are attracted towards the negatively charged metal plate, β⁻-particles are attracted towards the positively charged plate and γ-rays pass through undeflected.

● Particle tracks

The paths of particles of radiation were first shown up by the ionisation they produced in devices called **cloud chambers**. When air containing a vapour, alcohol, is cooled enough, saturation occurs. If ionising radiation passes through the air, further cooling causes the saturated vapour to condense on the ions created. The resulting white line of tiny liquid drops shows up as a track when illuminated.

In a **diffusion cloud chamber**, α-particles showed straight, thick tracks (Figure 49.7a). Very fast β-particles produced thin, straight tracks while slower ones gave short, twisted, thicker tracks (Figure 49.7b). Gamma-rays eject electrons from air molecules; the ejected electrons behaved like β⁻-particles in the cloud chamber and produced their own tracks spreading out from the γ-rays.

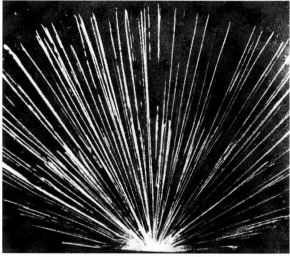

a α-particles

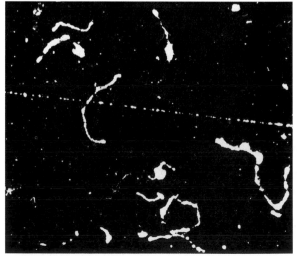

b Fast and slow β-particles

Figure 49.7 Tracks in a cloud chamber

The **bubble chamber**, in which the radiation leaves a trail of bubbles in liquid hydrogen, has now replaced the cloud chamber in research work. The higher density of atoms in the liquid gives better defined tracks, as shown in Figure 49.8, than obtained in a cloud chamber. A magnetic field is usually applied across the bubble chamber which causes charged particles to move in circular paths; the sign of the charge can be deduced from the way the path curves.

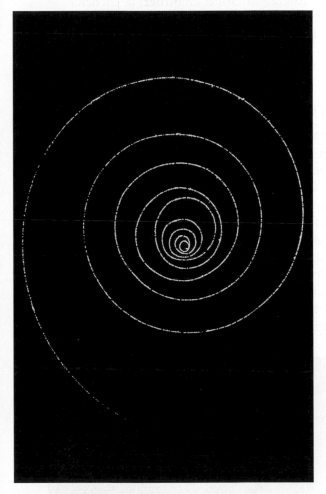

Figure 49.8 Charged particle track in a bubble chamber

● Radioactive decay

Radioactive atoms have unstable nuclei and, when they emit α-particles or β-particles, they **decay** into atoms of different elements that have more stable nuclei. These changes are spontaneous and cannot be controlled; also, it does not matter whether the material is pure or combined chemically with something else.

Half-life

The **rate of decay** is unaffected by temperature but every radioactive element has its own definite decay rate, expressed by its **half-life**. This is the **average time for half the atoms in a given sample to decay**. It is difficult to know when a substance has lost all its radioactivity, but the time for its activity to fall to half its value can be found more easily.

Decay curve

The average number of disintegrations (i.e. decaying atoms) per second of a sample is its **activity**. If it is measured at different times (e.g. by finding the count-rate using a GM tube and ratemeter), a decay curve of activity against time can be plotted. The ideal curve for one element (Figure 49.9) shows that the activity decreases by the same fraction in successive equal time intervals. It falls from 80 to 40 disintegrations per second in 10 minutes, from 40 to 20 in the next 10 minutes, from 20 to 10 in the third 10 minutes and so on. The half-life is 10 minutes.

Half-lives vary from millionths of a second to millions of years. For radium it is 1600 years.

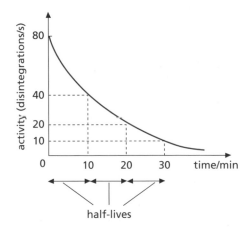

Figure 49.9 Decay curve

Experiment to find the half-life of thoron

The half-life of the α-emitting gas **thoron** can be found as shown in Figure 49.10. The thoron bottle is squeezed three or four times to transfer some thoron to the flask (Figure 49.10a). The clips are then closed, the bottle removed and the stopper replaced by a GM tube so that it seals the top (Figure 49.10b).

When the ratemeter reading has reached its maximum and started to fall, the count-rate is noted every 15 s for 2 minutes and then every 60 s for the next few minutes. (The GM tube is left in the flask for at least 1 hour until the radioactivity has decayed.)

A measure of the background radiation is obtained by recording the counts for a period (say 10 minutes) at a position well away from the thoron equipment. The count-rates in the thoron decay experiment are then corrected by subtracting the average background count-rate from each reading. A graph of the corrected count-rate against time is plotted and the half-life (52 s) estimated from it.

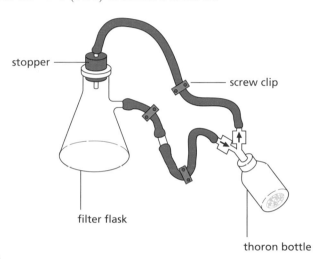

a

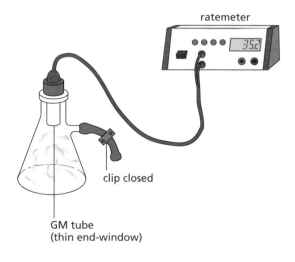

b

Figure 49.10

Random nature of decay

During the previous experiment it becomes evident that the count-rate varies irregularly: the loudspeaker of the ratemeter 'clicks' erratically, not at a steady rate. This is because radioactive decay is a **random** process, in that it is a matter of pure chance whether or not a particular atom will decay during a certain period of time. All we can say is that about half the atoms in a sample will decay during the half-life. We cannot say which atoms these will be, nor can we influence the process in any way. Radioactive emissions occur randomly over space and time.

● Uses of radioactivity

Radioactive substances, called **radioisotopes**, are now made in nuclear reactors and have many uses.

a) Thickness gauge

If a radioisotope is placed on one side of a moving sheet of material and a GM tube on the other, the count-rate decreases if the thickness increases. This technique is used to control automatically the thickness of paper, plastic and metal sheets during manufacture (Figure 49.11). Because of their range, β-emitters are suitable sources for monitoring the thickness of thin sheets but γ-emitters would be needed for thicker materials.

Flaws in a material can be detected in a similar way; the count-rate will increase where a flaw is present.

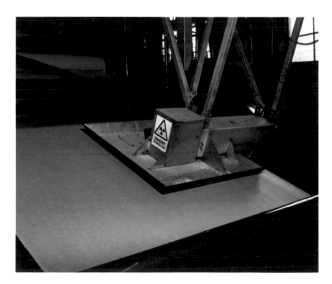

Figure 49.11 Quality control in the manufacture of paper using a radioactive gauge

b) Tracers

The progress of a small amount of a weak radioisotope injected into a system can be 'traced' by a GM tube or other detector. The method is used in medicine to detect brain tumours and internal bleeding, in agriculture to study the uptake of fertilisers by plants, and in industry to measure fluid flow in pipes.

A tracer should be chosen whose half-life matches the time needed for the experiment; the activity of the source is then low after it has been used and so will not pose an ongoing radiation threat. For medical purposes, where short exposures are preferable, the time needed to transfer the source from the production site to the patient also needs to be considered.

c) Radiotherapy

Gamma rays from strong cobalt radioisotopes are used in the treatment of cancer.

d) Sterilisation

Gamma rays are used to sterilise medical instruments by killing bacteria. They are also used to 'irradiate' certain foods, again killing bacteria to preserve the food for longer. They are safe to use as no radioactive material goes into the food.

e) Archaeology

A radioisotope of carbon present in the air, carbon-14, is taken in by living plants and trees along with non-radioactive carbon-12. When a tree dies no fresh carbon is taken in. So as the carbon-14 continues to decay, with a half-life of 5700 years, the amount of carbon-14 compared with the amount of carbon-12 becomes smaller. By measuring the residual radioactivity of carbon-containing material such as wood, linen or charcoal, the age of archaeological remains can be estimated within the range 1000 to 50 000 years (Figure 49.12). See Worked example 2, on p. 236.

The ages of rocks have been estimated in a similar way by measuring the ratio of the number of atoms of a radioactive element to those of its decay product in a sample. See Worked example 3, on p. 236.

Figure 49.12 The year of construction of this Viking ship has been estimated by radiocarbon techniques to be AD 800.

● Dangers and safety

We are continually exposed to radiation from a range of sources, both natural ('background') and artificial, as indicated in Figure 49.13.

(i) Cosmic rays (high-energy particles from outer space) are mostly absorbed by the atmosphere and produce radioactivity in the air we breathe, but some reach the Earth's surface.

(ii) Numerous homes, particularly in Scotland, are built from granite rocks that emit radioactive radon gas; this can collect in basements or well-insulated rooms if the ventilation is poor.

(iii) Radioactive potassium-40 is present in food and is absorbed by our bodies.

(iv) Various radioisotopes are used in certain medical procedures.

(v) Radiation is produced in the emissions from nuclear power stations and in fall-out from the testing of nuclear bombs; the latter produce strontium isotopes with long half-lives which are absorbed by bone.

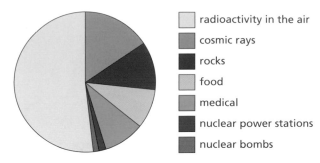

Figure 49.13 Radiation sources

We cannot avoid exposure to radiation in small doses but large doses can be dangerous to our health. The ionising effect produced by radiation causes damage to cells and tissues in our bodies and can also lead to the mutation of genes. The danger from α-particles is small, unless the source enters the body, but β- and γ-radiation can cause radiation burns (i.e. redness and sores on the skin) and delayed effects such as eye cataracts and cancer. Large exposures may lead to radiation sickness and death. The symbol used to warn of the presence of radioactive material is shown in Figure 49.14.

Figure 49.14 Radiation hazard sign

The increasing use of radioisotopes in medicine and industry has made it important to find ways of disposing of radioactive waste safely. One method is to enclose the waste in steel containers which are then buried in concrete bunkers; possible leakage is a cause of public concern, as water supplies could be contaminated allowing radioactive material to enter the food chain.

The weak sources used at school should always be:

● **lifted with forceps,**
● **held away from the eyes,** and
● **kept in their boxes when not in use**.

In industry, sources are handled by long tongs and transported in thick lead containers. Workers are protected by lead and concrete walls, and wear

radiation dose badges that keep a check on the amount of radiation they have been exposed to over a period (usually one month). The badge contains several windows which allow different types of radiation to fall onto a photographic film; when the film is developed it is darkest where the exposure to radiation was greatest.

● Worked examples

1 A radioactive source has a half-life of 20 minutes. What fraction is left after 1 hour?

After 20 minutes, fraction left = 1/2
After 40 minutes, fraction left = $1/2 \times 1/2 = 1/4$
After 60 minutes, fraction left = $1/2 \times 1/4 = 1/8$

2 Carbon-14 has a half-life of 5700 years. A 10 g sample of wood cut recently from a living tree has an activity of 160 counts/minute. A piece of charcoal taken from a prehistoric campsite also weighs 10 g but has an activity of 40 counts/minute. Estimate the age of the charcoal.

After 1×5700 years the activity will be $160/2 = 80$ counts per minute
After 2×5700 years the activity will be $80/2 = 40$ counts per minute
The age of the charcoal is $2 \times 5700 = 11\,400$ years

3 The ratio of the number of atoms of argon-40 to potassium-40 in a sample of radioactive rock is analysed to be 1 : 3. Assuming that there was no potassium in the rock originally and that argon-40 decays to potassium-40 with a half-life of 1500 million years, estimate the age of the rock.

Assume there were N atoms of argon-40 in the rock when it was formed.

After 1×1500 million years there will be $N/2$ atoms of argon left and $N - (N/2) = N/2$ atoms of potassium formed, giving an Ar : K ratio of 1 : 1.

After $2 \times 1500 = 3000$ million years, there would be $(N/2)/2 = N/4$ argon atoms left and $N - (N/4) = 3N/4$ potassium atoms formed, giving an Ar : K ratio of 1 : 3 as measured.
The rock must be about 3000 million years old.

Questions

1 Which type of radiation from radioactive materials
  a has a positive charge?
  b is the most penetrating?
  c is easily deflected by a magnetic field?
  d consists of waves?
  e causes the most intense ionisation?
  f has the shortest range in air?
  g has a negative charge?
  h is not deflected by an electric field?

2 In an experiment to find the half-life of radioactive iodine, the count-rate falls from 200 counts per second to 25 counts per second in 75 minutes. What is its half-life?

3 If the half-life of a radioactive gas is 2 minutes, then after 8 minutes the activity will have fallen to a fraction of its initial value. This fraction is
  A 1/4   B 1/6   C 1/8   D 1/16   E 1/32

Checklist

After studying this chapter you should be able to

- recall that the radiation emitted by a radioactive substance can be detected by its ionising effect,
- explain the principle of operation of a Geiger–Müller tube and a diffusion cloud chamber,
- recall the nature of α-, β- and γ-radiation,
- describe experiments to compare the range and penetrating power of α-, β- and γ-radiation in different materials,
- recall the ionising abilities of α-, β- and γ-radiation and relate them to their ranges,

- predict how α-, β- and γ-radiation will be deflected in magnetic and electric fields,

- define the term half-life,
- describe an experiment from which a radioactive decay curve can be obtained,
- show from a graph that radioactive decay processes have a constant half-life,
- solve simple problems on half-life,
- recall that radioactivity is (a) a random process, (b) due to nuclear instability, (c) independent of external conditions,
- recall some uses of radioactivity,
- describe sources of radiation,
- discuss the dangers of radioactivity and safety precautions necessary.

50 Atomic structure

- Nuclear atom
- Protons and neutrons
- Isotopes and nuclides
- Radioactive decay

- Nuclear stability
- Models of the atom
- Nuclear energy

The discoveries of the electron and of radioactivity seemed to indicate that atoms contained negatively and positively charged particles and were not indivisible as was previously thought. The questions then were 'How are the particles arranged inside an atom?' and 'How many are there in the atom of each element?'

An early theory, called the 'plum-pudding' model, regarded the atom as a positively charged sphere in which the negative electrons were distributed all over it (like currants in a pudding) and in sufficient numbers to make the atom electrically neutral. Doubts arose about this model.

● Nuclear atom

While investigating radioactivity, the physicist Rutherford noticed that not only could α-particles pass straight through very thin metal foil as if it weren't there but also that some were deflected from their initial direction. With the help of Geiger (of GM tube fame) and Marsden, Rutherford investigated this in detail at Manchester University using the arrangement in Figure 50.1. The fate of the α-particles after striking the gold foil was detected by the scintillations (flashes of light) they produced on a glass screen coated with zinc sulfide and fixed to a rotatable microscope.

They found that most of the α-particles were undeflected, some were scattered by appreciable angles and a few (about 1 in 8000) surprisingly 'bounced' back. To explain these results Rutherford proposed in 1911 a 'nuclear' model of the atom in which **all the positive charge and most of the mass of an atom** formed a dense core or **nucleus**, of very small size compared with the whole atom. The electrons surrounded the nucleus some distance away.

He derived a formula for the number of α-particles deflected at various angles, assuming that the electrostatic force of repulsion between the positive charge on an α-particle and the positive charge on the nucleus of a gold atom obeyed an inverse-square law (i.e. the force increases four times if the separation is halved). Geiger and Marsden's experimental results completely confirmed Rutherford's formula and supported the view that an atom is mostly empty space. In fact the nucleus and electrons occupy about one million millionth of the volume of an atom. Putting it another way, the nucleus is like a sugar lump in a very large hall and the electrons a swarm of flies.

Figure 50.2 shows the paths of three α-particles. Particle 1 is clear of all nuclei and passes straight through the gold atoms.
Particle 2 is deflected slightly.
Particle 3 approaches a gold nucleus so closely that it is violently repelled by it and 'rebounds', appearing to have had a head on 'collision'.

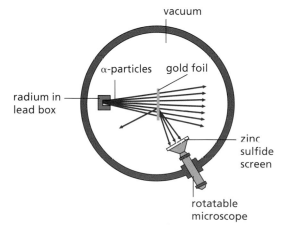

Figure 50.1 Geiger and Marsden's scattering experiment

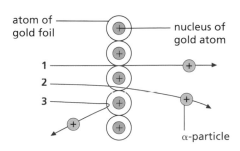

Figure 50.2 Electrostatic scattering of α-particles

● Protons and neutrons

We now believe as a result of other experiments, in some of which α and other high-speed particles were used as 'atomic probes', that atoms contain three basic particles – protons, neutrons and electrons.

A **proton** is a hydrogen atom minus an electron, i.e. a positive hydrogen ion. Its charge is equal in size but opposite in sign to that of an electron but its mass is about 2000 times greater.

A **neutron** is uncharged with almost the same mass as a proton.

Protons and neutrons are in the nucleus and are called **nucleons**. Together they account for the mass of the nucleus (and most of that of the atom); the protons account for its positive charge. These facts are summarised in Table 50.1.

Table 50.1

Particle	Relative mass	Charge	Location
proton	1836	+e	in nucleus
neutron	1839	+0	in nucleus
electron	1	−e	outside nucleus

In a neutral atom the number of protons equals the number of electrons surrounding the nucleus. Table 50.2 shows the particles in some atoms. Hydrogen is simplest with one proton and one electron. Next is the inert gas helium with two protons, two neutrons and two electrons. The soft white metal lithium has three protons and four neutrons.

Table 50.2

	Hydrogen	Helium	Lithium	Oxygen	Copper
protons	1	2	3	8	29
neutrons	0	2	4	8	34
electrons	1	2	3	8	29

The atomic or proton number Z of an atom is the number of protons in the nucleus.

The **atomic number** is also the number of electrons in the atom. The electrons determine the chemical properties of an atom and when the elements are arranged in order of atomic number in the Periodic Table, they fall into chemical families.

In general, $A = Z + N$, where N is the **neutron number** of the element.

Atomic **nuclei** are represented by symbols. Hydrogen is written as ^1_1H, helium as ^4_2He and lithium a ^7_3Li. In general atom X is written as ^A_ZX, where A is the nucleon number and Z the proton number.

The mass or nucleon number A of an atom is the number of nucleons in the nucleus.

● Isotopes and nuclides

Isotopes of an element are atoms that have the same number of protons but different numbers of neutrons. That is, their proton numbers are the same but not their nucleon numbers.

Isotopes have identical chemical properties since they have the same number of electrons and occupy the same place in the Periodic Table. (In Greek, *isos* means same and *topos* means place.)

Few elements consist of identical atoms; most are mixtures of isotopes. Chlorine has two isotopes; one has 17 protons and 18 neutrons (i.e. $Z = 17$, $A = 35$) and is written $^{35}_{17}Cl$, the other has 17 protons and 20 neutrons (i.e. $Z = 17$, $A = 37$) and is written $^{37}_{17}Cl$. They are present in ordinary chlorine in the ratio of three atoms of $^{35}_{17}Cl$ to one atom of $^{37}_{17}Cl$, giving chlorine an average atomic mass of 35.5.

Hydrogen has three isotopes: ^1_1H with one proton, **deuterium** ^2_1D with one proton and one neutron and **tritium** ^3_1T with one proton and two neutrons. Ordinary hydrogen consists 99.99 per cent of ^1_1H atoms. Water made from deuterium is called 'heavy water' (D_2O); it has a density of $1.108\,g/cm^3$, it freezes at $3.8\,°C$ and boils at $101.4\,°C$.

Each form of an element is called a **nuclide**. Nuclides with the same Z number but different A numbers are isotopes. Radioactive isotopes are termed **radioisotopes** or **radionuclides**; their nuclei are unstable.

● Radioactive decay

Radioactive atoms have unstable nuclei which change or 'decay' into atoms of a different element when they emit α- or β-particles. The decay is spontaneous and cannot be controlled; also it does not matter whether the material is pure or combined chemically with something else.

a) Alpha decay

An α-particle is a helium nucleus, having two protons and two neutrons, and when an atom decays by emission of an α-particle, its nucleon number decreases by four and its proton number by two. For example, when radium of nucleon number 226 and proton number 88 emits an α-particle, it decays to radon of nucleon number 222 and proton number 86. We can write:

$$^{226}_{88}\text{Ra} \rightarrow ^{222}_{86}\text{Rn} + ^{4}_{2}\text{He}$$

The values of A and Z must balance on both sides of the equation since nucleons and charge are conserved.

b) Beta decay

In β⁻ decay a neutron changes to a proton and an electron. The proton remains in the nucleus and the electron is emitted as a β⁻-particle. The new nucleus has the same nucleon number, but its proton number increases by one since it has one more proton. Radioactive carbon, called carbon-14, decays by β⁻ emission to nitrogen:

$$^{14}_{6}\text{C} \rightarrow ^{14}_{7}\text{N} + ^{0}_{-1}\text{e}$$

A particle called an **antineutrino** ($\bar{\nu}$), with no charge and negligible mass, is also emitted in β⁻ decay. Note that a β⁻ decay is often referred to as just a β decay.

Positrons are subatomic particles with the same mass as an electron but with opposite (positive) charge. They are emitted in some decay processes as β⁺-particles. Their tracks can be seen in bubble chamber photographs. The symbol for a positron is $^{0}_{+1}\text{e}$. In β⁺- decay a proton in a nucleus is converted to a neutron and a positron, for example in the reaction:

$$^{64}_{29}\text{Cu} \rightarrow ^{64}_{28}\text{N} + ^{0}_{+1}\text{e}$$

A **neutrino** (ν) is also emitted in β⁺ decay. Neutrinos are emitted from the Sun in large numbers, but they rarely interact with matter so are very difficult to detect. Antineutrinos and positrons are the 'antiparticles' of neutrinos and electrons, respectively. If a particle and its antiparticle collide, they annihilate each other, producing energy in the form of γ-rays.

c) Gamma emission

After emitting an α-particle, or β⁻- or β⁺-particles, some nuclei are left in an 'excited' state. Rearrangement of the protons and neutrons occurs and a burst of γ-rays is released.

● Nuclear stability

The stability of a nucleus depends on both the number of protons (Z) and the number of neutrons (N) it contains. Figure 50.3 is a plot of N against Z for all known nuclides. The blue band indicates the region over which stable nuclides occur; unstable nuclides occur outside this band. The continuous line, drawn through the centre of the band, is called the stability line.

It is found that for **stable nuclides**:

(i) $N = Z$ for the lightest,
(ii) $N > Z$ for the heaviest,
(iii) most nuclides have *even* N and Z, implying that the α-particle combination of two neutrons and two protons is likely to be particularly stable.

For **unstable nuclides**:

(i) disintegration tends to produce new nuclides nearer the stability line and continues until a stable nuclide is formed,
(ii) a nuclide above the stability line decays by β⁻ emission (a neutron changes to a proton and electron) so that the N/Z ratio decreases,
(iii) a nuclide below the stability line decays by β⁺ emission (a proton changes to a neutron and positron) so that the N/Z ratio increases,
(iv) nuclei with more than 82 protons usually emit an α-particle when they decay.

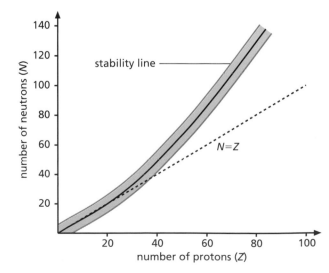

Figure 50.3 Stability of nuclei

● Models of the atom

Rutherford–Bohr model

Shortly after Rutherford proposed his nuclear model of the atom, Bohr, a Danish physicist, developed it to explain how an atom emits light. He suggested that the electrons circled the nucleus at high speed, being kept in certain orbits by the electrostatic attraction of the nucleus for them. He pictured atoms as miniature solar systems. Figure 50.4 shows the model for three elements.

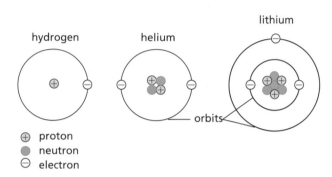

proton
neutron
electron

Figure 50.4 Electron orbits

Normally the electrons remain in their orbits but if the atom is given energy, for example by being heated, electrons may jump to an outer orbit. The atom is then said to be **excited**. Very soon afterwards the electrons return to an inner orbit and, as they do, they emit energy in the form of bursts of electromagnetic radiation (called **photons**), such as infrared light, ultraviolet or X-rays (Figure 50.5). The wavelength of the radiation emitted depends on the two orbits between which the electrons jump. If an atom gains enough energy for an electron to escape altogether, the atom becomes an ion and the energy needed to achieve this is called the **ionisation energy** of the atom.

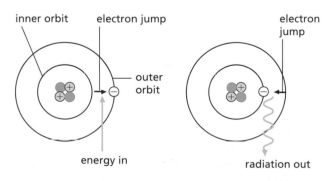

Figure 50.5 Bohr's explanation of energy changes in an atom

Schrödinger model

Although it remains useful for some purposes, the Rutherford–Bohr model was replaced by a mathematical model developed by Erwin Schrödinger, which is not easy to picture. The best we can do, without using advanced mathematics, is to say that the atom consists of a nucleus surrounded by a hazy cloud of electrons. Regions of the atom where the mathematics predicts that electrons are more likely to be found are represented by denser shading (Figure 50.6).

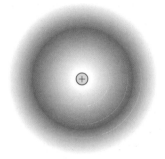

Figure 50.6 Electron cloud

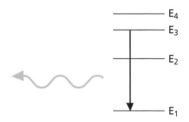

Figure 50.7 Energy levels of an atom

This theory does away with the idea of electrons moving in definite orbits and replaces them by **energy levels** that are different for each element. When an electron 'jumps' from one level, say E_3 in Figure 50.7, to a lower one E_1, a photon of electromagnetic radiation is emitted with energy equal to the difference in energy of the two levels. The frequency (and wavelength) of the radiation emitted by an atom is thus dependent on the arrangement of energy levels. For an atom emitting visible light, the resulting **spectrum** (produced for example by a prism) is a series of coloured **lines** that is unique to each element. Sodium vapour in a gas discharge tube (such as a yellow street light) gives two adjacent yellow–orange lines (Figure 50.8a). Light from the Sun is due to energy changes in many different atoms and the resulting spectrum is a **continuous** one with all colours (see Figure 50.8b).

Figure 50.8a Line spectrum due to energy changes in sodium

Figure 50.8b A continuous spectrum

● Nuclear energy

a) $E = mc^2$

Einstein predicted that if the energy of a body changes by an amount E, its mass changes by an amount m given by the equation

$$E = mc^2$$

where c is the speed of light (3×10^8 m/s). The implication is that any reaction in which there is a decrease of mass, called a **mass defect**, is a source of energy. The energy and mass changes in physical and chemical changes are very small; those in some nuclear reactions, Such as radioactive decay, are millions of times greater. It appears that mass (matter) is a very concentrated form of energy.

b) Fission

The heavy metal uranium is a mixture of isotopes of which $^{235}_{92}\mathrm{U}$, called uranium-235, is the most important. Some atoms of this isotope decay quite naturally, emitting high-speed neutrons. If one of these hits the nucleus of a neighbouring uranium-235 atom (being uncharged the neutron is not repelled by thc nucleus), this may break (**fission** of the nucleus) into two nearly equal radioactive nuclei, often of barium and krypton, with the production of two or three more neutrons:

$$^{235}_{92}\mathrm{U} + {}^{1}_{0}\mathrm{n} \rightarrow {}^{144}_{56}\mathrm{Ba} + {}^{90}_{36}\mathrm{Kr} + 2{}^{1}_{0}\mathrm{n}$$

neutron   fission fragments   neutrons

The mass defect is large and appears mostly as k.e. of the fission fragments. These fly apart at great speed,

colliding with surrounding atoms and raising their average k.e., i.e. their temperature, so producing heat.

If the fission neutrons split other uranium-235 nuclei, a **chain reaction** is set up (Figure 50.9). In practice some fission neutrons are lost by escaping from the surface of the uranium before this happens. The ratio of those causing fission to those escaping increases as the mass of uranium-235 increases. This must exceed a certain **critical** value to sustain the chain reaction.

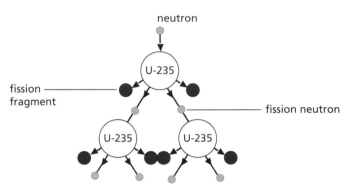

Figure 50.9 Chain reaction

c) Nuclear reactor

In a nuclear power station heat from a nuclear reactor produces the steam for the turbines. Figure 50.10 is a simplified diagram of one type of reactor.

The chain reaction occurs at a steady rate which is controlled by inserting or withdrawing neutron-absorbing rods of boron among the uranium rods. The graphite core is called the **moderator** and slows down the fission neutrons; fission of uranium-235 occurs more readily with slow than with fast neutrons. Carbon dioxide gas is pumped through the core and carries off heat to the **heat exchanger** where steam is produced. The concrete shield gives workers protection from γ-rays and escaping neutrons. The radioactive fission fragments must be removed periodically if the nuclear fuel is to be used efficiently.

In an **atomic bomb**, an increasing uncontrolled chain reaction occurs when two pieces of uranium-235 come together and exceed the critical mass.

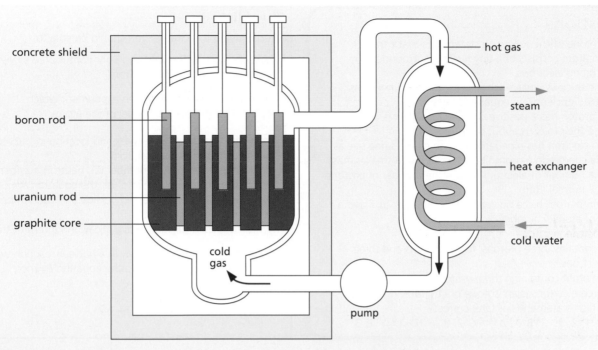

Figure 50.10 Nuclear reactor

d) Fusion

If light nuclei join together to make heavier ones, this can also lead to a loss of mass and, as a result, the release of energy. Such a reaction has been achieved in the **hydrogen bomb**. At present, research is being done on the controlled **fusion** of isotopes of hydrogen (deuterium and tritium) to give helium.

$$\underset{\text{deuterium}}{{}_{1}^{2}\text{H}} \quad + \quad \underset{\text{tritium}}{{}_{1}^{3}\text{H}} \quad \rightarrow \quad \underset{\text{helium}}{{}_{2}^{4}\text{He}} \quad + \quad \underset{\text{neutron}}{{}_{0}^{1}\text{n}}$$

Fusion can only occur if the reacting nuclei have enough energy to overcome their mutual electrostatic repulsion. This can happen if they are raised to a very high temperature (over 100 million °C) so that they collide at very high speeds. If fusion occurs, the energy released is enough to keep the reaction going; since heat is required, it is called **thermonuclear** fusion.

The source of the Sun's energy is nuclear fusion. The temperature in the Sun is high enough for the conversion of hydrogen into helium to occur, in a sequence of thermonuclear fusion reactions known as the 'hydrogen burning' sequence.

$$ {}_{1}^{1}\text{H} + {}_{1}^{1}\text{H} \rightarrow {}_{1}^{2}\text{H} + \text{positron}\ ({}_{0}^{1}\text{e}) + \text{neutrino}\ (\nu) $$

$$ {}_{1}^{1}\text{H} + {}_{1}^{2}\text{H} \rightarrow {}_{2}^{3}\text{He} + \gamma - \text{ray} $$

$$ {}_{2}^{3}\text{He} + {}_{2}^{3}\text{He} \rightarrow {}_{2}^{4}\text{He} + {}_{1}^{1}\text{H} + {}_{1}^{1}\text{H} $$

Each of these fusion reactions results in a loss of mass and a release of energy. Overall, tremendous amounts of energy are created that help to maintain the very high temperature of the Sun.

Questions

1 Which one of the following statements is *not* true?
  A An atom consists of a tiny nucleus surrounded by orbiting electrons.
  B The nucleus always contains protons and neutrons, called nucleons, in equal numbers.
  C A proton has a positive charge, a neutron is uncharged and their mass is about the same.
  D An electron has a negative charge of the same size as the charge on a proton but it has a much smaller mass.
  E The number of electrons equals the number of protons in a normal atom.

2 A lithium atom has a nucleon (mass) number of 7 and a proton (atomic) number of 3.
  1 Its symbol is $^{7}_{4}$Li.
  2 It contains three protons, four neutrons and three electrons.
  3 An atom containing three protons, three neutrons and three electrons is an isotope of lithium.
     Which statement(s) is (are) correct?
     A 1, 2, 3   B 1, 2   C 2, 3   D 1   E 3

Checklist

After studying this chapter you should be able to

• describe how Rutherford and Bohr contributed to views about the structure of the atom,

• describe the Geiger–Marsden experiment which established the nuclear model of the atom,

• recall the charge, relative mass and location in the atom of protons, neutrons and electrons,

• define the terms proton number (Z), neutron number (N) and nucleon number (A), and use the equation $A = Z + N$,

• explain the terms isotope and nuclide and use symbols to represent them, e.g. $^{35}_{17}$Cl,

• write equations for radioactive decay and interpret them,

• connect the release of energy in a nuclear reaction with a change of mass according to the equation $E = mc^2$,

• outline the process of fission,

• outline the process of fusion.

Revision questions

General physics
Measurements and motion

1   Which are the basic SI units of mass, length and time?
    A kilogram, kilometre, second
    B gram, centimetre, minute
    C kilogram, centimetre, second
    D gram, centimetre, second
    E kilogram, metre, second

2   Density can be calculated from the expression
    A mass/volume
    B mass × volume
    C volume/mass
    D weight/area
    E area × weight

3   Which of the following properties are the same for an object on Earth and on the Moon?
    1 weight   2 mass   3 density
    Use the answer code:
    A 1, 2, 3   B 1, 2   C 2, 3   D 1   E 3

4 a The smallest division marked on a metre rule is 1 mm. A student measures a length with the ruler and records it as 0.835 m. Is he justified in giving three significant figures?
  b The SI unit of density is
     A kg m   B kg/m²   C kg m³
     D kg/m   E kg/m³

Forces and momentum

5   A 3 kg mass falls with its terminal velocity. Which of the combinations A to E gives its weight, the air resistance and the resultant force acting on it?

	Weight	Air resistance	Resultant force
A	0.3 N down	zero	zero
B	3 N down	3 N up	3 N up
C	10 N down	10 N up	10 N down
D	30 N down	30 N up	zero
E	300 N down	zero	300 N down

6   A boy whirls a ball at the end of a string round his head in a horizontal circle, centre O. He lets go of the string when the ball is at X in the diagram. In which direction does the ball fly off?
    A 1   B 2   C 3   D 4   E 5

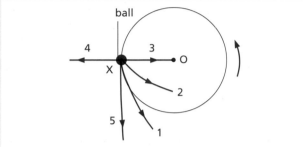

Energy, work, power and pressure

7   The work done by a force is
    1 calculated by multiplying the force by the distance moved in the direction of the force
    2 measured in joules
    3 the amount of the energy changed.
    Which statement(s) is (are) correct?
    A 1, 2, 3   B 1, 2   C 2, 3   D 1   E 3

8   The main energy change occurring in the device named is
    1 electric lamp   electrical to heat and light
    2 battery   chemical to electrical
    3 pile driver   k.e. to p.e.
    Which statement(s) is (are) correct?
    A 1, 2, 3   B 1, 2   C 2, 3   D 1   E 3

9   The efficiency of a machine which raises a load of 200 N through 2 m when an effort of 100 N moves 8 m is
    A 0.5%   B 5%   C 50%
    D 60%   E 80%

10 Which one of the following statements is not true?
    A Pressure is the force acting on unit area.
    B Pressure is calculated from force/area.
    C The SI unit of pressure is the pascal (Pa) which equals 1 newton per square metre (1 N/m²).
    D The greater the area over which a force acts the greater is the pressure.
    E Force = pressure × area.

11 A stone of mass 2 kg is dropped from a height of 4 m. Neglecting air resistance, the kinetic energy (k.e.) of the stone in joules just before it hits the ground is
A 6   B 8   C 16   D 80   E 160

12 An object of mass 2 kg is fired vertically upwards with a k.e. of 100 J. Neglecting air resistance, which of the numbers in **A** to **E** below is
a the velocity in m/s with which it is fired,
b the height in m to which it will rise?
A 5   B 10   C 20   D 100   E 200

13 An object has k.e. of 10 J at a certain instant. If it is acted on by an opposing force of 5 N, which of the numbers **A** to **E** below is the furthest distance it travels in metres before coming to rest?
A 2   B 5   C 10   D 20   E 50

2 Thermal physics

Thermal properties and temperature

14 If the piston in the diagram is pulled out of the cylinder from position X to position Y, without changing the temperature of the air enclosed, the air pressure in the cylinder is
A reduced to a quarter
B reduced to a third
C the same
D trebled
E quadrupled.

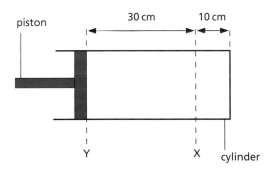

15 Which one of the following statements is *not* true?
A Temperature tells us how hot an object is.
B Temperature is measured by a thermometer which uses some property of matter (e.g. the expansion of mercury) that changes continuously with temperature.

C Heat flows naturally from an object at a lower temperature to one at a higher temperature.
D The molecules of an object move faster when its temperature rises.
E Temperature is measured in °C, heat is measured in joules.

16 The pressure exerted by a gas in a container
1 is due to the molecules of the gas bombarding the walls of the container
2 decreases if the gas is cooled
3 increases if the volume of the container increases.
Which statement(s) is (are) correct?
A 1, 2, 3   B 1, 2   C 2, 3   D 1   E 3

17 A drink is cooled more by ice at 0 °C than by the same mass of water at 0 °C because ice
A floats on the drink
B has a smaller specific heat capacity
C gives out latent heat to the drink as it melts
D absorbs latent heat from the drink to melt
E is a solid.

Thermal processes

18 Which of the following statements is/are true?
1 In cold weather the wooden handle of a saucepan feels warmer than the metal pan because wood is a better conductor of heat.
2 Convection occurs when there is a change of density in parts of a fluid.
3 Conduction and convection cannot occur in a vacuum.
A 1, 2, 3   B 1, 2   C 2, 3   D 1   E 3

19 Which one of the following statements is *not* true?
A Energy from the Sun reaches the Earth by radiation only.
B A dull black surface is a good absorber of radiation.
C A shiny white surface is a good emitter of radiation.
D The best heat insulation is provided by a vacuum.
E A vacuum flask is designed to reduce heat loss or gain by conduction, convection and radiation.

3 Properties of waves
General wave properties

20 In the transverse wave shown below distances are in centimetres. Which pair of entries A to E is correct?

	A	B	C	D	E
Amplitude	2	4	4	8	8
Wavelength	4	4	8	8	12

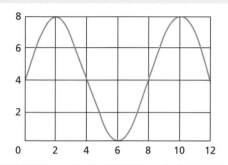

21 When a water wave goes from deep to shallow water, the changes (if any) in its speed, wavelength and frequency are

	Speed	Wavelength	Frequency
A	greater	greater	the same
B	greater	less	less
C	the same	less	greater
D	less	the same	less
E	less	less	the same

22 When the straight water waves in the diagram pass through the narrow gap in the barrier they are diffracted. What changes (if any) occur in
a the shape of the waves,
b the speed of the waves,
c the wavelength?

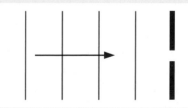

23 The diagram below shows the complete electromagnetic spectrum.

radio waves	microwaves	A	visible light	ultraviolet	B	γ rays

a Name the radiation found at
(i) A,
(ii) B.
b State which of the radiations marked on the diagram would have
(i) the lowest frequency,
(ii) the shortest wavelength.

24 The wave travelling along the spring in the diagram is produced by someone moving end X of the spring to and fro in the directions shown by the arrows.
a Is the wave longitudinal or transverse?
b What is the region called where the coils of the spring are **(i)** closer together, **(ii)** further apart, than normal?

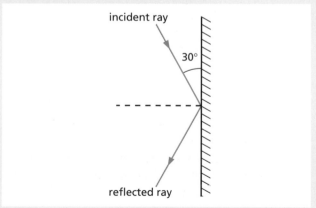

Light
25 In the diagram a ray of light is shown reflected at a plane mirror. What is
a the angle of incidence,
b the angle the reflected ray makes with the mirror?

26 In the diagram below a ray of light IO changes direction as it enters glass from air.
  a What name is given to this effect?
  b Which line is the normal?
  c Is the ray bent towards or away from the normal in the glass?
  d What is the value of the angle of incidence in air?
  e What is the value of the angle of refraction in glass?

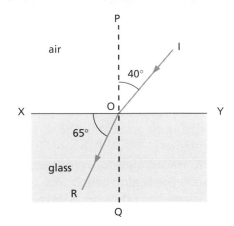

27 In the diagram, which of the rays **A** to **E** is most likely to represent the ray emerging from the parallel-sided sheet of glass?

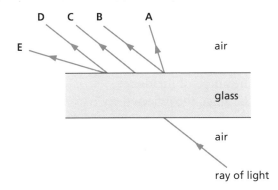

28 A narrow beam of white light is shown passing through a glass prism and forming a spectrum on a screen.
  a What is the effect called?
  b Which colour of light appears at (i) A, (ii) B?

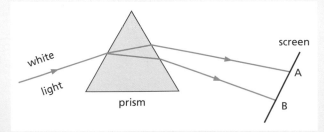

29 When using a magnifying glass to see a small object
  1 an upright image is seen
  2 the object should be less than one focal length away
  3 a real image is seen.
  Which statement(s) is (are) correct?
  A 1, 2, 3   **B** 1, 2   **C** 2, 3   **D** 1   **E** 3

Sound

30 If a note played on a piano has the same pitch as one played on a guitar, they have the same
  A frequency   **B** amplitude
  C quality   **D** loudness
  E harmonics.

31 The waveforms of two notes P and Q are shown below. Which one of the statements **A** to **E** is true?

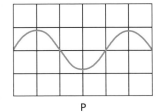
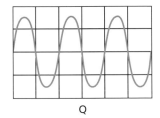

  P   Q

  A P has a higher pitch than Q and is not so loud.
  B P has a higher pitch than Q and is louder.
  C P and Q have the same pitch and loudness.
  D P has a lower pitch than Q and is not so loud.
  E P has a lower pitch than Q and is louder.

32 Examples of transverse waves are
  1 water waves in a ripple tank
  2 all electromagnetic waves
  3 sound waves.
  Which statement(s) is (are) correct?
  A 1, 2, 3   **B** 1, 2   **C** 2, 3   **D** 1   **E** 3

4 Electricity and magnetism
Simple phenomena of magnetism

33 Which one of the following statements about the diagram below is *not* true?

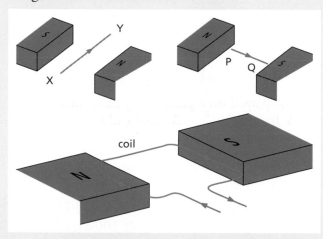

A If a current is passed through the wire XY, a vertically upwards force acts on it.
B If a current is passed through the wire PQ, it does not experience a force.
C If a current is passed through the coil, it rotates clockwise.
D If the coil had more turns and carried a larger current, the turning effect would be greater.
E In a moving-coil loudspeaker a coil moves between the poles of a strong magnet.

Electrical quantities and circuits

34 For the circuit below calculate
  a the total resistance,
  b the current in each resistor,
  c the p.d. across each resistor.

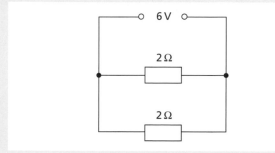

35 Repeat question 33 for the circuit below.

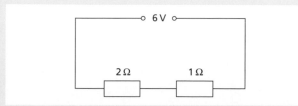

36 An electric kettle for use on a 230 V supply is rated at 3000 W. For safe working, the cable supplying it should be able to carry at least
  A 2 A **B** 5 A **C** 10 A **D** 15 A **E** 30 A

37 Which one of the following statements is *not* true?
  A In a house circuit, lamps are wired in parallel.
  B Switches, fuses and circuit breakers should be placed in the neutral wire.
  C An electric fire has its earth wire connected to the metal case to prevent the user receiving a shock.
  D When connecting a three-core cable to a 13 A three-pin plug the brown wire goes to the live pin.
  E The cost of operating three 100 W lamps for 10 hours at 10p per unit is 30p.

38 Which of the units **A** to **E** could be used to measure
  a electric charge,
  b electric current,
  c p.d.,
  d energy,
  e power?
  A ampere **B** joule **C** volt **D** watt
  E coulomb

39 Which one of the following statements about the transistor circuit shown below is *not* true?

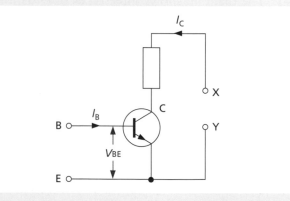

  A The collector current I_C is zero until base current I_B flows.
  B I_B is zero until the base–emitter p.d. V_{BE} is +0.6 V.
  C A small I_B can switch on and control a large I_C.
  D When used as an amplifier the input is connected across B and E.
  E X must be connected to supply the – terminal and Y to the + terminal.

Electromagnetic effects

40 A magnet is pushed, N pole first, into a coil as in the diagram below. Which one of the following statements **A** to **E** is *not* true?

A A p.d. is induced in the coil and causes a current through the galvanometer.

B The induced p.d. increases if the magnet is pushed in faster and/or the coil has more turns.

C Mechanical energy is changed to electrical energy.

D The coil tends to move to the right because the induced current makes face X a N pole which is repelled by the N pole of the magnet.

E The effect produced is called electrostatic induction.

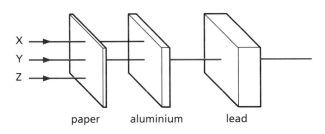

5 Atomic physics

41 The diagram shows three types of radiation, X, Y and Z.

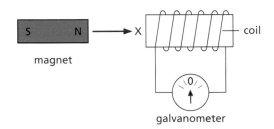

Which of the columns **A** to **E** correctly names the radiations X, Y and Z?

	A	B	C	D	E
X	alpha	beta	gamma	gamma	beta
Y	beta	alpha	alpha	beta	gamma
Z	gamma	gamma	beta	alpha	alpha

42 The graph shows the decay curve of a radioactive substance.

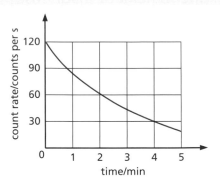

What is its half-life in minutes?

A 1   **B** 2   **C** 3   **D** 4   **E** 5

43 A radioactive source which has a half-life of 1 hour gives a count-rate of 100 counts per second at the start of an experiment and 25 counts per second at the end. The time taken by the experiment was, in hours,

A 1   **B** 2   **C** 3   **D** 4   **E** 5

44 Which symbol **A** to **E** below is used in equations for nuclear reactions to represent

a an alpha particle,

b a beta particle,

c a neutron,

d an electron?

A $_{-1}^{0}e$ **B** $_{0}^{1}n$ **C** $_{2}^{4}He$ **D** $_{-1}^{1}e$ **E** $_{1}^{1}n$

45 a Radon $_{86}^{220}Rn$ decays by emitting an alpha particle to form an element whose symbol is

A $_{85}^{216}At$ **B** $_{86}^{216}Rn$ **C** $_{84}^{218}Po$

D $_{84}^{216}Po$ **E** $_{85}^{217}At$

b Thorium $_{90}^{234}Th$ decays by emitting a beta particle to form an element whose symbol is

A $_{90}^{235}Th$ **B** $_{89}^{230}Ac$ **C** $_{89}^{234}Ac$

D $_{88}^{232}Ra$ **E** $_{91}^{234}Pa$

Cambridge IGCSE exam questions

1 General physics

Measurements and motion

1 a (i) The two diagrams show the dimensions of a rectangular block being measured using a ruler. They are not shown full size. Use the scales shown to find the length and the width of the block, giving your answers in cm. [2]

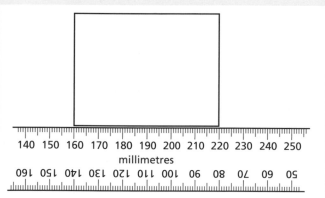

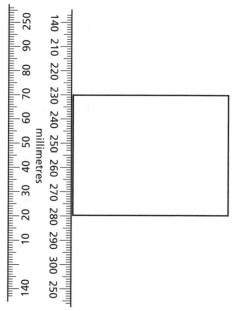

(ii) When the block was made, it was cut from a piece of metal 2.0 cm thick. Calculate the volume of the block. [2]

b Another block has a volume of 20 cm³. The diagram shows the reading when the block is placed on a balance.

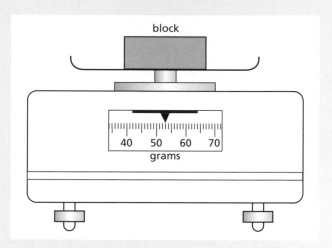

Find the density of this block. [4]

[Total: 8]

(Cambridge IGCSE Physics 0625 Paper 21 Q1 November 2010)

2 An engineering machine has a piston which is going up and down approximately 75 times per minute.

Describe carefully how a stopwatch may be used to find accurately the time for one up-and-down cycle of the piston. [4]

[Total: 4]

(Cambridge IGCSE Physics 0625 Paper 31 Q1 June 2009)

3 Imagine that you live beside a busy road. One of your neighbours thinks that many of the vehicles are travelling faster than the speed limit for the road.

You decide to check this by measuring the speeds of some of the vehicles.

a Which two quantities will you need to measure in order to find the speed of a vehicle, and which instruments would you use to measure them?

Quantity measured	Instrument used

[4]

b State the equation you would use to calculate the speed of the vehicle. If you use symbols, state what your symbols mean. [1]

c One lorry travels from your town to another town. The lorry reaches a top speed of 90 km/h, but its average speed between the towns is only 66 km/h.

(i) Why is the average speed less than the top speed? [1]

(ii) The journey between the towns takes 20 minutes.
Calculate the distance between the towns. [3]

[Total: 9]

(Cambridge IGCSE Physics 0625 Paper 21 Q1 June 2010)

4 The top graph shows the distance/time graph for a girl's bicycle ride and the bottom graph gives the axes for the corresponding speed/time graph.

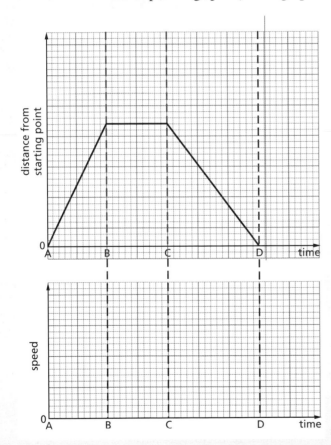

a Look at the distance/time graph that has been drawn for you.

(i) Answer the following questions for the time interval AB.

1 What is happening to the distance from the starting point? [2]

2 What can you say about the speed of the bicycle? [1]

(ii) On a copy of the speed/time axes on the bottom graph, draw a thick line that could show the speed during AB. [1]

b On your copy of the speed/time axes

(i) draw a thick line that could show the speed during BC, [1]

(ii) draw a thick line that could show the speed during CD. [2]

c How far from her starting point is the girl when she has finished her ride? [1]

[Total: 8]

(Cambridge IGCSE Physics 0625 Paper 02 Q3 November 2009)

5 In a training session, a racing cyclist's journey is in three stages.

Stage 1 He accelerates uniformly from rest to 12 m/s in 20 s.

Stage 2 He cycles at 12 m/s for a distance of 4800 m.

Stage 3 He decelerates uniformly to rest.

The whole journey takes 500 s.

a Calculate the time taken for stage 2. [2]

b On a copy of the grid below, draw a speed/time graph of the cyclist's ride. [3]

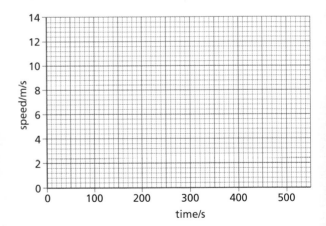

c Show that the total distance travelled by the cyclist is 5400 m. [4]

d Calculate the average speed of the cyclist. [2]

[Total: 11]

(Cambridge IGCSE Physics 0625 Paper 02 Q2 June 2007)

6 A large plastic ball is dropped from the top of a tall building. The diagram shows the speed/time graph for the falling ball until it hits the ground.

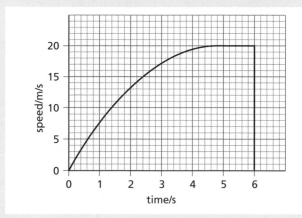

a From the graph estimate,
  (i) the time during which the ball is travelling with terminal velocity, [1]
  (ii) the time during which the ball is accelerating, [1]
  (iii) the distance fallen while the ball is travelling with terminal velocity, [2]
  (iv) the height of the building. [2]
b Explain, in terms of the forces acting on the ball, why
  (i) the acceleration of the ball decreases, [3]
  (ii) the ball reaches terminal velocity. [2]

[Total: 11]

(Cambridge IGCSE Physics 0625 Paper 03 Q1 November 2007)

Forces and momentum

7 A student investigated the stretching of a spring by hanging various weights from it and measuring the corresponding extensions. The results are shown in the table below.

Weight/N	0	1	2	3	4	5
Extension/mm	0	21	40	51	82	103

a On a copy of the grid, plot the points from these results. Do not draw a line through the points yet. [2]

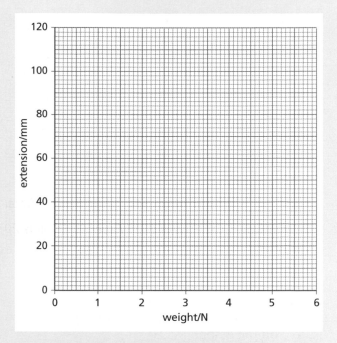

b The student appears to have made an error in recording one of the results. Which result is this? [1]
c Ignoring the incorrect result, draw the best straight line through the remaining points. [1]
d State and explain whether this spring is obeying Hooke's law. [2]
e Describe how the graph might be shaped if the student continued to add several more weights to the spring. [1]
f The student estimates that if he hangs a 45 N load on the spring, the extension will be 920 mm.
Explain why this estimate may be unrealistic. [1]

[Total: 8]

(Cambridge IGCSE Physics 0625 Paper 31 Q3 November 2009)

8 In an experiment, forces are applied to a spring as shown in the diagram. The results of this experiment are shown on the graph.

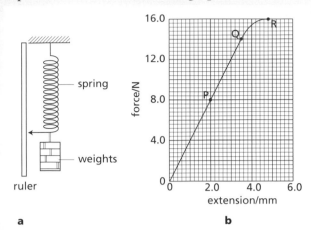

a   **b**

a What is the name given to the point marked Q on the graph? [1]

b For the part OP of the graph, the spring obeys Hooke's law. State what this means. [1]

c The spring is stretched until the force and extension are shown by the point R on the graph. Compare how the spring stretches, as shown by the part of the graph OQ, with that shown by QR. [1]

d The part OP of the graph shows the spring stretching according to the expression

$$F = kx$$

Use values from the graph to calculate the value of k. [2]

[Total: 5]

(Cambridge IGCSE Physics 0625 Paper 03 Q2 November 2006)

9 An object of weight W is suspended by two ropes from a beam, as shown in the diagram. The tensions in the ropes are 50.0 N and 86.6 N, as shown.

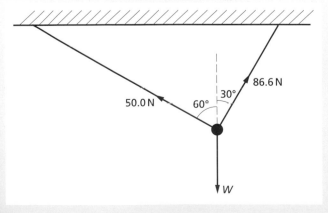

a On graph paper, draw a scale diagram to find the resultant of the two tensions. Use a scale of 1.0 cm = 10 N. Clearly label the resultant. [3]

b From your diagram, find the value of the resultant. [1]

c State the direction in which the resultant is acting. [1]

d State the value of W. [1]

[Total: 6]

(Cambridge IGCSE Physics 0625 Paper 31 Q1 November 2010)

10 The diagram shows a circular metal disc of mass 200 g, freely pivoted at its centre.

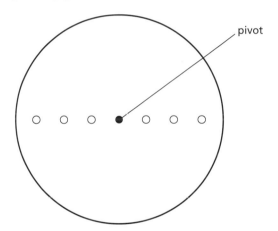

Masses of 100 g, 200 g, 300 g, 400 g, 500 g and 600 g are available, but only one of each value. These may be hung with string from any of the holes. There are three small holes on each side of the centre, one at 4.0 cm from the pivot, one at 8.0 cm from the pivot and one at 12.0 cm from the pivot.

The apparatus is to be used to show that there is no net moment of force acting on a body when it is in equilibrium.

a On a copy of the diagram, draw in **two different** value masses hanging from appropriate holes. The values of the masses should be chosen so that there is no net moment. Alongside the masses chosen, write down their values. [2]

b Explain how you would test that your chosen masses give no net moment to the disc. [1]

c Calculate the moments about the pivot due to the two masses chosen. [2]

d Calculate the force on the pivot when the two masses chosen are hanging from the disc. [2]

[Total: 7]

(Cambridge IGCSE Physics 0625 Paper 31 Q2 November 2008)

11 A piece of stiff cardboard is stuck to a plank of wood by means of two sticky-tape 'hinges'. This is shown in the diagram.

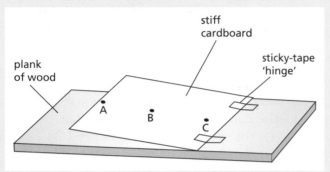

a The cardboard is lifted as shown, using a force applied either at A or B or C.
  (i) On a copy of the diagram, draw the force in the position where its value will be as small as possible. [2]
  (ii) Explain why the position you have chosen in **a(i)** results in the smallest force. [1]
b Initially, the cardboard is flat on the plank of wood. A box of matches is placed on it. The cardboard is then slowly raised at the left-hand edge, as shown in the diagram below.

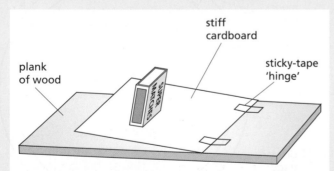

State the condition for the box of matches to fall over. [2]
c The box of matches is opened, as shown in the diagram below. The procedure in **b** is repeated.

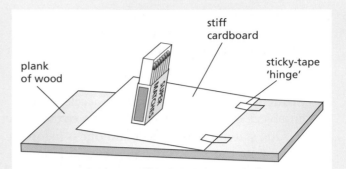

  (i) Copy and complete the sentence below, using either the words 'greater than' or 'the same as' or 'less than'.

  When the box of matches is open, the angle through which the cardboard can be lifted before the box of matches falls is .... the angle before the closed box of matches falls. [1]
  (ii) Give a reason for your answer to **c(i)**. [1]

[Total: 7]

(Cambridge IGCSE Physics 0625 Paper 02 Q3 June 07)

12 a State the two factors on which the turning effect of a force depends. [2]
b Forces F_1 and F_2 are applied vertically downwards at the ends of a beam resting on a pivot P. The beam has weight W.

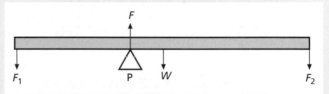

  (i) Copy and complete the statements about the two requirements for the beam to be in equilibrium.
  1 There must be no resultant . . .
  2 There must be no resultant . . . [2]
  (ii) The beam is in equilibrium. F is the force exerted on the beam by the pivot P. Copy and complete the following equation about the forces on the beam.
  $$F =$$ [1]
  (iii) Which one of the four forces on the beam does **not** exert a moment about P? [1]

[Total: 6]

(Cambridge IGCSE Physics 0625 Paper 02 Q5 November 2006)

13 Two students make the statements about acceleration that are given below.
Student A: For a given mass the acceleration of an object is proportional to the resultant force applied to the object.
Student B: For a given force the acceleration of an object is proportional to the mass of the object.
   a One statement is correct and one is incorrect. Rewrite the incorrect statement, making changes so that it is now correct. [1]
   b State the equation which links acceleration *a*, resultant force *F* and mass *m*. [1]
   c Describe what happens to the motion of a moving object when
      (i) there is no resultant force acting on it, [1]
      (ii) a resultant force is applied to it in the opposite direction to the motion, [1]
      (iii) a resultant force is applied to it in a perpendicular direction to the motion. [1]

[Total: 5]

(Cambridge IGCSE Physics 0625 Paper 31 Q3 June 2010)

14 A car travels around a circular track at constant speed.
   a Why is it incorrect to describe the circular motion as having constant velocity? [1]
   b A force is required to maintain the circular motion.
      (i) Explain why a force is required. [2]
      (ii) In which direction does this force act? [1]
      (iii) Suggest what provides this force. [1]

[Total: 5]

(Cambridge IGCSE Physics 0625 Paper 31 Q2 November 2010)

15 The diagram shows apparatus used to find a relationship between the force applied to a trolley and the acceleration caused by the force.

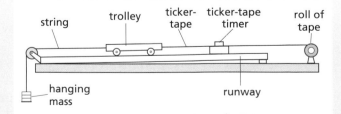

For each mass, hung as shown, the acceleration of the trolley is determined from the tape. Some of the results are given in the table below.

Weight of the hanging mass/N	Acceleration of the trolley/m/s²
0.20	0.25
0.40	0.50
0.70	
0.80	1.0

   a **(i)** Explain why the trolley accelerates. [2]
      (ii) Suggest why the runway has a slight slope as shown. [1]
   b Calculate the mass of the trolley, assuming that the accelerating force is equal to the weight of the hanging mass. [2]
   c Calculate the value missing from the table. Show your working. [2]
   d In one experiment, the hanging mass has a weight of 0.4 N and the trolley starts from rest. Use data from the table to calculate
      (i) the speed of the trolley after 1.2 s, [2]
      (ii) the distance travelled by the trolley in 1.2 s. [2]

[Total: 11]

(Cambridge IGCSE Physics 0625 Paper 31 Q1 November 2008)

16 The diagram shows a model car moving clockwise around a horizontal circular track.

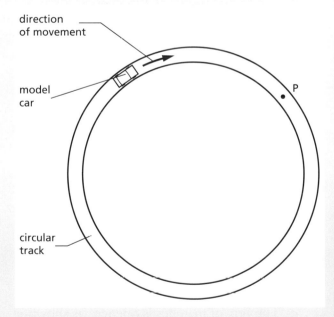

   a A force acts on the car to keep it moving in a circle.
      (i) Draw an arrow on a copy of the diagram to show the direction of this force. [1]

(ii) The speed of the car increases. State what happens to the magnitude of this force. [1]

b (i) The car travels too quickly and leaves the track at P. On your copy of the diagram, draw an arrow to show the direction of travel after it has left the track. [1]

(ii) In terms of the forces acting on the car, suggest why it left the track at P. [2]

c The car, starting from rest, completes one lap of the track in 10 s. Its motion is shown graphically in the graph below.

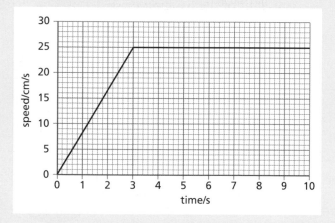

(i) Describe the motion between 3.0 s and 10.0 s after the car has started. [1]

(ii) Use the graph to calculate the circumference of the track. [2]

(iii) Calculate the increase in speed per second during the time 0 to 3.0 s. [2]

[Total: 10]

(Cambridge IGCSE Physics 0625 Paper 03 Q1 June 2007)

Energy, work, power and pressure

17 a The diagram represents the energy into and out of a machine.

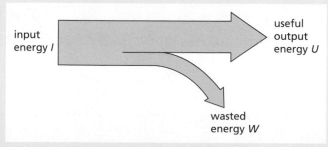

Write down the equation linking *I*, *U* and *W*. [1]

b An electric motor and a pulley in a warehouse are being used to lift a packing case of goods from the ground up to a higher level. This is shown in the diagram.

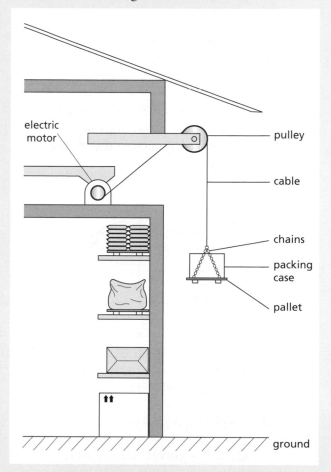

The packing case of goods, the chains and the pallet together weigh 850 N.

(i) State the value of the tension force in the cable when the load is being lifted at a steady speed. [1]

(ii) When the load is just leaving the floor, why is the force larger than your answer to **b(i)**? [1]

(iii) The warehouse manager wishes to calculate the useful work done when the load is lifted from the ground to the higher level. Which quantity, other than the weight, does he need to measure? [1]

(iv) Which further quantity does the manager need to know, in order to calculate the power required to lift the load? [1]

c How does the electrical energy supplied to the electric motor compare with the increase in energy of the load? Answer by copying and completing the sentence.

The electrical energy supplied to the motor is .... the increase in energy of the load. [1]

[Total: 6]

(Cambridge IGCSE Physics 0625 Paper 21 Q3 June 2010)

18 A car of mass 900 kg is travelling at a steady speed of 30 m/s against a resistive force of 2000 N, as illustrated in the diagram.

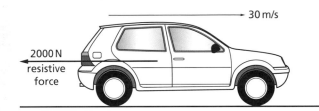

a Calculate the kinetic energy of the car. [2]
b Calculate the energy used in 1.0 s against the resistive force. [2]
c What is the minimum power that the car engine has to deliver to the wheels? [1]
d What form of energy is in the fuel, used by the engine to drive the car? [1]
e State why the energy in the fuel is converted at a greater rate than you have calculated in c. [1]

[Total: 7]

(Cambridge IGCSE Physics 0625 Paper 31 Q2 June 2010)

19 a Name three different energy resources used to obtain energy directly from water (not steam). [3]
b Choose one of the energy resources you have named in a and write a **brief** description of how the energy is converted to electrical energy. [3]

[Total: 6]

(Cambridge IGCSE Physics 0625 Paper 21 Q3 November 2010)

20 The diagram shows a manometer, containing mercury, being used to monitor the pressure of a gas supply.

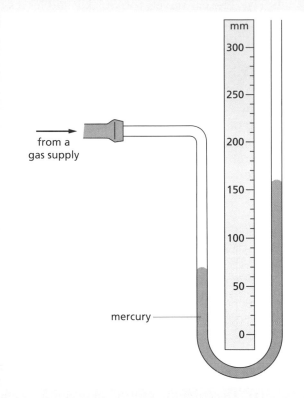

a Using the scale on the diagram, find the vertical difference between the two mercury levels. [1]
b What is the value of the excess pressure of the gas supply, measured in millimetres of mercury? [1]
c The atmospheric pressure is 750 mm of mercury.
Calculate the actual pressure of the gas supply. [1]
d The gas pressure now decreases by 20 mm of mercury. On a copy of the diagram, mark the new positions of the two mercury levels. [2]

[Total: 5]

(Cambridge IGCSE Physics 0625 Paper 02 Q4 June 2009)

21 The diagram shows a design for remotely operating an electrical switch using air pressure.

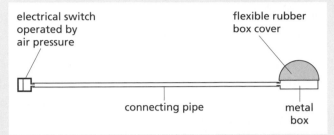

The metal box and the pipe contain air at normal atmospheric pressure and the switch is off. When the pressure in the metal box and pipe is raised to 1.5 times atmospheric pressure by pressing down on the flexible rubber box cover, the switch comes on.

a Explain in terms of pressure and volume how the switch is made to come on. [2]

b Normal atmospheric pressure is 1.0×10^5 Pa. At this pressure, the volume of the box and pipe is 60 cm³.
Calculate the **reduction** in volume that must occur for the switch to be on. [3]

c Explain, in terms of air particles, why the switch may operate, without the rubber cover being squashed, when there is a large rise in temperature. [2]

[Total: 7]

(Cambridge IGCSE Physics 0625 Paper 31 Q4 June 2008)

22 The diagram shows two mercury barometers standing side-by-side. The right-hand diagram is incomplete. The space labelled X is a vacuum.

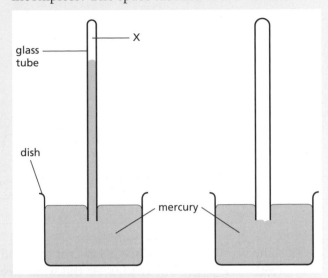

a On a copy of the left-hand barometer, carefully mark the distance that would have to be measured in order to find the value of the atmospheric pressure. [2]

b A small quantity of air is introduced into X.
  (i) State what happens to the mercury level in the tube. [1]
  (ii) In terms of the behaviour of the air molecules, explain your answer to **b(i)**. [2]

c The space above the mercury in the right-hand barometer is a vacuum.
On a copy of the right-hand diagram, mark the level of the mercury surface in the tube. [1]

d The left-hand tube now has air above the mercury; the right-hand tube has a vacuum. complete the table below, using words chosen from the following list, to indicate the effect of changing the external conditions.
  rises   falls   stays the same

change	effect on the level of the mercury in the left-hand tube	effect on the level of the mercury in the right-hand tube
atmospheric pressure rises		
temperature rises		

[4]

[Total: 10]

(Cambridge IGCSE Physics 0625 Paper 02 Q6 November 2008)

23 A wind turbine has blades, which sweep out an area of diameter 25 m as shown in the diagram.

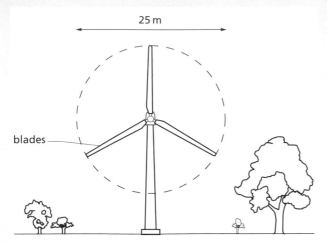

25 m

blades

a The wind is blowing directly towards the wind turbine at a speed of 12 m/s. At this wind speed, 7500 kg of air passes every second through the circular area swept out by the blades.

   (i) Calculate the kinetic energy of the air travelling at 12 m/s, which passes through the circular area in 1 second.   [3]

   (ii) The turbine converts 10% of the kinetic energy of the wind to electrical energy. Calculate the electrical power output of the turbine. State any equation that you use.   [3]

b On another day, the wind speed is half that in **a**.

   (i) Calculate the mass of air passing through the circular area per second on this day. [1]

   (ii) Calculate the power output of the wind turbine on the second day as a fraction of that on the first day.   [3]

[Total: 10]

(Cambridge IGCSE Physics 0625 Paper 31 Q5 June 2009)

2 Thermal physics
Simple kinetic molecular model of matter

24 The whole of a sealed, empty, dusty room is kept at a constant temperature of 15 °C. Light shines into the room through a small outside window. An observer points a TV camera with a magnifying lens into the room through a second small window, set in an inside wall at right angles to the outside wall.

Dust particles in the room show up on the TV monitor screen as tiny specks of light.

a Draw a diagram to show the motion of one of the specks of light over a short period of time.   [1]

b After a period of one hour the specks are still observed, showing that the dust particles have not fallen to the floor. Explain why the dust particles have not fallen to the floor. You may draw a labelled diagram to help your explanation.   [2]

c On another day, the temperature of the room is only 5 °C. All other conditions are the same and the specks of light are again observed. Suggest any differences that you would expect in the movement of the specks when the temperature is 5 °C, compared to before.   [1]

[Total: 4]

(Cambridge IGCSE Physics 0625 Paper 31 Q4 November 2008)

25 a Here is a list of descriptions of molecules in matter.

Description	Solid	Gas
free to move around from place to place		
can only vibrate about a fixed position		
closely packed		
relatively far apart		
almost no force between molecules		
strong forces are involved between molecules		

Copy the table and in the columns alongside the descriptions, put ticks next to those which apply to the molecules in
(i) a solid,   **(ii)** a gas.   [4]

b The water in a puddle of rainwater is evaporating.
Describe what happens to the molecules when the water evaporates.   [2]

[Total: 6]

(Cambridge IGCSE Physics 0625 Paper 02 Q5 June 2007)

Thermal properties and temperature

26 A certain substance is in the solid state at a temperature of −36 °C. It is heated at a constant rate for 32 minutes. The record of its temperature is given in the table at the bottom of the page.
a State what is meant by the term *latent heat*. [2]
b State a time at which the energy is being supplied as latent heat of fusion.   [1]
c Explain the energy changes undergone by the molecules of a substance during the period when latent heat of vaporisation is being supplied.   [2]
d (i) The rate of heating is 2.0 kW.
Calculate how much energy is supplied to the substance during the period 18–22 minutes.   [2]

(ii) The specific heat capacity of the substance is 1760 J/(kg°C). Use the information in the table for the period 18–22 minutes to calculate the mass of the substance being heated.   [3]

[Total: 10]

(Cambridge IGCSE Physics 0625 Paper 31 Q5 June 2010)

27 Three wires and a meter are used to construct a thermocouple for measuring the surface temperature of a pipe carrying hot liquid, as shown in the diagram.

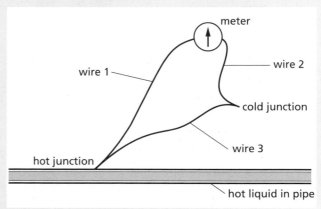

a Copper wire and constantan wire are used in the construction of the thermocouple. State which metal might be used for
wire 1 ....
wire 2 ....
wire 3 ....   [1]
b State what type of meter is used.   [1]
c State one particular advantage of thermocouples for measuring temperature. [1]

[Total: 3]

(Cambridge IGCSE Physics 0625 Paper 31 Q7 November 2009)

Time/min	0	1	2	6	10	14	18	22	24	26	28	30	32
Temperature/°C	−36	−16	−9	−9	−9	−9	32	75	101	121	121	121	121

28 a State what is meant by *specific heat capacity*. [2]

b Water has a very high specific heat capacity. Suggest why this might be a disadvantage when using water for cooking. [1]

c The diagram illustrates an experiment to measure the specific heat capacity of some metal.

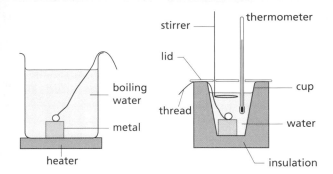

The piece of metal is heated in boiling water until it has reached the temperature of the water. It is then transferred rapidly to some water in a well-insulated cup. A very sensitive thermometer is used to measure the initial and final temperatures of the water in the cup.

specific heat capacity of water = 4200 J/(kg K)

The readings from the experiment are as follows.

mass of metal = 0.050 kg

mass of water in cup = 0.200 kg

initial temperature of water in cup = 21.1 °C

final temperature of water in cup = 22.9 °C

(i) Calculate the temperature rise of the water in the cup and the temperature fall of the piece of metal. [1]

(ii) Calculate the thermal energy gained by the water in the cup. State the equation that you use. [3]

(iii) Assume that only the water gained thermal energy from the piece of metal. Making use of your answers to **c(i)** and **c(ii)**, calculate the value of the specific heat capacity of the metal. Give your answer to three significant figures. [2]

(iv) Suggest one reason why the experiment might not have given a correct value for the specific heat capacity of the metal. [1]

[Total: 10]

(Cambridge IGCSE Physics 0625 Paper 31 Q9 November 2009)

29 a The thermometer shown below is calibrated at two fixed points, and the space between these is divided into equal divisions.

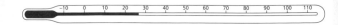

A thermometer is being calibrated with the Celsius scale.

(i) 1 Write down another name for the lower fixed point. [1]

     2 How is this temperature achieved? [2]

     3 What is the temperature of this fixed point? [1]

(ii) 1 Write down another name for the upper fixed point. [1]

     2 How is this temperature achieved? [2]

     3 What is the temperature of this fixed point? [2]

b A block of copper and a block of aluminium have identical masses. They both start at room temperature and are given equal quantities of heat. When the heating is stopped, the aluminium has a lower temperature than the copper.

Fill in the missing words in the sentence below, to explain this temperature difference.

The aluminium block has a smaller temperature rise than the copper block because the aluminium block has a larger

....

than the copper block. [1]

[Total: 10]

(Cambridge IGCSE Physics 0625 Paper 02 Q8 June 2008)

30 The diagram shows apparatus that could be used to determine the specific latent heat of fusion of ice.

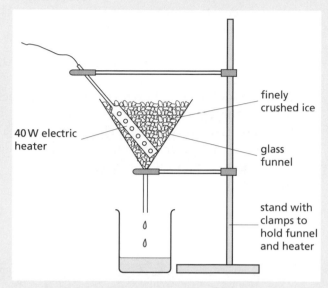

finely crushed ice

40 W electric heater

glass funnel

stand with clamps to hold funnel and heater

a In order to obtain as accurate a result as possible, state why it is necessary to
  (i) wait until water is dripping into the beaker at a constant rate before taking readings, [1]
  (ii) use finely crushed ice rather than large pieces. [1]
b The power of the heater and the time for which water is collected are known. Write down all the other readings that are needed to obtain a value for the specific latent heat of fusion of ice. [2]
c Using a 40 W heater, 16.3 g of ice is melted in 2.0 minutes. The heater is then switched off. In a further 2.0 minutes, 2.1 g of ice is melted. Calculate the value of the specific latent heat of fusion of ice from these results. [4]

[Total: 8]

(Cambridge IGCSE Physics 0625 Paper 31 Q5 November 2008)

31 The diagram shows a student's attempt to estimate the specific latent heat of fusion of ice by adding ice at 0 °C to water at 20 °C. The water is stirred continuously as ice is slowly added until the temperature of the water is 0 °C and all the added ice has melted.

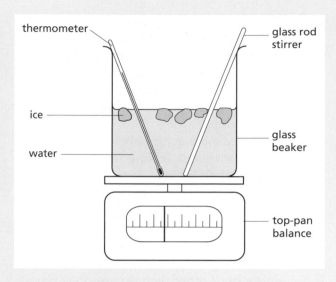

thermometer

glass rod stirrer

ice

water

glass beaker

top-pan balance

a Three mass readings are taken. A description of the first reading is given.
Write down descriptions of the other two.

reading 1: the mass of the beaker + stirrer + thermometer
reading 2: ....
reading 3: .... [2]
b Write down word equations which the student could use to find
  (i) the heat lost by the water as it cools from 20 °C to 0 °C, [1]
  (ii) the heat gained by the melting ice. [1]
c The student calculates that the water loses 12 800 J and that the mass of ice melted is 30 g. Calculate a value for the specific latent heat of fusion of ice. [2]
d Suggest two reasons why this value is only an approximate value. [2]

[Total: 8]

(Cambridge IGCSE Physics 0625 Paper 03 Q4 June 2007)

32 The diagram shows a liquid-in-glass thermometer.

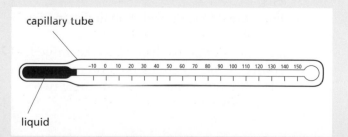

capillary tube

liquid

a The thermometer is used for measuring temperatures in school laboratory experiments. State the units in which the temperatures are measured. [1]

b On a copy of the diagram, mark where the liquid thread will reach when the thermometer is placed in
  (i) pure melting ice (label this point ICE), [1]
  (ii) steam above boiling water (label this point STEAM). [1]

c A liquid-in-glass thermometer makes use of the expansion of a liquid to measure temperature. Other thermometers make use of other properties that vary with temperature. In a copy of the table below, write in two properties, other than expansion of a liquid, that can be used to measure temperature.

example	expansion	OF	a liquid
1.		OF	
2.		OF	

[2]

[Total: 5]

(Cambridge IGCSE Physics 0625 Paper 02 Q5 November 2007)

3 Properties of waves

Light

33 The diagram shows how an image is formed by a converging lens.

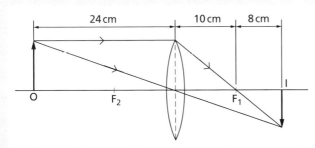

a State the value of the focal length of the lens. [1]

b The object O is moved a small distance to the left.
State two things that happen to the image I. [2]

c Points F_1 and F_2 are marked on the diagram.
  (i) State the name we give to these two points. [1]

(ii) On a copy of the diagram, draw the ray from the top of the object which passes through F_2. Continue your ray until it meets the image. [4]

[Total: 8]

(Cambridge IGCSE Physics 0625 Paper 21 Q8 June 2010)

34 In an optics lesson, a Physics student traces the paths of three rays of light near the boundary between medium A and air. The student uses a protractor to measure the various angles. The diagrams below illustrate the three measurements.

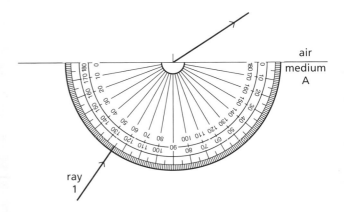

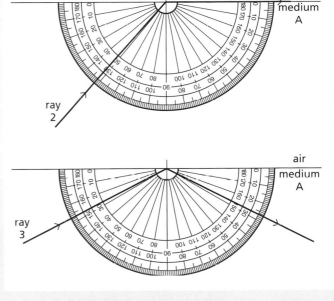

a State which is the optically denser medium, A or air, and how you can tell this. [1]

b State in which medium the light travels the faster, and how you know this. [1]

c State the critical angle of medium A. [1]

d State the full name for what is happening to ray 3 in the third diagram. [1]

e The refractive index of medium A is 1.49. Calculate the value of the angle of refraction of ray 1, showing all your working. [2]

f The speed of light in air is 3.0×10^8 m/s. Calculate the speed of light in medium A, showing all your working. [2]

[Total: 8]

(Cambridge IGCSE Physics 0625 Paper 31 Q8 June 2009)

35 The diagram shows an experiment in which an image is being formed on a card by a lens and a plane mirror.

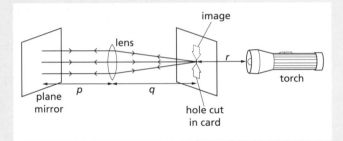

The card and the mirror are shown angled, so that you can see what is happening. In a real experiment they are each roughly perpendicular to the line joining the torch bulb and the centre of the lens.

a State which of the three marked distances, p, q and r, is the focal length of the lens. [1]

b On a copy of the diagram clearly mark a principal focus of the lens, using the letter F. [1]

c Which two features describe the image formed on the card?

erect

inverted

real

virtual [2]

d What can be said about the size of the image, compared with the size of the object? [1]

e In the experiment, the plane mirror is perpendicular to the beam of light. State what, if anything, happens to the image on the card if
(i) the plane mirror is moved slightly to the left, [1]
(ii) the lens is moved slightly to the left. [1]

[Total: 7]

(Cambridge IGCSE Physics 0625 Paper 02 Q7 November 2009)

36 A woman stands so that she is 1.0 m from a mirror mounted on a wall, as shown below.

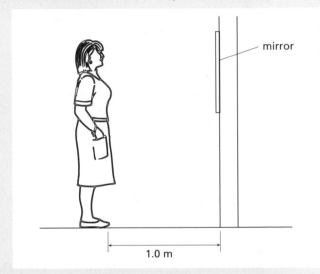

a Copy the diagram and carefully draw
(i) a clear dot to show the position of the image of her eye,
(ii) the normal to the mirror at the bottom edge of the mirror,
(iii) a ray from her toes to the bottom edge of the mirror and then reflected from the mirror. [5]

b Explain why the woman cannot see the reflection of her toes. [1]

c (i) How far is the woman from her image?
(ii) How far must the woman walk, and in what direction, before the distance between her and her image is 6.0 m? [4]

[Total: 10]

(Cambridge IGCSE Physics 0625 Paper 02 Q6 November 2006)

37 The diagram shows a ray of light, from the top of an object PQ, passing through two glass prisms.

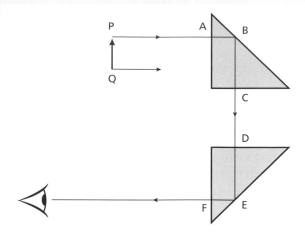

a Copy the sketch and complete the path through the two prisms of the ray shown leaving Q. [1]

b A person looking into the lower prism, at the position indicated by the eye symbol, sees an image of PQ. State the properties of this image. [2]

c Explain why there is no change in direction of the ray from P at points A, C, D and F. [1]

d The speed of light as it travels from P to A is 3×10^8 m/s and the refractive index of the prism glass is 1.5. Calculate the speed of light in the prism. [2]

e Explain why the ray AB reflects through 90° at B and does not pass out of the prism at B. [2]

[Total: 8]

(Cambridge IGCSE Physics 0625 Paper 03 Q6 November 2006)

38 a The sketch shows two rays of light from a point O on an object. These rays are incident on a plane mirror.

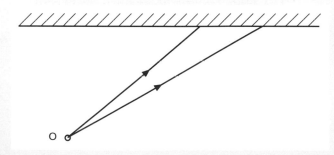

(i) Copy the diagram and continue the paths of the two rays after they reach the mirror. Hence locate the image of the object O. Label the image I.

(ii) Describe the nature of the image I. [4]

b The diagram below is drawn to scale. It shows an object PQ and a convex lens.

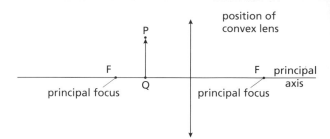

(i) Copy the diagram and draw two rays from the top of the object P that pass through the lens. Use these rays to locate the top of the image. Label this point T.

(ii) Draw an eye symbol to show the position from which the image T should be viewed. [4]

[Total: 8]

(Cambridge IGCSE Physics 0625 Paper 03 Q7 November 2005)

Sound

39 The diagram shows a workman hammering a metal post into the ground. Some distance away is a vertical cliff.

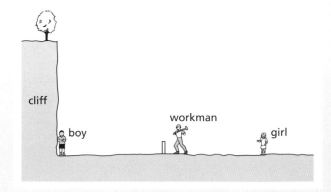

a A boy is standing at the foot of the cliff. The speed of sound in air is 330 m/s. It takes 1.5 s for the sound of the hammer hitting the post to reach the boy.

(i) What does the boy hear after he sees each strike of the hammer on the post? [1]

(ii) Calculate the distance between the post and the boy. [3]

b A girl is also watching the workman. She is standing the same distance behind the post as the boy is in front of it. She hears two separate sounds after each strike of the hammer on the post.

(i) Why does she hear **two** sounds? [2]

(ii) How long after the hammer strike does the girl hear each of these sounds?

girl hears first sound after ....s

girl hears second sound after .... s [2]

[Total: 8]

(Cambridge IGCSE Physics 0625 Paper 21 Q8 November 2010)

40 The trace shows the waveform of the note from a bell. A grid is given to help you take measurements.

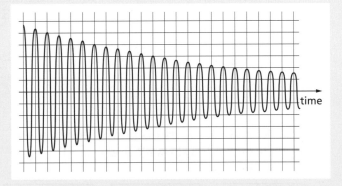

a (i) State what, if anything, is happening to the loudness of the note. [1]

(ii) State how you deduced your answer to **a(i)**. [1]

b (i) State what, if anything, is happening to the frequency of the note. [1]

(ii) State how you deduced your answer to **b(i)**. [1]

c (i) How many oscillations does it take for the amplitude of the wave to decrease to half its initial value? [1]

(ii) The wave has a frequency of 300 Hz.

1 What is meant by a frequency of 300 Hz? [1]

2 How long does 1 cycle of the wave take? [1]

3 How long does it take for the amplitude to decrease to half its initial value? [2]

d A student says that the sound waves, which travelled through the air from the bell, were longitudinal waves, and that the air molecules moved repeatedly closer together and then further apart.

(i) Is the student correct in saying that the sound waves are longitudinal?

(ii) Is the student correct about the movement of the air molecules?

(iii) The student gives light as another example of longitudinal waves. Is this correct? [2]

[Total: 11]

(Cambridge IGCSE Physics 0625 Paper 02 Q6 June 2009)

41 The diagram shows a student standing midway between a bell tower and a steep mountainside.

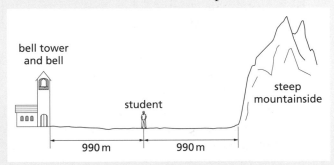

The bell rings once, but the student hears two rings separated by a short time interval.

a Explain why the student hears two rings. [2]

b State which of the sounds is louder, and why. [2]

c Sound in that region travels at 330 m/s.

(i) Calculate the time interval between the bell ringing and the student hearing it for the **first** time. [2]

(ii) Calculate the time interval between the bell ringing and the student hearing it for the **second** time. [1]

(iii) Calculate the time interval between the two sounds. [1]

[Total: 8]

(Cambridge IGCSE Physics 0625 Paper 02 Q8 November 2009)

4 Electricity and magnetism
Simple phenomena of magnetism

42 a Four rods are shown in the diagram.

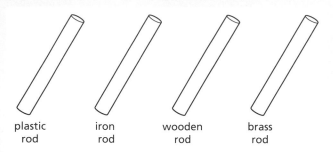

plastic   iron   wooden   brass
rod   rod   rod   rod

State which of these could be held in the hand at one end and be
  (i) magnetised by stroking it with a magnet, [1]
  (ii) charged by stroking it with a dry cloth. [1]
b Magnets A and B below are repelling each other.

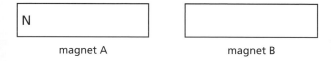

magnet A   magnet B

The north pole has been labelled on magnet A. On a copy of the diagram, label the other three poles. [1]
c Charged rods C and D below are attracting each other.

rod C   rod D

On a copy of the diagram, show the charge on rod D. [1]
d A plotting compass with its needle pointing north is shown below.

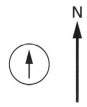

A brass rod is positioned in an east–west direction. A plotting compass is put at each end of the brass rod, as shown below.

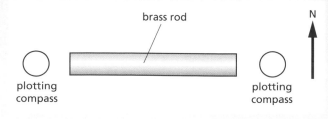

On a copy of the diagram, mark the position of the pointer on each of the two plotting compasses. [2]

[Total: 6]

(Cambridge IGCSE Physics 0625 Paper 02 Q8 June 2009)

43 a An iron rod is placed next to a bar magnet, as shown in the diagram.

iron rod

  (i) On a copy of the diagram above, mark clearly the north pole and the south pole that are induced in the iron rod. [1]
  (ii) What happens to the magnet and the rod? Tick the correct option below.
  nothing ☐
  they attract ☐
  they repel ☐ [1]
b A second bar magnet is now placed next to the iron rod, as shown below.

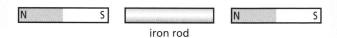

iron rod

  (i) On a copy of the diagram above, mark clearly the magnetic poles induced in the iron rod. [1]
  (ii) What happens to the iron rod and the second magnet?
  nothing
  they attract
  they repel [1]

c The iron rod is removed, leaving the two magnets, as shown below.

What happens to the two magnets?
nothing
they attract
they repel [1]

d The second magnet is removed and replaced by a charged plastic rod, as shown below.

charged
plastic rod

What happens to the magnet and the plastic rod?
nothing
they attract
they repel [1]

[Total: 6]

(Cambridge IGCSE Physics 0625 Paper 02 Q8 November 2008)

Electrical quantities and circuits

44 a A warning on the packaging of a light switch purchased from an electrical store reads

> Safety warning
> This push-button switch is not suitable for use in a washroom.
> Lights in washrooms should be operated by pull-cord switches.

  (i) Explain why it might be dangerous to use a push-button switch in a washroom. [2]
  (ii) Why is it safe to use a pull-cord switch in a washroom? [1]
b An electric heater, sold in the electrical store, has a current of 8 A when it is working normally. The cable fitted to the heater has a maximum safe current of 12 A.
  Which of the following fuses would be most suitable to use in the plug fitted to the cable of the heater?

  5 A   10 A   13 A   20 A

[1]

c The cable for connecting an electric cooker is much thicker than the cable on a table lamp.
  (i) Why do cookers need a much thicker cable? [1]
  (ii) What would happen if a thin cable were used for wiring a cooker to the supply? [1]

[Total: 6]

(Cambridge IGCSE Physics 0625 Paper 21 Q9 June 2010)

45 In the diagram, A and B are two conductors on insulating stands. Both A and B were initially uncharged.

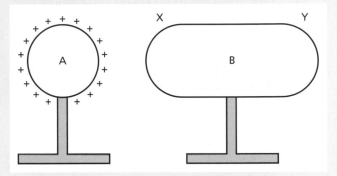

a Conductor A is given the positive charge shown on the diagram.
  (i) On a copy of the diagram, mark the signs of the charges induced at end X and at end Y of conductor B. [1]
  (ii) Explain how these charges are induced. [3]
  (iii) Explain why the charges at X and at Y are equal in magnitude. [1]
b B is now connected to earth by a length of wire. Explain what happens, if anything, to
  (i) the charge at X, [1]
  (ii) the charge at Y. [2]

[Total: 8]

(Cambridge IGCSE Physics 0625 Paper 31 Q9 November 2010)

46 The diagram shows a simple circuit.

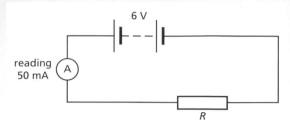

a What is the value of
   (i) the e.m.f. of the battery,
   (ii) the current in the circuit? [2]
b Calculate the resistance R of the resistor. [3]
c State how the circuit could be changed to
   (i) halve the current in the circuit, [2]
   (ii) reduce the current to zero. [1]
d A student wishes to include a switch in the circuit, but mistakenly connects it as shown below.

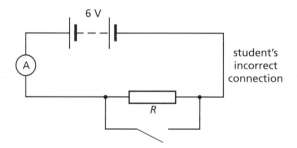

   (i) Comment on the size of the current in the circuit if the student closes the switch. [1]
   (ii) What effect would this current have on the circuit? [2]

[Total: 11]

(Cambridge IGCSE Physics 0625 Paper 02 Q9 June 2009)

47 The diagram shows a series circuit.

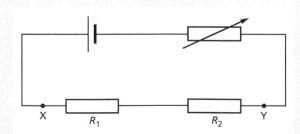

Resistance $R_1 = 25\,\Omega$ and resistance $R_2 = 35\,\Omega$. The cell has zero resistance.

a Calculate the combined resistance of R_1 and R_2. [2]
b On a copy of the diagram, use the correct circuit symbol to draw a voltmeter connected to measure the potential difference between X and Y. [1]
c The variable resistor is set to zero resistance. The voltmeter reads 1.5 V.
   (i) Calculate the current in the circuit. [4]
   (ii) State the value of the potential difference across the cell. [1]
d The resistance of the variable resistor is increased.
   (i) What happens to the current in the circuit? Tick the correct option below.
   increases ☐
   stays the same ☐
   decreases ☐ [1]
   (ii) What happens to the voltmeter reading?
   increases
   stays the same
   decreases [1]
   (iii) State the resistance of the variable resistor when the voltmeter reads 0.75 V. [1]

[Total: 11]

(Cambridge IGCSE Physics 0625 Paper 02 Q10 June 2008)

48 a Draw the symbol for a NOR gate. [1]
b Describe the action of a NOR gate in terms of its inputs and output. [2]
c A chemical process requires heating at low pressure to work correctly.
   When the heater is working, the output of a temperature sensor is high. When the pressure is low enough, a pressure sensor has a low output. Both outputs are fed into a NOR gate. A high output from the gate switches on an indicator lamp.
   (i) Explain why the indicator lamp is off when the process is working correctly. [1]
   (ii) State whether the lamp is on or off in the following situations.
   1 The pressure is low enough, but the heater stops working.
   2 The heater is working, but the pressure rises too high. [2]

[Total: 6]

(Cambridge IGCSE Physics 0625 Paper 31 Q10 June 2008)

49 a The circuit shows two resistors connected to a 6 V battery.

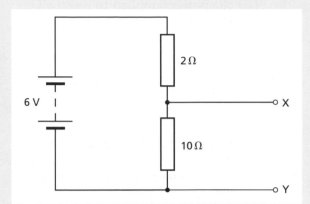

(i) What name do we use to describe this way of connecting resistors? [1]

(ii) Calculate the combined resistance of the two resistors. [1]

(iii) Calculate the current in the circuit. [4]

(iv) Use your answer to **a(iii)** to calculate the potential difference across the 10 Ω resistor. [2]

(v) State the potential difference between terminals X and Y. [1]

b The circuit shown is similar to the circuit above, but it uses a resistor AB with a sliding contact.

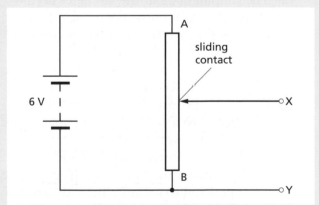

(i) State the potential difference between X and Y when the sliding contact is at

1 end A of the resistor, ... V

2 end B of the resistor. .... V [2]

(ii) The sliding contact of the resistor AB is moved so that the potential difference between X and Y is 5 V. On a copy of the circuit mark with the letter C the position of the sliding contact. [1]

[Total: 12]

(Cambridge IGCSE Physics 0625 Paper 02 Q9 June 2007)

50 The diagram shows part of a low-voltage lighting circuit containing five identical lamps.

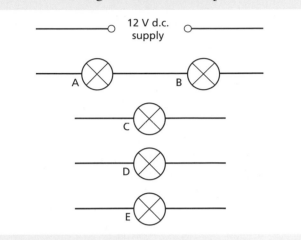

a Copy and complete the circuit, by the addition of components as necessary, so that

(i) the total current from the supply can be measured,

(ii) the brightness of lamp E only can be varied,

(iii) lamps C and D may be switched on and off together whilst lamps A, B and E remain on. [4]

b All five lamps are marked 12 V, 36 W. Assume that the resistance of each lamp is the same fixed value regardless of how it is connected in the circuit.

Calculate

(i) the current in one lamp when operating at normal brightness, [1]

(ii) the resistance of one lamp when operating at normal brightness, [1]

(iii) the combined resistance of two lamps connected in parallel with the 12 V supply, [1]

(iv) the energy used by one lamp in 30 s when operating at normal brightness. [1]

c The whole circuit is switched on. Explain why the brightness of lamps A and B is much less than that of one lamp operating at normal brightness. [2]

[Total: 10]

(Cambridge IGCSE Physics 0625 Paper 03 Q8 June 2007)

51 The diagram shows two electrical circuits.
The batteries in circuit 1 and circuit 2 are identical.

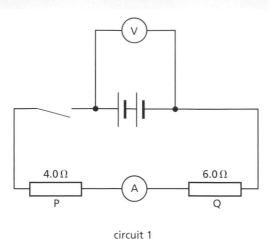

circuit 1

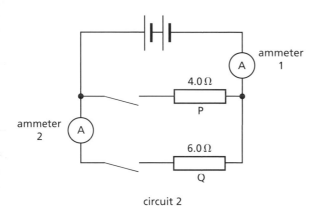

circuit 2

a Put ticks in a copy of the table below to describe the connections of the two resistors P and Q.

	Series	Parallel
circuit 1		
circuit 2			[1]

b The resistors P and Q are used as small electrical heaters. State two advantages of connecting them as shown in circuit 2. [2]
c In circuit 1, the ammeter reads 1.2 A when the switch is closed. Calculate the reading of the voltmeter in this circuit. [2]
d The two switches in circuit 2 are closed. Calculate the combined resistance of the two resistors in this circuit. [2]
e When the switches are closed in circuit 2, ammeter 1 reads 5 A and ammeter 2 reads 2 A.

Calculate
(i) the current in resistor P, [1]
(ii) the power supplied to resistor Q, [1]
(iii) the energy transformed in resistor Q in 300 s. [1]

[Total: 10]

(Cambridge IGCSE Physics 0625 Paper 03 Q8 November 2007)

52 The diagram shows an electric circuit.

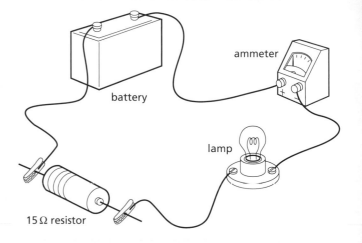

a The lamp lights, but the ammeter needle moves the wrong way. What change should be made so that the ammeter works correctly? [1]
b What does an ammeter measure? [1]
c Draw a circuit diagram of the circuit in the diagram, using correct circuit symbols. [2]
d (i) Name the instrument that would be needed to measure the potential difference (p.d.) across the 15 Ω resistor.
(ii) Using the correct symbol, add this instrument to your circuit diagram in **c**, in a position to measure the p.d. across the 15 Ω resistor. [2]
e The potential difference across the 15 Ω resistor is 6 V.
Calculate the current in the resistor. [3]
f Without any further calculation, state the value of the current in the lamp. [1]
g Another 15 Ω resistor is connected in parallel with the 15 Ω resistor that is already in the circuit.
(i) What is the combined resistance of the two 15 Ω resistors in parallel?
30 Ω, 15 Ω, 7.5 Ω or zero?

(ii) State what effect, if any, adding this extra resistor has on the current in the lamp. [2]

[Total: 12]

(Cambridge IGCSE Physics 0625 Paper 02 Q12 November 2006)

Electromagnetic effects

53 Alternating current electricity is delivered at 22 000 V to a pair of transmission lines. The transmission lines carry the electricity to the customer at the receiving end, where the potential difference is *V*. This is shown in the diagram. Each transmission line has a resistance of 3 Ω.

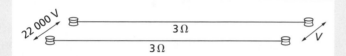

a The a.c. generator actually generates at a much lower voltage than 22 000 V.
  (i) Suggest how the voltage is increased to 22 000 V. [1]
  (ii) State one advantage of delivering electrical energy at high voltage. [1]
b The power delivered by the generator is 55 kW.
  Calculate the current in the transmission lines. [2]
c Calculate the rate of loss of energy from one of the 3 Ω transmission lines. [2]
d Calculate the voltage drop across one of the transmission lines. [2]
e Calculate the potential difference *V* at the receiving end of the transmission lines. [2]

[Total: 10]

(Cambridge IGCSE Physics 0625 Paper 31 Q10 November 2009)

54 a The diagram illustrates the left-hand rule, which helps when describing the force on a current-carrying conductor in a magnetic field.

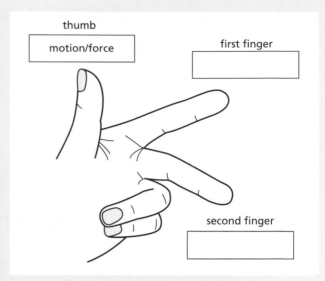

One direction has been labelled for you. In each of the other two boxes, write the name of the quantity that direction represents. [1]
b The diagram below shows a simple d.c. motor connected to a battery and a switch.

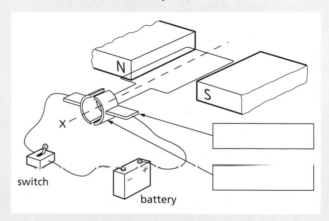

  (i) On a copy of the diagram, write in each of the boxes the name of the part of the motor to which the arrow is pointing. [2]
  (ii) State which way the coil of the motor will rotate when the switch is closed, when viewed from the position X. [1]
  (iii) State two things which could be done to increase the speed of rotation of the coil. [2]

[Total: 6]

(Cambridge IGCSE Physics 0625 Paper 31 Q9 June 2010)

55 The diagram shows a transformer.

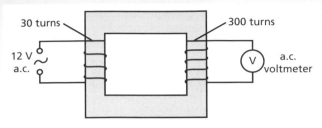

a (i) On a copy of the diagram, clearly label the core of the transformer. [1]
  (ii) Name a suitable material from which the core could be made. [1]
  (iii) State the purpose of the core. [1]
b Calculate the reading on the voltmeter. [3]

[Total: 6]

(Cambridge IGCSE Physics 0625 Paper 02 Q10 November 2009)

56 a An experimenter uses a length of wire ABC in an attempt to demonstrate electromagnetic induction. The wire is connected to a sensitive millivoltmeter G as shown in the diagram.

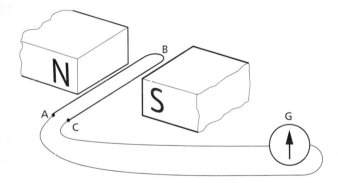

Using the arrangement in the diagram, the experimenter finds that she does not obtain the expected deflection on G when she moves the wire ABC down through the magnetic field.
  (i) Explain why there is no deflection shown on G. [2]
  (ii) What change should be made in order to observe a deflection on G? [1]
b Name one device that makes use of electromagnetic induction. [1]

[Total: 4]

(Cambridge IGCSE Physics 0625 Paper 02 Q11 June 2008)

57 The diagram shows apparatus used to investigate electromagnetic effects around straight wires.

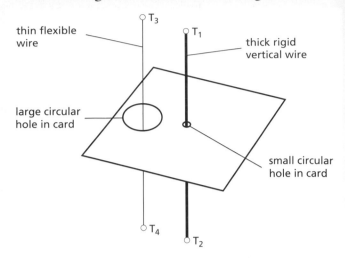

The diagram below is a view looking down on the apparatus shown above.

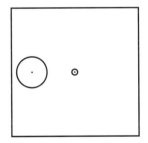

a A battery is connected to T_1 and T_2 so that there is a current vertically down the thick wire. On a copy of the diagram of the view looking down, draw three magnetic field lines and indicate, with arrows, the direction of all three. [2]
b Using a variable resistor, the p.d. between terminals T_1 and T_2 is gradually reduced. State the effect, if any, that this will have on
  (i) the strength of the magnetic field, [1]
  (ii) the direction of the magnetic field. [1]
c The battery is now connected to terminals T_3 and T_4, as well as to terminals T_1 and T_2, so that there is a current down both wires. This causes the flexible wire to move.
  (i) Explain why the flexible wire moves. [2]
  (ii) State the direction of the movement of the flexible wire. [1]

(iii) The battery is replaced by one that delivers a smaller current. State the effect that this will have on the force acting on the flexible wire. [1]

[Total: 8]

(Cambridge IGCSE Physics 0625 Paper 31 Q9 June 2008)

58 The circuit in the diagram shows an electromagnetic relay being used to switch an electric motor on and off. The relay coil has a much greater resistance than the potential divider.

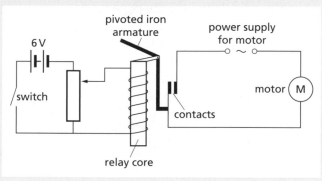

a The relay operates when there is a potential difference of 3 V across the coil. On a copy of the diagram, mark the position of the slider of the potential divider when the relay just operates. [1]

b Describe how the relay closes the contacts in the motor circuit. [3]

[Total: 4]

(Cambridge IGCSE Physics 0625 Paper 02 Q10 November 2008)

59 A coil of insulated wire is connected in series with a battery, a resistor and a switch as shown below.

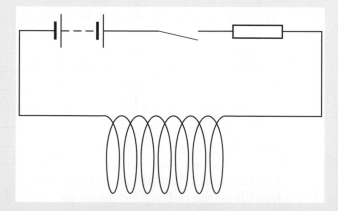

a The switch is closed and the current in the coil creates a magnetic field.

(i) On a copy of the diagram, draw the shape of the magnetic field, both inside and outside the coil. [4]

(ii) A glass bar, an iron bar and a Perspex bar are placed in turn inside the coil. Which one makes the field stronger? [1]

b Two thin iron rods are placed inside the coil as shown below. The switch is then closed.

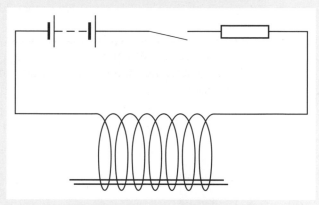

The iron rods move apart. Suggest why this happens. [3]

[Total: 8]

(Cambridge IGCSE Physics 0625 Paper 02 Q10 November 2007)

60 Electromagnetic induction may be demonstrated using a magnet, a solenoid and other necessary apparatus.

a Explain what is meant by *electromagnetic induction*. [2]

b Draw a labelled diagram of the apparatus set up so that electromagnetic induction may be demonstrated. [2]

c Describe how you would use the apparatus to demonstrate electromagnetic induction. [2]

d State two ways of increasing the magnitude of the induced e.m.f. in this experiment. [2]

[Total: 8]

(Cambridge IGCSE Physics 0625 Paper 03 Q9 November 2007)

5 Atomic physics

61 Here is a list of different types of radiation.
alpha (α), beta (β), gamma (γ), infra-red, radio, ultra-violet, visible, X-rays

a List all those radiations in the list which are **not** electromagnetic radiations. [2]

b Which radiation is the most penetrating? [1]

c Which radiation has the longest wavelength? [1]

d Which radiation consists of particles that are the same as ^{4}He nuclei? [1]

[Total: 5]

(Cambridge IGCSE Physics 0625 Paper 21 Q5 November 2010)

62 Emissions from a radioactive source pass through a hole in a lead screen and into a magnetic field, as shown in the diagram.

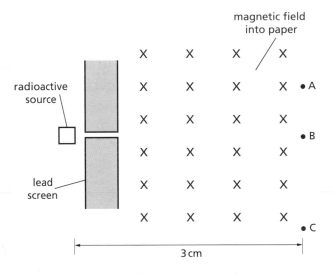

Radiation detectors are placed at A, B and C. They give the following readings:

A	B	C
32 counts/min	543 counts/min	396 counts/min

The radioactive source is then completely removed, and the readings become:

A	B	C
33 counts/min	30 counts/min	31 counts/min

a Explain why there are still counts being recorded at A, B and C, even when the radioactive source has been removed, and give the reason for them being slightly different. [2]

b From the data given, deduce the type of emission being detected, if any, at A, at B and at C when the radiation source is present.

State the reasons for your answers.

detector at A .... [2]

detector at B .... [3]

detector at C .... [3]

[Total: 10]

(Cambridge IGCSE Physics 0625 Paper 31 Q10 November 2010)

63 A beam of ionising radiation, containing α-particles, β-particles and γ-rays, is travelling left to right across the page. A magnetic field acts perpendicularly into the page.

a In a copy of the table below, tick the boxes that describe the deflection of each of the types of radiation as it passes through the magnetic field. One row has been completed to help you. [3]

	not deflected	deflected towards top of page	deflected towards bottom of page	large deflection	small deflection
α-particles		✓			✓
β-particles					
γ-rays					

b An electric field is now applied, in the same region as the magnetic field and at the same time as the magnetic field. What is the direction of the electric field in order to cancel out the deflection of the α-particles? [2]

[Total: 5]

(Cambridge IGCSE Physics 0625 Paper 31 Q11 June 2009)

64 **a** The table shows how the activity of a sample of a radioactive substance changes with time.

Time /minutes	Activity /counts/s
0	128
30	58
60	25
90	11
120	5

Use the data in the table to estimate the half-life of the radioactive substance. [2]

b The half-lives of various substances are given below.

radon-220   55 seconds
iodine-128   25 minutes
radon-222   3.8 days
strontium-90   28 years

(i) If the radioactive substance in **a** is one of these four, which one is it? [1]

(ii) A sample of each of these substances is obtained. Which sample will have the greatest proportion of decayed nuclei by the end of one year, and why? [2]

[Total: 5]

(Cambridge IGCSE Physics 0625 Paper 02 Q12 June 2008)

65 a Chlorine has two isotopes, one of nucleon number 35 and one of nucleon number 37. The proton number of chlorine is 17. The table refers to neutral atoms of chlorine. Copy and complete the table.

	Nucleon number 35	Nucleon number 37
number of protons		
number of neutrons		
number of electrons		

[3]

b Some isotopes are radioactive. State the three types of radiation that may be emitted from radioactive isotopes. [1]

c (i) State one practical use of a radioactive isotope. [1]

(ii) Outline how it is used. [1]

[Total: 6]

(Cambridge IGCSE Physics 0625 Paper 31 Q11 June 2008)

66 The nucleus of one of the different nuclides of polonium can be represented by the symbol $^{218}_{84}$Po.

a State the proton number of this nuclide. [1]
b State the nucleon number of this nuclide. [1]
c The nucleus decays according to the following equation.

$$^{218}_{84}\text{Po} \rightarrow \,^{214}_{82}\text{Pb} + \text{emitted particle}$$

(i) State the proton number of the emitted particle. [1]

(ii) State the nucleon number of the emitted particle. [1]

(iii) Name the emitted particle. Choose from the following:

α-particle   ☐

β-particle   ☐

neutron   ☐

proton   ☐   [1]

[Total: 5]

(Cambridge IGCSE Physics 0625 Paper 02 Q12 November 2008)

67 The diagram shows the paths of three α-particles moving towards a thin gold foil.

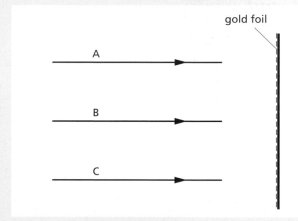

Particle A is moving directly towards a gold nucleus.
Particle B is moving along a line which passes close to a gold nucleus.
Particle C is moving along a line which does not pass close to a gold nucleus.

a On a copy of the diagram, complete the paths of the α-particles A, B and C. [3]

b State how the results of such an experiment, using large numbers of α-particles, provides evidence for the existence of nuclei in gold atoms. [3]

[Total: 6]

(Cambridge IGCSE Physics 0625 Paper 03 Q11 June 2007)

68 The activity of a sample of radioactive material is determined every 10 minutes for an hour. The results are shown in the table.

Time / minutes	0	10	20	30	40	50	60
Activity / counts/s	461	332	229	162	106	81	51

a From the figures in the table, estimate the half-life of the radioactive material. [1]

b A second experiment is carried out with another sample of the same material. At the start of the experiment, this sample has twice the number of atoms as the first sample. Suggest what values might be obtained for
   (i) the activity at the start of the second experiment, [1]
   (ii) the half-life of the material in the second experiment. [1]

c Name one type of particle that the material might be emitting in order to cause this activity. [1]

[Total: 4]

(Cambridge IGCSE Physics 0625 Paper 02 Q11 November 2007)

69 A beam of cathode rays is travelling in a direction perpendicularly out of the page. The beam is surrounded by four metal plates P_1, P_2, P_3 and P_4 as shown in the diagram.
The beam is shown as the dot at the centre.

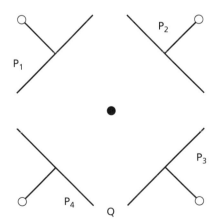

a Cathode rays are produced by thermionic emission.
What is the name of the particles which make up cathode rays? [1]

b A potential difference is applied between P_1 and P_3, with P_1 positive with respect to P_3. State what happens to the beam of cathode rays. [2]

c The potential difference in **b** is removed. Suggest how the beam of cathode rays can now be deflected down the page towards Q. [2]

d Cathode rays are invisible. State one way to detect them. [1]

[Total: 6]

(Cambridge IGCSE Physics 0625 Paper 02 Q12 November 2007)

70 The diagram shows an experiment to test the absorption of β-particles by thin sheets of aluminium.
Ten sheets are available, each 0.5 mm thick.

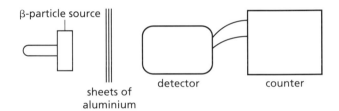

a Describe how the experiment is carried out, stating the readings that should be taken. [4]

b State the results that you would expect to obtain. [2]

[Total: 6]

(Cambridge IGCSE Physics 0625 Paper 03 Q11 November 2007)

Mathematics for physics

USE THIS SECTION AS THE NEED ARISES

● Solving physics problems

When tackling physics problems using mathematical equations it is suggested that you **do not substitute numerical values until you have obtained the expression in symbols which gives the answer.** That is, work in symbols until you have solved the problem and only then insert the numbers in the expression to get the final result.

This has two advantages. First, it reduces the chance of errors in the arithmetic (and in copying down). Second, you write less since a symbol is usually a single letter whereas a numerical value is often a string of figures.

Adopting this 'symbolic' procedure frequently requires you to change round an equation first. The next two sections and the questions that follow them are intended to give you practice in doing this and then substituting numerical values to get the answer.

● Equations – type 1

In the equation $x = a/b$, the subject is x. To change it we **multiply or divide both** sides of the equation by the same quantity.

To change the subject to a
We have

$$x = \frac{a}{b}$$

If we multiply both sides by b, the equation will still be true.

$$\therefore \qquad x \times b = \frac{a}{b} \times b$$

The b's on the right-hand side cancel

$$\therefore \qquad b \times x = \frac{a}{b} \times b = a$$

and

$$a = b \times x$$

To change the subject to b
We have

$$x = \frac{a}{b}$$

Multiplying both sides by b as before, we get

$$a = b \times x$$

Dividing both sides by x:

$$\frac{a}{x} = \frac{b \times x}{x} = \frac{b \times x}{x} = b$$

$$\therefore \qquad b = \frac{a}{x}$$

Note that the **reciprocal** of x is $1/x$.
Can you show that

$$\frac{1}{x} = \frac{b}{a}?$$

Now try the following questions using these ideas.

Questions

1 What is the value of x if
  a $2x = 6$ b $3x = 15$ c $3x = 8$
  d $\dfrac{x}{2} = 10$ e $\dfrac{x}{3} = 4$ f $\dfrac{2x}{3} = 4$
  g $\dfrac{4}{x} = 2$ h $\dfrac{9}{x} = 3$ i $\dfrac{x}{6} = \dfrac{4}{3}$

2 Change the subject to
  a f in $v = f\lambda$ b λ in $v = f\lambda$
  c I in $V = IR$ d R in $V = IR$
  e m in $d = \dfrac{m}{V}$ f V in $d = \dfrac{m}{V}$
  g s in $v = \dfrac{s}{t}$ h t in $v = \dfrac{s}{t}$

3 Change the subject to
  a I^2 in $P = I^2R$ b I in $P = I^2R$
  c a in $s = \dfrac{1}{2}at^2$ d t^2 in $s = \dfrac{1}{2}at^2$
  e t in $s = \dfrac{1}{2}at^2$ f v in $\dfrac{1}{2}mv^2 = mgh$
  g y in $\lambda = \dfrac{ay}{D}$ h ρ in $R = \dfrac{\rho l}{A}$

4 By replacing (substituting) find the value of $v = f\lambda$ if
  a $f = 5$ and $\lambda = 2$ b $f = 3.4$ and $\lambda = 10$
  c $f = 1/4$ and $\lambda = 8/3$ d $f = 3/5$ and $\lambda = 1/6$
  e $f = 100$ and $\lambda = 0.1$ f $f = 3 \times 10^5$ and $\lambda = 10^3$

5 By changing the subject and replacing find
  a f in $v = f\lambda$, if $v = 3.0 \times 10^8$ and $\lambda = 1.5 \times 10^3$
  b h in $p = 10hd$, if $p = 10^5$ and $d = 10^3$
  c a in $n = a/b$, if $n = 4/3$ and $b = 6$
  d b in $n = a/b$, if $n = 1.5$ and $a = 3.0 \times 10^8$
  e F in $p = F/A$ if $p = 100$ and $A = 0.2$
  f s in $v = s/t$, if $v = 1500$ and $t = 0.2$

● Equations – type 2

To change the subject in the equation $x = a + by$ we **add or subtract the same quantity from each side**. We may also have to divide or multiply as in type 1. Suppose we wish to change the subject to y in

$$x = a + by$$

Subtracting a from both sides,

$$x - a = a + by - a = by$$

Dividing both sides by b,

$$\frac{x - a}{b} = \frac{by}{b} = y$$

$$\therefore \qquad y = \frac{x - a}{b}$$

Questions

6 What is the value of x if
  a $x + 1 = 5$ b $2x + 3 = 7$ c $x - 2 = 3$

  d $2(x - 3) = 10$ e $\frac{x}{2} - \frac{1}{3} = 0$ f $\frac{x}{3} + \frac{1}{4} = 0$

  g $2x + \frac{5}{3} + 6$ h $7 - \frac{x}{4} = 11$ i $\frac{3}{x} + 2 = 5$

7 By changing the subject and replacing, find the value of a in $v = u + at$ if
  a $v = 20$, $u = 10$ and $t = 2$
  b $v = 50$, $u = 20$ and $t = 0.5$
  c $v = 5/0.2$, $u = 2/0.2$ and $t = 0.2$
8 Change the subject in $v^2 = u^2 + 2as$ to a.

● Proportion (or variation)

One of the most important mathematical operations in physics is finding the relation between two sets of measurements.

a) Direct proportion

Suppose that in an experiment two sets of readings are obtained for the quantities x and y as in Table M1 (units omitted).

Table M1

x	1	2	3	4
y	2	4	6	8

We see that when x is doubled, y doubles; when x is trebled, y trebles; when x is halved, y halves; and so

on. There is a one-to-one correspondence between each value of x and the corresponding value of y.

We say that y is **directly proportional** to x, or y **varies directly** as x. In symbols

$$y \propto x$$

Also, the ratio of one to the other, e.g. y to x, is always the same, i.e. it has a constant value which in this case is 2. Hence

$$\frac{y}{x} = \text{a constant} = 2$$

The constant, called the **constant of proportionality** or **constant of variation**, is given a symbol, e.g. k, and the relation (or law) between y and x is then summed up by the equation

$$\frac{y}{x} = k \quad \text{or} \quad y = kx$$

Notes

1 In practice, because of inevitable experimental errors, the readings seldom show the relation so clearly as here.

2 If instead of using numerical values for x and y we use letters, e.g. x_1, x_2, x_3, etc., and y_1, y_2, y_3, etc., then we can also say

$$\frac{y_1}{x_1} = \frac{y_2}{x_2} = \frac{y_3}{x_3} = \quad \dots \quad = k$$

or

$$y_1 = kx_1, \ y_2 = kx_2, \ y_3 = kx_3, \dots$$

b) Inverse proportion

Two sets of readings for the quantities p and V are given in Table M2 (units omitted).

Table M2

p	3	4	6	12
V	4	3	2	1

There is again a one-to-one correspondence between each value of p and the corresponding value of V, but when p is doubled, V is halved, when p is trebled, V has one-third its previous value, and so on.

We say that V is **inversely proportional** to p, or V **varies inversely** as p, i.e.

$$V \propto \frac{1}{p}$$

Also, the **product** $p \times V$ is always the same ($=12$ in this case) and we write

$$V = \frac{k}{p} \quad \text{or} \quad pV = k$$

where k is the constant of proportionality or variation and equals 12 in this case.

Using letters for values of p and V we can also say

$$p_1 V_1 = p_2 V_2 = p_3 V_3 = \ldots = k$$

● Graphs

Another useful way of finding the relation between two quantities is by a graph.

a) Straight line graphs

When the readings in Table M1 are used to plot a graph of y against x, a **continuous** line joining the points is **a straight line passing through the origin** as in Figure M1. Such a graph shows there is direct proportionality between the quantities plotted, i.e. $y \propto x$. But note that the line must go through the origin.

A graph of p against V using the readings in Table M2 is a curve, as in Figure M2. However if we plot p against $1/V$ (Table M3) (or V against $1/p$) we get a straight line through the origin, showing that $p \propto V$, as in Figure M3 (or $V \propto 1/p$).

Table M3

p	V	1/V
3	4	0.25
4	3	0.33
6	2	0.50
12	1	1.00

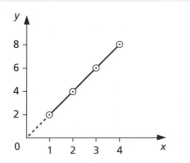

Figure M1

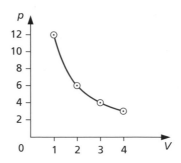

Figure M2

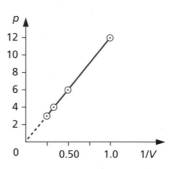

Figure M3

b) Slope or gradient

The slope or gradient of a straight line graph equals the constant of proportionality. In Figure M1, the slope is $y/x = 2$; in Figure M3 it is $p/(1/V) = 12$.

In practice, points plotted from actual measurements may not lie exactly on a straight line due to experimental errors. The 'best straight line' is then drawn 'through' them so that they are equally distributed about it. This automatically averages the results. Any points that are well off the line stand out and may be investigated further.

c) Variables

As we have seen, graphs are used to show the relationship between two physical quantities. In an experiment to investigate how potential difference, V, varies with the current, I, a graph can be drawn of V/V values plotted against the values of I/A. This will reveal how the potential difference depends upon the current (see Figure M4).

In the experiment there are two **variables**. The quantity I is varied and the value for V is dependent upon the value for I. So V is called the **dependent variable** and I is called the **independent variable**.

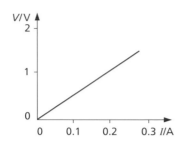

Figure M4

Note that in Figure M4 each axis is labelled with the quantity *and* the unit. Also note that there is a scale along each axis. The statement *V/V against I/A* means that *V/V*, the dependent variable, is plotted along the *y*-axis and the independent variable *I* is plotted along the *x*-axis (see Figure M5).

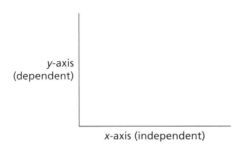

Figure M5

Questions

9 In an experiment different masses were hung from the end of a spring held in a stand and the extensions produced were as shown below.

Mass/g	100	150	200	300	350	500	600
Extension/cm	1.9	3.1	4.0	6.1	6.9	10.0	12.2

  a Plot a graph of extension along the vertical (*y*) axis against mass along the horizontal (*x*) axis.
  b What is the relation between extension and mass? Give a reason for your answer.

10 Pairs of readings of the quantities *m* and *v* are given below.

m	0.25	1.5	2.5	3.5
v	20	40	56	72

  a Plot a graph of *m* along the vertical axis and *v* along the horizontal axis.
  b Is *m* directly proportional to *v*? Explain your answer.
  c Use the graph to find *v* when $m = 1$.

11 The distances *s* (in metres) travelled by a car at various times *t* (in seconds) are shown below.

s/m	0	2	8	18	32	50
t/m	0	1	2	3	4	5

Draw graphs of
  a *s* against *t*,
  b *s* against t^2.
What can you conclude?

d) Practical points

(i) The axes should be labelled giving the quantities being plotted and their units, e.g. *I/A* meaning current in amperes.

(ii) If possible the origin of both scales should be on the paper and the scales chosen so that the points are spread out along the graph. It is good practice to draw a large graph.

(iii) The scale should be easy to use. A scale based on multiples of 10 or 5 is ideal. Do not use a scale based on a multiple of 3; such scales are very difficult to use.

(iv) Mark the points ⊙ or ×.

Further experimental investigations

Stretching of a rubber band

Set up the equipment as shown in Chapter 6 (Figure 6.3, p. 25) but replace the spring with a thick rubber band. Draw up a table in which to record stretching force/N, scale reading/mm and total extension/mm. Take readings for increasing loads on the hanger.

Plot a graph with stretching force/N along the x-axis and extension/mm along the y-axis. Draw the best straight line through your points; are your results consistent with Hooke's law for all loads?

(If weights and a hanger are not available you could use coins (all similar) in a paper cup instead; in this case the stretching force would be proportional to the number of coins used.)

Toppling

The stability of a body can be investigated using a 1 litre drinks carton or can as shown in Figure E1a. When the carton is tilted so that the centre of mass moves outside the base, the carton will topple over.

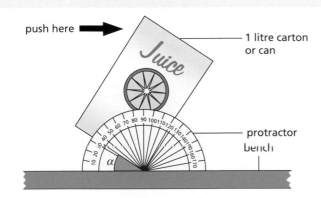

a

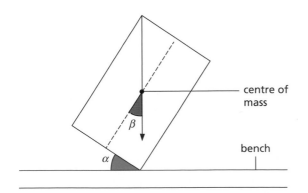

b

Figure E1

(i) Attach a protractor to the bench with Blu-tack. Fill a carton with water and gently push it at the top so that it tilts. Measure the maximum angle, α, that the carton can be tilted through without toppling; repeat your measurement several times and obtain an average value for α.

(ii) Draw a full-size diagram of the face of the carton; mark the centre of mass on the face and measure the angle β between the long side and a diagonal as shown in Figure E1b; how do your values for α and β compare?

(iii) Repeat part (i) with the carton half full, a quarter full and empty. Draw up a table of your results as shown below.

Liquid volume/litres	α_1/°	α_2/°	α_3/°	Average α/°
1.0				
0.5				
0.25				
0.0				
0.5 (frozen)				

Where is the centre of mass of an empty carton? Plot a graph with volume/litres on the y-axis and α/° on the x-axis. What angle of topple would you expect if the carton was one third full of water? How does changing the position of the centre of mass affect the stability of the carton?

(iv) Put a half-full carton in the freezer; when the water is fully frozen repeat part (i); add your results to the table.
Will the carton be more or less stable when the water has melted? How are the centre of mass and the angle of topple changed by freezing the water?

(v) Turn a full carton on its side and repeat steps (i) and (ii). Is the carton more or less stable than when upright? Explain why.

Summarise the factors that influence the stability of a body.

Cooling and evaporation

For this experiment you will need two heat sensors connected to a datalogger and computer. Use some cotton thread to tie a piece of tissue paper loosely over one of the heat sensors. Insert both heat sensors into a beaker of hot water and wait until they reach a constant temperature.

Experiment 1: With the datalogger running, remove the heat sensors from the water and quickly dry the sensor that is not covered by tissue paper. Hang each sensor on a retort stand and allow each to cool to room temperature. Use the computer to record a graph of temperature (on the y-axis) versus time (on the x-axis) for each sensor – these are 'cooling curves'.

Discuss the general shape of the cooling curves – when do the bulbs cool most rapidly? How do the cooling curves differ for the 'wet' compared with the 'dry' heat sensor? Which sensor reaches the lower temperature – can you explain why?

Experiment 2: Repeat the first experiment but this time hang the sensors in a draught to cool. An artificial draught can be produced by an electric cooling fan. Compare the cooling curves recorded by the computer with those obtained when there was no draught (Experiment 1). Comment on how the rate of cooling and the lowest temperature reached have changed for each sensor and try to explain your results.

From your findings, summarise the factors that affect the rate at which an object cools.

(If dataloggers and computers are not available this experiment could be done with mercury thermometers and a 'team' of students to help record temperatures manually every 15 seconds!)

Variation of the resistance of a wire with length

Several different lengths of resistance wire (constantan SWG 34 is suitable) are needed, in addition to the equipment shown in Chapter 38 (Figure 38.6, p. 168).

Cut the following lengths (l) of resistance wire: 20 cm, 40 cm, 60 cm, 80 cm and 100 cm. Wind each wire into a coil, ensuring that adjacent turns do not touch if the wire is not insulated. Set up the circuit shown in Figure 38.6 with the shortest coil in position R. (Set the rheostat near the midway position.) Draw up a table in which to record l, I, V and R for each coil. Determine R ($= V/I$) from your readings; repeat the measurements and calculation of R for each coil.

Draw a graph with average R values on the y-axis and l values on the x-axis. Is it consistent with the relation $R = \rho l$? Calculate the slope of the graph.

Measure the diameter of the constantan wire with a micrometer screw gauge and determine a value for the resistivity, ρ, of the wire.

Practical test questions

1 In this experiment, you are to investigate the stretching of springs. You have been provided with the apparatus shown in Figure P1.

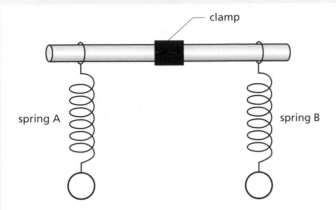

Figure P1

a (i) Measure the length l_A of spring **A**.
  (ii) On a copy of Figure P1 show clearly where you decided to start and end the length measurement l_A.
  (iii) Hang the 200 g mass on spring **A**. Measure the new length l of the spring.
  (iv) Calculate the extension e_A of spring **A** using the equation $e_A = (l - l_A)$. [3]
b (i) Measure the length l_B of spring **B**.
  (ii) Hang the 200 g mass on spring **B**. Measure the new length l of the spring.
  (iii) Calculate the extension e_B of spring **B** using the equation $e_B = (l - l_B)$. [2]
c Use the small length of wooden rod provided to hang the 400 g mass midway between the springs as shown in Figure P2.

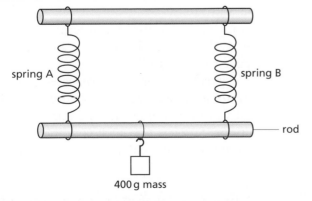

Figure P2

(i) Measure the new lengths of each of the springs.
(ii) Calculate the extension of each spring using the appropriate equation from parts **a** and **b**.
(iii) Calculate the average of these two extensions e_{av}. Show your working. [2]
d Theory suggests that

$$\frac{(e_a + e_b)}{2} = e_{av}$$

State whether your results support this theory and justify your answer with reference to the results. [2]
e Describe briefly one precaution that you took to obtain accurate length measurements. [1]

[Total 10]

(Cambridge IGCSE Physics 0625 Paper 51 Q1 June 2010)

2 In this experiment, you will investigate the effect of the length of resistance wire in a circuit on the potential difference across a lamp. The circuit has been set up for you.
a Figure P3 shows the circuit without the voltmeter.
Draw on a copy of the circuit diagram the voltmeter as it is connected in the circuit. [2]

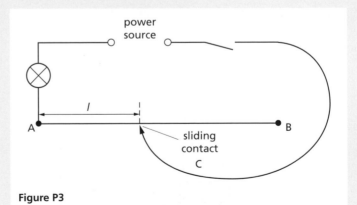

Figure P3

b (i) Switch on and place the sliding contact **C** on the resistance wire at a distance $l = 0.150$ m from end **A**. Record the value of l and the potential difference V across the lamp in the table. Switch off.

(ii) Repeat step (i) using the following values of *l*:
0.350 m, 0.550 m, 0.750 m and 0.950 m. Record all the values of *l* and *V* in a copy of the table.

l/m	V/V	V/l

(iii) For each pair of readings in the table calculate and record in the table the value of *V/l*.
(iv) Complete the table by writing in the unit for *V/l*. [5]

c A student suggests that the potential difference *V* across the lamp is directly proportional to the length *l* of resistance wire in the circuit. State whether or not you agree with this suggestion and justify your answer by reference to your results. [2]

d State one precaution that you would take in order to obtain accurate readings in this experiment. [1]

[Total 10]

(Cambridge IGCSE Physics 0625 Paper 51 Q3 June 2010)

3 In this experiment you will investigate the rate of heating and cooling of a thermometer bulb. Carry out the following instructions referring to Figure P4. You are provided with a beaker of hot water.

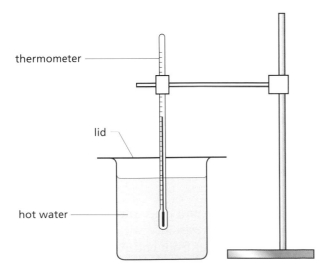

thermometer

lid

hot water

Figure P4

a Record the room temperature θ_r. [1]
b (i) Place the thermometer into the water as shown in Figure P4. When the temperature shown on the thermometer stops rising, record the temperature θ in a copy of Table A at time $t = 0$ s.
(ii) Remove the thermometer from the beaker of water and immediately start the stopclock. Record in Table A the temperature shown on the thermometer as it cools in the air. Take readings at 30 s intervals from $t = 30$ s until you have a total of seven values up to time $t = 180$ s. [2]
c (i) Set the stopclock back to zero. With the thermometer still out of the beaker, record in a copy of Table B the temperature θ shown on the thermometer at time $t = 0$ s.
(ii) Replace the thermometer in the beaker of hot water as shown in Figure P4 and immediately start the stopclock. Record in Table B the temperature shown by the thermometer at 10 s intervals until you have a total of seven values up to time $t = 60$ s.

Table A		Table B	
t/	θ/	t/	θ/

[2]

d Copy and complete the column headings in both tables. [1]
e Estimate the time that would be taken in part **b** for the thermometer to cool from the reading at time $t = 0$ s to room temperature θ_r. [1]
f State in which table the rate of temperature change is the greater. Justify your answer by reference to your readings. [1]
g If this experiment were to be repeated in order to determine an average temperature for each time, it would be important to control the conditions. Suggest two such conditions that should be controlled. [2]

[Total: 10]

(Cambridge IGCSE Physics 0625 Paper 51 Q2 November 2010)

4 In this experiment you will investigate reflection of light through a transparent block.
Carry out the following instructions referring to Figure P5.

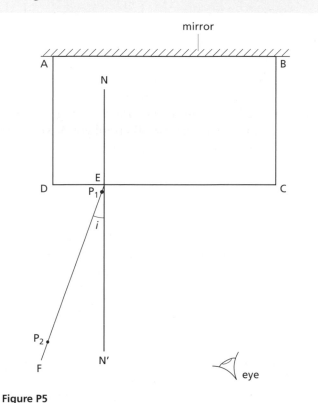

Figure P5

a Place the transparent block, largest face down, on the ray-trace sheet supplied. The block should be on the top half of the paper. Draw the outline of the block and label it **ABCD**.

b Remove the block and draw the normal **NN'** to side **CD** so that the normal is 2.0 cm from **D**. Label the point **E** where **NN'** crosses **CD**.

c Draw the line **EF** at an angle of incidence $i = 20°$ as shown in Figure P5.

d Place the paper on the pinboard. Stand the plane mirror vertically and in contact with face **AB** of the block as shown in Figure P5.

e Push two pins P_1 and P_2 into line **EF**. Pin P_1 should be about 1 cm from the block and pin P_2 some distance from the block.

f Replace the block and observe the images of P_1 and P_2 through side **CD** of the block from the direction indicated by the eye in Figure P5 so that the images of P_1 and P_2 appear one behind the other.

Push two pins P_3 and P_4 into the surface, between your eye and the block, so that P_3, P_4 and the images of P_1 and P_2, seen through the block, appear in line.
Mark the positions of P_1, P_2, P_3 and P_4. Remove the block.

g Continue the line joining the positions of P_1 and P_2 so that it crosses **CD** and extends as far as side **AB**.

h Draw a line joining the positions of P_3 and P_4. Continue the line so that it crosses **CD** and extends as far as side **AB**. Label the point **G** where this line crosses the line from P_1 and P_2.

i Remove the pins, block and mirror from the ray trace sheet. Measure the acute angle θ between the lines meeting at **G**. [1]

j Calculate the difference $(\theta - 2i)$. [1]

k Repeat steps **c** to **j** using an angle of incidence $i = 30°$. [1]

l Theory suggests that $\theta = 2i$. State whether your result supports the theory and justify your answer by reference to your results. [2]

[5]

[Total: 10]

(Cambridge IGCSE Physics 0625 Paper 51 Q4 November 2010)

5 In this experiment, you are to make two sets of measurements as accurately as you can in order to determine the density of glass.
Carry out the following instructions referring to Figure P6.

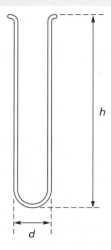

Figure P6

Method 1

a (i) Use the two blocks of wood and the rule to measure the external diameter d of the test tube in cm.

(ii) Draw a labelled diagram to show how you used the blocks of wood and the rule to find, as accurately as possible, a value for the external diameter of the test tube.

(iii) Measure the height h of the test tube in cm.

(iv) Calculate the external volume V_e of the test tube using the equation

$$V_e = \frac{\pi d^2 h}{4}$$ [3]

b Use the balance provided to measure the mass m_1 of the test tube. [1]

c (i) Completely fill the test tube with water. Pour the water into the measuring cylinder and record the volume V_i of the water.

(ii) Calculate the density ρ of the glass using the equation [1]

$$\rho = \frac{m_1}{(V_e - V_i)}$$

Method 2

d (i) Pour water into the measuring cylinder up to about the 175 cm^3 mark. Record this volume V_1.

(ii) Carefully lower the test tube, open end uppermost, into the measuring cylinder so that it floats. Record the new volume reading V_2 from the measuring cylinder.

(iii) Calculate the difference in volumes $(V_2 - V_1)$.

(iv) Calculate the mass m_2 of the test tube using the equation $m_2 = k(V_2 - V_1)$ where $k = 1.0 \text{ g/cm}^3$. [3]

e (i) Use the wooden rod to push the test tube, open end uppermost, down to the bottom of the measuring cylinder so that the test tube is full of water and below the surface. Remove the wooden rod. Record the new volume reading V_3 from the measuring cylinder.

(ii) Calculate the density ρ of the glass using the equation

$$\rho = \frac{m_1}{(V_3 - V_1)}$$ [2]

[Total: 10]

(Cambridge IGCSE Physics 0625 Paper 51 Q1 June 2009)

6 In this experiment, you are to determine the focal length of a converging lens. Carry out the following instructions referring to Figure P7.

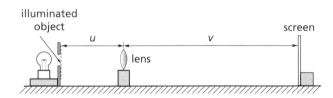

Figure P7

a Place the lens so that its centre is a distance $u = 25.0 \text{ cm}$ from the illuminated object.

b In a copy of the table record the distance u in cm from the centre of the lens to the illuminated object, as shown in Figure P7.

c Place the screen close to the lens. Move the screen away from the lens until a focused image of the object is seen on the screen.

d Measure and record in your table the distance v in cm from the centre of the lens to the screen.

u/cm	v/cm	f/cm

e Calculate and record in your table the focal length f of the lens using the equation

$$f = \frac{uv}{(u + v)}$$ [5]

f Place the lens so that its centre is 45.0 cm from the illuminated object.

g Repeat steps **b** to **e**.

h Calculate the average value of the focal length. [3]

i State and briefly explain one precaution you took in order to obtain reliable measurements. [2]

[Total: 10]

(Cambridge IGCSE Physics 0625 Paper 51 Q4 June 2009)

7 In this experiment, you are to investigate the period of oscillation of a simple pendulum. Carry out the following instructions referring to Figure P8 and Figure P9.
The pendulum has been set up for you. Do not adjust the position of the clamp supporting the pendulum.

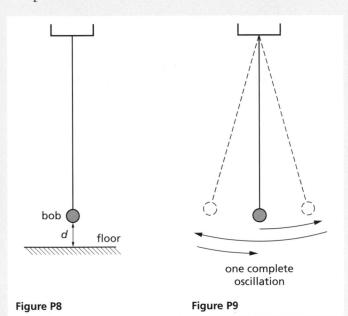

Figure P8   **Figure P9**

a Measure and record in a copy of the table the vertical distance d from the floor to the bottom of the pendulum bob.

b Displace the pendulum bob slightly and release it so that it swings. Measure and record in your table the time t for 20 complete oscillations of the pendulum (see Figure P9).

c Calculate the period T of the pendulum. The period is the time for one complete oscillation. Record the value of T in the table.

d Without changing the position of the clamp supporting the pendulum, adjust the length until the vertical distance d from the floor to the bottom of the pendulum bob is about 20 cm. Measure and record in the table the actual value of d to the nearest 0.1 cm. Repeat steps **b** and **c**.

c Repeat step **d** using d values of about 30 cm, 40 cm and 50 cm.

d/cm	t/s	T/s

[4]

f Plot a graph of T/s (y-axis) against d/cm (x-axis). [5]

g State whether or not your graph shows that T is directly proportional to d. Justify your statement by reference to the graph. [1]

[Total: 10]

(Cambridge IGCSE Physics 0625 Paper 51 Q1 November 2009)

8 In this experiment, you are to compare the combined resistance of lamps arranged in series and in parallel.
Carry out the following instructions, referring to Figure P10 and Figure P11.
The circuit shown in Figure P10 has been set up for you.

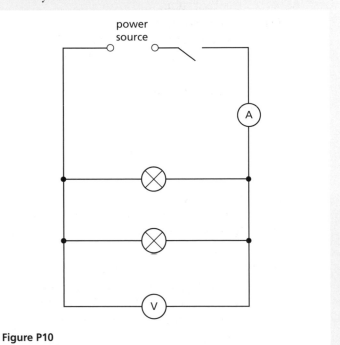

Figure P10

a Switch on. Measure and record in a copy of the table the current I in the circuit and the p.d. V across the two lamps. Switch off.

b Calculate the combined resistance R of the two lamps using the equation

$$R = \frac{V}{I}$$

Record this value of R in your table.

	V/	I/	R/
Figure P10			
Figure P11			

[4]

c Complete the column headings in the table.
d Disconnect the lamps and the voltmeter. Set up the circuit shown in Figure P11.

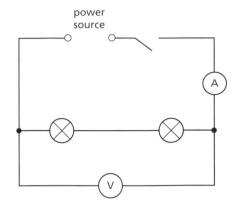

Figure P11

e Switch on. Measure and record in the table the current I in the circuit and the p.d. V across the two lamps. Switch off.
f Calculate the combined resistance R of the two lamps using the equation

$$R = \frac{V}{I}$$

Record this value of R in the table.
g Using the values of resistance obtained in **b** and **f**, calculate the ratio y of the resistances using the equation

$$y = \frac{\text{resistance of lamps in series}}{\text{resistance of lamps in parallel}}$$ [3]

h **(i)** Figure P12a shows a circuit including two motors A and B.
Draw a diagram of the circuit using standard circuit symbols. The circuit symbol for a motor is shown in Figure P12b.

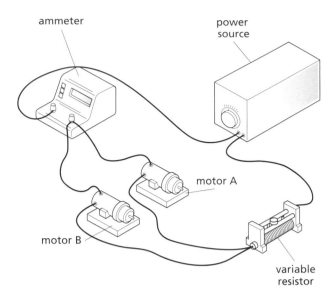

Figure P12a

Figure P12b

(ii) An engineer wishes to measure the voltage across motor A.
On a copy of Figure P12a mark with the letters X and Y where the engineer should connect the voltmeter.
(iii) State the purpose of the variable resistor. [3]

[Total: 10]

(Cambridge IGCSE Physics 0625 Paper 51 Q3 November 2009)

Alternative to practical test questions

1 The IGCSE class is investigating the cooling of water. Figure P13 shows the apparatus used.

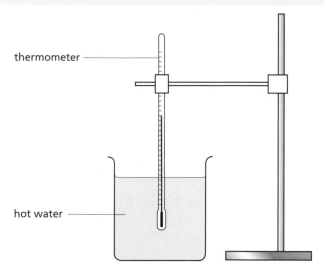

Figure P13

Hot water is poured into the beaker and temperature readings are taken as the water cools.
The table shows the readings taken by one student.

t/s	θ/°C
0	85
30	78
60	74
90	71
120	69
150	67
300	63

a (i) Using the information in the table, calculate the temperature change T_1 of the water in the first 150 s.
(ii) Using the information in the table, calculate the temperature change T_2 of the water in the final 150 s. [3]
b Plot a graph of θ/°C (y-axis) against t/s (x-axis) for the first 150 s. [5]
c During the experiment the rate of temperature change decreases.
(i) Describe briefly how the results that you have calculated in part **a** show this trend.

(ii) Describe briefly how the graph line shows this trend. [2]

[Total: 10]

(Cambridge IGCSE Physics 0625 Paper 61 Q2 June 2010)

2 The IGCSE class is investigating the current in a circuit when different resistors are connected in the circuit.
The circuit is shown in Figure P14. The circuit contains a resistor **X**, and there is a gap in the circuit between points **A** and **B** that is used for adding extra resistors to the circuit.

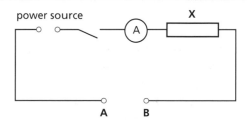

Figure P14

a A student connects points **A** and **B** together, switches on and measures the current I_0 in the circuit.
The reading is shown on the ammeter in Figure P15.
Write down the ammeter reading. [1]

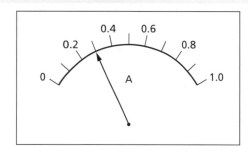

Figure P15

b The student connects a 3.3 Ω resistor between points **A** and **B**, switches on and records the current *I*. He repeats the procedure with a 4.7 Ω resistor and then a 6.8 Ω resistor.
Finally he connects the 3.3 Ω resistor and the 6.8 Ω resistor in series between points **A** and **B**, and records the current *I*.

(i) Complete the column headings in a copy of the table. [1]

R/	V/
3.3	0.23
4.7	0.21
6.8	0.18
	0.15

(ii) Write the combined resistance of the 3.3 Ω resistor and the 6.8 Ω resistor in series in the space in the resistance column of the table. [1]

c Theory suggests that the current will be 0.5 I_0 when the total resistance in the circuit is twice the value of the resistance of resistor **X**. Use the readings in the table, and the value of I_0 from **a**, to estimate the resistance of resistor **X**. [2]

d On a copy of Figure P14 draw two resistors in parallel connected between **A** and **B** and also a voltmeter connected to measure the potential difference across resistor **X**. [3]

[Total: 8]

(Cambridge IGCSE Physics 0625 Paper 61 Q3 November 2010)

3 The IGCSE class is investigating the reflection of light by a mirror as seen through a transparent block. Figure P16 shows a student's ray-trace sheet.

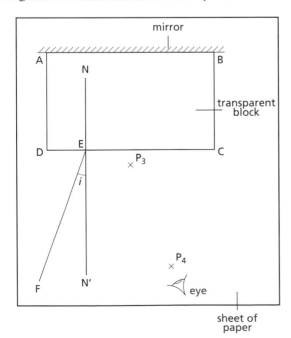

Figure P16

a A student draws the outline of the transparent block **ABCD** on the ray-trace sheet. He draws the normal **NN'** to side **CD**. He draws the incident ray **EF** at an angle of incidence $i = 20°$. He pushes two pins P_1 and P_2 into line **EF** and places the block on the sheet of paper. He then observes the images of P_1 and P_2 through side **CD** of the block from the direction indicated by the eye in Figure P16 so that the images of P_1 and P_2 appear one behind the other. He pushes two pins P_3 and P_4 into the surface, between his eye and the block, so that P_3, P_4 and the images of P_1 and P_2, seen through the block, appear in line. (The plane mirror along side **AB** of the block reflects the light.)

The positions of P_3 and P_4 are marked on Figure P16.

(i) Make a copy of Figure P16. On line **EF**, mark with neat crosses (×) suitable positions for the pins P_1 and P_2.

(ii) Continue the line **EF** so that it crosses **CD** and extends as far as side **AB**.

(iii) Draw a line joining the positions of P_4 and P_3. Continue the line so that it crosses **CD** and extends as far as side **AB**. Label the point **G** where this line crosses the line from P_1 and P_2. [4]

(iv) Measure the acute angle θ between the lines meeting at **G**.

(v) Calculate the difference $(\theta - 2i)$. [2]

b The student repeats the procedure using an angle of incidence $i = 30°$ and records the value of θ as 62°.

(i) Calculate the difference $(\theta - 2i)$.

(ii) Theory suggests that $\theta = 2i$. State whether the results support the theory and justify your answer by reference to the results. [3]

c To place the pins as accurately as possible, the student views the bases of the pins. Explain briefly why viewing the bases of the pins, rather than the tops of the pins, improves the accuracy of the experiment. [1]

[Total: 10]

(Cambridge IGCSE Physics 0625 Paper 61 Q4 November 2010)

4 An IGCSE student is investigating moments using a simple balancing experiment.
He uses a pivot on a bench as shown in Figure P17. First, the student balances the metre rule, without loads, on the pivot. He finds that it does not balance at the 50.0 cm mark, as he expects, but it balances at the 49.7 cm mark.
Load **Q** is a metal cylinder with diameter a little larger than the width of the metre rule, so that it covers the markings on the rule. Load **Q** is placed carefully on the balanced metre rule with its centre at the 84.2 cm mark. The rule does not slip on the pivot.

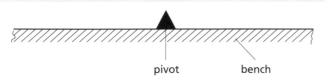

pivot   bench

Figure P17

a Draw on a copy of Figure P17 the metre rule with load **Q** on it. [2]
b Explain, using a labelled diagram, how the student would ensure that the metre rule reading at the centre of **Q** is 84.2 cm. [2]
c Calculate the distance between the pivot and the centre of load **Q**. [1]

[Total: 5]

(Cambridge IGCSE Physics 0625 Paper 61 Q5 June 2009)

5 The IGCSE class is investigating the period of oscillation of a simple pendulum. Figure P18 shows the set-up.

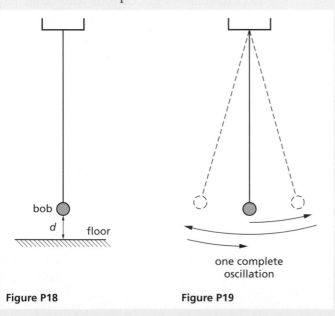

bob

d   floor

one complete
oscillation

Figure P18   **Figure P19**

a (i) On Figure P18, measure the vertical distance *d* from the floor to the bottom of the pendulum bob.
(ii) Figure P18 is drawn one twentieth actual size. Calculate the actual distance *x* from the floor to the bottom of the pendulum bob. Enter this value in the top row of a copy of the table.
The students displace the pendulum bob slightly and release it so that it swings. They measure and record in the table the time *t* for 20 complete oscillations of the pendulum (see Figure P19).

x/cm	t/s	T/s	T²/s²
	20.0		
20.0	19.0		
30.0	17.9		
40.0	16.8		
50.0	15.5		

[4]

b (i) Copy the table and calculate the period *T* of the pendulum for each set of readings. The period is the time for one complete oscillation. Enter the values in the table.
(ii) Calculate the values of T^2. Enter the T^2 values in the table.
c Use your values from the table to plot a graph of T^2/s^2 (*y*-axis) against *x*/cm (*x*-axis). Draw the best-fit line. [5]
d State whether or not your graph shows that T^2 is directly proportional to *x*. Justify your statement by reference to the graph. [1]

[Total: 10]

(Cambridge IGCSE Physics 0625 Paper 61 Q1 November 2009)

6   An IGCSE student is carrying out an optics experiment.
The experiment involves using a lens to focus the image of an illuminated object onto a screen.
a   Copy and complete Figure P20 to show the apparatus you would use. Include a metre rule to measure the distances between the object and the lens and between the lens and the screen. The illuminated object is drawn for you.   [3]

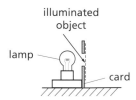

Figure P20

b   State two precautions that you would take to obtain accurate results in this experiment.   [2]

[Total: 5]

(Cambridge IGCSE Physics 0625 Paper 61 Q5 November 2009)

7   The IGCSE class is comparing the combined resistance of resistors in different circuit arrangements. The first circuit is shown in Figure P21.

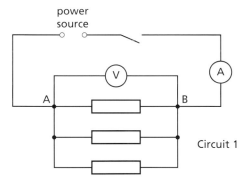

Figure P21

a   The current *I* in the circuit and the p.d. *V* across the three resistors are measured and recorded. Three more circuit arrangements are used. For each arrangement, a student disconnects the resistors and then reconnects them between points **A** and **B** as shown in Figures P22–24.

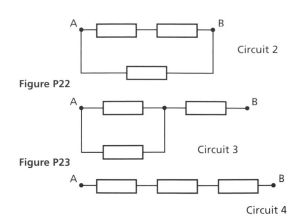

Figure P22

Figure P23

Figure P24

The voltage and current readings are shown in the table.

Circuit	V/	I/	R/
1	1.87	1.68	
2	1.84	0.84	
3	1.87	0.37	
4	1.91	0.20	

(i)   Copy and complete the column headings for each of the *V*, *I* and *R* columns of the table.
(ii)   For each circuit, calculate the combined resistance *R* of the three resistors using the equation

$$R = \frac{V}{I}$$

Record these values of *R* in your table.   [3]

b   Theory suggests that, if all three resistors have the same resistance under all conditions, the combined resistance in circuit 1 will be one half of the combined resistance in circuit 2.
(i)   State whether, within the limits of experimental accuracy, your results support this theory. Justify your answer by reference to the results.
(ii)   Suggest one precaution you could take to ensure that the readings are as accurate as possible.   [3]

[Total: 6]

(Cambridge IGCSE Physics 0625 Paper 61 Q2 June 2008)

8 The IGCSE class is investigating the change in temperature of hot water as cold water is added to the hot water.

A student measures and records the temperature θ of the hot water before adding any of the cold water available.

He then pours 20 cm³ of the cold water into the beaker containing the hot water. He measures and records the temperature θ of the mixture of hot and cold water.

He repeats this procedure four times until he has added a total of 100 cm³ of cold water.

The temperature readings are shown in the table. V is the volume of cold water added.

V/	θ/
0	82
	68
	58
	50
	45
	42

a (i) Copy and complete the column headings in the table.

(ii) Enter the values for the volume of cold water added. [2]

b Use the data in the table to plot a graph of temperature (*y*-axis) against volume (*x*-axis). Draw the best-fit curve. [4]

c During this experiment, some heat is lost from the hot water to the surroundings. Also, each time the cold water is added, it is added in quite large volumes and at random times. Suggest two improvements you could make to the procedure to give a graph that more accurately shows the pattern of temperature change of the hot water, due to addition of cold water alone. [2]

[Total: 8]

(Cambridge IGCSE Physics 0625 Paper 61 Q3 November 2008)

9 a The table shows some measurements taken by three IGCSE students. The second column shows the values recorded by the three students. For each quantity, underline the value most likely to be correct. The first one is done for you.

Quantity measured	Recorded values
The mass of a wooden metre rule	0.112 kg 1.12 kg 11.2 kg
The weight of an empty 250 cm³ glass beaker	0.7 N 7.0 N 70 N
The volume of one sheet of this paper	0.6 cm³ 6.0 cm³ 60 cm³
The time taken for one swing of a simple pendulum of length 0.5 m	0.14 s 1.4 s 14 s
The pressure exerted on the ground by a student standing on one foot	0.4 N/cm² 4.0 N/cm² 40 N/cm²

[4]

b (i) A student is to find the value of the resistance of a wire by experiment. Potential difference V and current I can be recorded. The resistance is then calculated using the equation

$$R = \frac{V}{I}$$

The student knows that an increase in temperature will affect the resistance of the wire.

Assuming that variations in room temperature will not have a significant effect, suggest two ways by which the student could minimise temperature increases in the wire during the experiment. [2]

(ii) Name the circuit component that the student could use to control the current. [1]

[Total: 7]

(Cambridge IGCSE Physics 0625 Paper 61 Q5 November 2008)

10 The IGCSE class is investigating the resistance of a wire. The circuit is as shown in Figure P25.

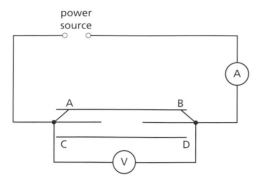

Figure P25

a A student uses the switches to connect the wire **AB** into the circuit and records the p.d. *V* across the wire between **A** and **B**. He also records the current *I* in the wire.
The student then repeats the measurements using the wire **CD** in place of wire **AB**. The readings are shown in the table.

Wire	V/	I/	R/
AB	1.9	0.24	
CD	1.9	0.96	

[3]

(i) Calculate the resistance *R* of each wire, using the equation $R = V/I$.
Record the values in a copy of the table.
(ii) Complete the column headings in your table.
b The two wires **AB** and **CD** are made of the same material and are of the same length. The diameter of wire **CD** is twice the diameter of wire **AB**.
(i) Look at the results in the table. Below are four possible relationships between *R* and the diameter *d* of the wire. Which relationship best matches the results?

R is proportional to *d*

R is proportional to $1/d$

R is proportional to d^2

R is proportional to $1/d^2$

(ii) Explain briefly how the results support your answer in part **b(i)**.   [2]

c Following this experiment, the student wishes to investigate whether two lamps in parallel with each other have a smaller combined resistance than the two lamps in series. Draw one circuit diagram showing
(i) two lamps in parallel with each other connected to a power source,
(ii) an ammeter to measure the total current in the circuit,
(iii) a voltmeter to measure the potential difference across the two lamps.   [3]

[Total: 8]

(Cambridge IGCSE Physics 0625 Paper 61 Q3 June 2007)

11 a An IGCSE student is investigating the differences in density of small pieces of different rocks. She is using an electronic balance to measure the mass of each sample and using the 'displacement method' to determine the volume of each sample. Figure P26 shows the displacement method.

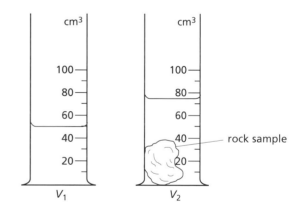

Figure P26

(i) Write down the volume shown in each measuring cylinder.
(ii) Calculate the volume *V* of the rock sample.
(iii) Calculate the density of sample **A** using the equation

$$\text{density} = \frac{m}{V}$$

where the mass *m* of the sample of rock is 109 g.   [4]

b The table shows the readings that the student obtains for samples of rocks **B** and **C**. Copy and complete the table by

    (i) inserting the appropriate column headings with units,

    (ii) calculating the densities using the equation

$$\text{density} = \frac{m}{V}$$

Sample	*m*/g			*V*/	Density/
B	193	84	50	34	
C	130	93	50	43	

[4]

c Explain briefly how you would determine the density of sand grains. [1]

[Total: 9]

(Cambridge IGCSE Physics 0625 Paper 61 Q5 November 2007)

Answers

Higher level questions are marked with *. The questions, example answers, marks awarded and/or comments that appear in this book were written by the authors. In examination the way marks would be awarded to answers like these may be different. Cambridge International Examinations bears no responsibility for the example answers to questions taken from its past question papers which are contained in this publication.

General physics

Measurements and motion

1 Measurements
1 a 10
  b 40
  c 5
  d 67
  e 1000
2 a 3.00
  b 5.50
  c 8.70
  d 0.43
  e 0.1
3 a 1.0×10^5; 3.5×10^3;
    4.28×10^8; 5.04×10^2;
    2.7056×10^4
  b 1000; 2 000 000; 69 200;
    134; 1 000 000 000
4 a 1×10^{-3}; 7×10^{-5};
    1×10^{-7}; 5×10^{-5}
  b 5×10^{-1}; 8.4×10^{-2};
    3.6×10^{-4}; 1.04×10^{-3}
5 10 mm
6 a Two
  b Three
  c Four
  d Two
7 24 cm³
8 40 cm³; 5
9 80
10 a 250 cm³
   b 72 cm³
11 a 53.3 mm
   b 95.8 mm
12 a 2.31 mm
   b 14.97 mm
13 a Metre, kilogram,
     second
   b Different number of
     significant figures
   c (i) πr^2
     (ii) $\frac{4}{3}\pi r^3$
     (iii) $\pi r^2 h$

2 Speed, velocity and acceleration
1 a 20 m/s
  b 6.25 m/s
2 a 15 m/s
  b 900 m
3 2 m/s²
4 50 s
5 a 6 m/s
  b 14 m/s
6 4 s
7 a Uniform acceleration
  b 75 cm/s²
8 a 1 s
  b (i) 10 cm/tentick²
    (ii) 50 cm/s per tentick
    (iii) 250 cm/s²
  c 0
9 A
10 E

3 Graphs of equations
1 a 60 km
  b 5 hours
  c 12 km/h
  d 2
  e $1\frac{1}{2}$ hours
  f 60 km/$3\frac{1}{2}$ h = 17 km/h
  g Steepest line: EF
2 a 100 m
  b 20 m/s
  c Slows down
3 a $\frac{5}{4}$ m/s²
  b (i) 10 m
    (ii) 45 m
  c 22 s
4 a (i) OA, BC: accelerating;
    (ii) DE: decelerating;
    (iii) AB, CD: uniform velocity
  b OA: $a = +80$ km/h²;
    AB: $v = 80$ km/h;
    BC: $a = +40$ km/h²;
    CD: $v = 100$ km/h;
    DE: $a = -200$ km/h²

  c OA 40 km; AB 160 km;
    BC (5 + 40) = 45 km;
    CD 100 km; DE 25 km
  d 370 km
  e 74 km/h
5 a Uniform velocity
  b 600 m
  c 20 m/s

4 Falling bodies
1 a (i) 10 m/s
    (ii) 20 m/s
    (iii) 30 m/s
    (iv) 50 m/s
  b (i) 5 m
    (ii) 20 m
    (iii) 45 m
    (iv) 125 m
2 3 s; 45 m

5 Density
1 a (i) 0.5 g
    (ii) 1 g
    (iii) 5 g
  b (i) 10 g/cm³
    (ii) 3 kg/m³
  c (i) 2.0 cm³
    (ii) 5.0 cm³
2 a 8.0 g/cm³
  b 8.0×10^3 kg/m³
3 15 000 kg
4 130 kg
5 1.1 g/cm³
6 Density of ice is less than density
  of water

Forces and momentum

6 Weight and stretching
1 a 1 N
  b 50 N
  c 0.50 N
2 a 120 N
  b 20 N
3 a 2000 N/m
  b 50 N/m
4 A

ANSWERS

7 Adding forces
1 40 N
2 50 N
3 25 N
4 50 N at an angle of 53° to the 30 N force
5 a 7 N
  b 13 N

8 Force and acceleration
1 D
2 20 N
3 a 5000 N
  b 15 m/s^2
4 a 4 m/s^2
  b 2 N
5 a 0.5 m/s^2
  b 2.5 m/s
  c 25 m
6 a 1000 N
  b 160 N
7 a 5000 N
  b 20 000 N; 40 m/s^2
8 a (i) Weight
    (ii) Air resistance
  b Falls at constant velocity (terminal velocity)

9 Circular motion
1 Force is greater than string can bear
2 a Sideways friction between tyres and road
  b (i) Larger
    (ii) Smaller
    (iii) Larger
3 Slicks allow greater speed in dry conditions but in wet conditions treads provide frictional force to prevent skidding
4 5000 s (83 min)

10 Moments and levers
1 E
2 (i) C
  (ii) A
  (iii) B

11 Centres of mass
1 a B
  b A
  c C
2 Tips to right

12 Momentum
1 a 50 kg m/s
  b 2 kg m/s
  c 100 kg m/s
2 2 m/s
3 4 m/s
4 0.5 m/s
5 2.5 m/s
6 a 40 kg m/s
  b 80 kg m/s
  c 20 kg m/s^2
  d 20 N
7 2.5 m/s

Energy, work, power and pressure

13 Energy transfer
1 a Electrical to sound
  b Sound to electrical
  c k.e. to p.e.
  d Electrical to light (and heat)
  e Chemical to electrical to light and heat
2 A chemical; B heat; C kinetic; D electrical
3 180 J
4 1.5 × 10^5 J
5 a 150 J
  b 150 J
  c 10 W
6 500 W
7 a (300/1000) × 100 = 30%
  b Heat
  c Warms surroundings
8 a Electricity transferred to k.e. and heat
  b Electricity transferred to heat
  c Electricity transferred to sound
9 3.5 kW

14 Kinetic and potential energy
1 a 2 J
  b 160 J
  c 100 000 = 10^5 J
2 a 20 m/s
  b (i) 150 J
    (ii) 300 J
3 a 1.8 J
  b 1.8 J
  c 6 m/s
  d 1.25 J
  e 5 m/s
4 3.5 × 10^9 W = 3500 MW

15 Energy sources
1 a 2%
  b Water
  c Cannot be used up
  d Solar, wind
  e All energy ends up as heat which is difficult to use and there is only a limited supply of non-renewable sources
2 Renewable, non-polluting (i.e. no CO_2, SO_2 or dangerous waste), low initial building cost of station to house energy converters, low running costs, high energy density, reliable, allows output to be readily adjusted to varying energy demands

16 Pressure and liquid pressure
1 a (i) 25 Pa
    (ii) 0.50 Pa
    (iii) 100 Pa
  b 30 N
2 a 100 Pa
  b 200 N
3 a A liquid is nearly incompressible
  b A liquid transfers the pressure applied to it
4 1 150 000 Pa (1.15 × 10^6 Pa) (ignoring air pressure)
5 a Vacuum
  b Atmospheric pressure
  c 740 mmHg
  d Becomes less; atmospheric pressure lower

6 E
7 B

Thermal physics
Simple kinetic molecular model of matter

17 Molecules
1 B
2 a Air is readily compressed
  b Steel is not easily compressed

18 The gas laws
1 a $15\,cm^3$
  b $6\,cm^3$

Thermal properties and temperature

19 Expansion of solids, liquids and gases
2 Aluminium
3 B
4 A

20 Thermometers
1 a $1530\,°C$
  b $19\,°C$
  c $0\,°C$
  d $-12\,°C$
  e $37\,°C$
2 C
3 a Property must change continuously with temperature
  b Volume of a liquid, resistance, pressure of a gas
  c (i) Platinum resistance
    (ii) Thermocouple
    (iii) Alcohol

21 Specific heat capacity
1 $15\,000\,J$, $1500\,J/°C$
2 A = $2000\,J/(kg\,°C)$;
  B = $200\,J/(kg\,°C)$;
  C = $1000\,J/(kg\,°C)$
3 Specific heat capacity of jam is higher than that of pastry so it cools more slowly

22 Specific latent heat
1 a $3400\,J$
  b $6800\,J$
2 a $5 \times 340 + 5 \times 4.2 \times 50$
    $= 2750\,J$
  b $1700\,J$
3 $680\,s$
4 a $0\,°C$
  b $45\,g$
5 a $9200\,J$
  b $25\,100\,J$
6 $157\,g$
7 a Ice has a high specific latent heat of fusion
  b Water has a high specific latent heat of vaporisation
8 Heat drawn from the water when it evaporates
9 Heat drawn from the milk when the water evaporates
10 $1200\,J$

Thermal processes

23 Conduction and convection
1 a Newspaper is a poor conductor of heat
  b The fur would trap more air, which is a good insulator, and so keep wearer warmer
  c Holes in a string vest trap air, which is a poor conductor, next to the skin
3 a If small amounts of hot water are to be drawn off frequently it may not be necessary to heat the whole tank
  b If large amounts of hot water are needed it will be necessary to heat the whole tank
4 Metal is a better conductor of heat than rubber

24 Radiation
1 Black surfaces absorb radiation better than white ones so the ice on the black sections of the canopy melts faster than on the white sections

2 a The Earth radiates energy back into space
  b Clouds reduce the amount of energy radiated into space, keeping the ground warmer

Properties of waves
General wave properties

25 Mechanical waves
1 a $1\,cm$
  b $1\,Hz$
  c $1\,cm/s$
2 A, C
3 a Speed of ripple depends on depth of water
  b AB since ripples travel more slowly towards it, therefore water shallower in this direction
4 a Trough
  b (i) $3.0\,mm$
    (ii) $15\,mm/s$
    (iii) $5\,Hz$

Light

26 Light rays
1 Larger, less bright
2 a Four images
  b Brighter but blurred
3 C
4 Before; sound travels slower than light

27 Reflection of light
1 a $40°$
  c $40°$, $50°$, 50
  d Parallel
2 A
3 Top half

28 Plane mirrors
1 B
2 D
3 $4\,m$ towards mirror
4 B

29 Refraction of light
3 $250\,000\,km/s$
4 C
6 E
7 A

30 Total internal reflection

1 a Angle of incidence = 0
  b Angle of incidence > critical angle
3 Periscope, binoculars
4 a Ray passes into air and is refracted away from the normal
  b Total internal reflection occurs in water
5 48.6°

31 Lenses

1 Parallel
2 a Converging
  c Image 9 cm from lens, 3 cm high
3 Distance from lens:
  a beyond 2F
  b 2F
  c between F and 2F
  d nearer than F
4 Towards
5 a 4 cm
  b 8 cm behind lens, virtual, $m = 2$
6 A: converging $f = 10$ cm
  B: converging $f = 5$ cm

32 Electromagnetic radiation

1 a 0.7 μm
  b 0.4 μm
2 a B
  b D
3 a Ultraviolet
  b Microwaves
  c Gamma rays
  d Infrared
  e Infrared/microwaves
  f X-rays
4 a 3 m
  b 2×10^{-4} s
5 E

Sound

33 Sound waves

1 1650 m (about 1 mile)
2 a $2 \times 160 = 320$ m/s

b $240/(3/4) = 320$ m/s
c 320 m
3 a Reflection, refraction, diffraction, interference
  b Vibrations are perpendicular to rather than along the direction of travel of the wave; longitudinal
4 b (i) 1.0 m
    (ii) 2.0 m

Electricity and magnetism

Simple phenomena of magnetism

34 Magnetic fields

1 C

Electrical quantities and circuits

35 Static electricity

1 D
2 Electrons are transferred from the cloth to the polythene
3 C

36 Electric current

1 a 5 C
  b 50 C
  c 1500 C
2 a 5 A
  b 0.5 A
  c 2 A
3 B
4 C
5 All read 0.25 A

37 Potential difference

1 a 12 J
  b 60 J
  c 240 J
2 a 6 V
  b (i) 2 J
    (ii) 6 J
3 B
4 b Very bright
  c Normal brightness
  d No light
  e Brighter than normal

f Normal brightness
5 a 6 V
  b 360 J
6 $x = 18$, $y = 2$, $z = 8$

38 Resistance

1 3 Ω
2 20 V
3 C
4 A = 3 V; B = 3 V; C = 6 V
5 2 Ω
6 a 15 Ω
  b 1.5 Ω
7 D
8 a (i) ohm's law
    (ii) 2 Ω
9 B

39 Capacitors

2 a (i) Maximum
    (ii) Zero
  b (i) Maximum
    (ii) Zero

40 Electric power

1 a 100 J
  b 500 J
  c 6000 J
2 a 24 W
  b 3 J/s
3 C
4 2.99 kW
5 Fuse is in live wire in a but not in b
7 a 3 A
  b 13 A
  c 13 A
8 40p
9 a (i) 2 kW
    (ii) 60 W
    (iii) 850 W
  b 4 A

41 Electronic systems

1 b L_1 lights, L_2 does not
  c L_1 and L_2 light
  d L_1 lights, L_2 does not
2 a $V_1 = V_2 = 3$ V
  b $V_1 = 1$ V, $V_2 = 5$ V
  c $V_1 = 4$ V, $V_2 = 2$ V

42 Digital electronics
1 **A** AND
  B OR
  C NAND
  D NOR
2 **A** OR
  B NOT
  C NAND
  D NOR
  E AND

Electromagnetic effects

43 Generators
1 **a** A: slip rings, B: brushes
  b Increase the number of turns on the coil, the strength of the magnet and the speed of rotation of the coil.
2 The galvanometer needle swings alternately in one direction and then the other as the rod vibrates. This is due to a p.d. being induced in the metal rod when it cuts the magnetic field lines; current flows in alternate directions round the circuit as the rod moves up or down

44 Transformers
2 B
3 **a** 24
  b 1.9 A
4 B

45 Electromagnets
1 **a** North
  b East
2 S
3 **a** To complete the circuits to the battery negative
  b One contains the starter switch and relay coil; the other contains the relay contacts and starter motor
  c Carries much larger current to starter motor
  d Allows wires to starter switch to be thin since they only carry the small current needed to energise the relay

46 Electric motors
1 E
2 Clockwise
3 E

47 Electric meters
3 **a** 0–5 V, 0–10 V
  b 0.1 V
  c 0–5 V
  d Above the 4
  e Parallax error introduced

48 Electrons
1 **a** A −ve, B +ve
  b Down
2 **a** 1.6×10^{-16} J
  b 1.9×10^7 m/s

Atomic physics

49 Radioactivity
1 **a** α
  b γ
  c β
  d γ
  e α
  f α
  g β−
  h γ
2 25 minutes
3 D

50 Atomic structure
1 B
2 C (symbol is ^7_3Li)

● Revision questions
1 E
2 A
3 C
4* **a** Yes, 1 mm = 0.001 m
  b E
5 D
6 E
7 A
8 B
9 C
10 D
11 D
12 **a** B
  b A
13 A
14 A
15 C
16 B
17 D
18 C
19 C
20 C
21 E
22 **a** Become circular
  b No change
  c No change
23 **a** (i) Infrared
     (ii) X-rays
  b (i) Radio
     (ii) γ-rays
24 **a** Longitudinal
  b (i) Compression
     (ii) Rarefaction
25 **a** 60°
  b 30°
26 **a** Refraction
  b POQ
  c Towards
  d 40°
  e 90−65 = 25°
27 C
28 **a** Dispersion
  b (i) Red
     (ii) Violet
29 B
30 A
31 D
32 B
33 C
34 **a** 1 Ω
  b 3 A
  c 6 V
35 **a** 3 Ω
  b 2 A
  c 4 V across 2; 2 V across 1
36 D
37 B
38 **a** E
  b A
  c C
  d B
  e D
39 E

40 E
41 E
42 B
43 B
44 a C
   b A
   c B
   d A
45 a D
   b E

Cambridge IGCSE exam questions

1 General physics

Measurements and motion

1 a (i) 6 cm and 5 cm
   (ii) 60 cm³
   b 2.65 g/cm³
2* Time 10 cycles and calculate the average

3 a

Distance	Tape measure
Time	Stopwatch

   b Speed = distance/time
   c (i) Some distances at slower speeds
   (ii) 22 km
4 a (i) 1 Increasing
   2 Constant
   c Zero distance
5 a 400 s
   d 10.8 m/s
6* a (i) 1.6 s
   (ii) 4.2 s
   (iii) 32 m
   (iv) 70–95 m (area under graph)
   b (i) Weight of ball down, air resistance up
   (ii) Up force = down force

Forces and momentum

7* b 3 N reading
   d Straight line through the origin shows Hooke's law
   e Graph curves

f Exceeded elastic limit
8* a Limit of proportionality
   b Force proportional to extension
   c OQ extension proportional to force
     QR extension/unit force greater
   d 4.0 N/mm
9* b 98 N–102 N
   c Vertically upwards
   d 98 N–102 N
10* c Mass × distance
11 a (i) At A
   (ii) Greatest distance from the hinge
   b When centre of mass is outside base
   c (i) Less than
   (ii) Centre of mass of matchbox has been raised
12 a Force, perpendicular distance from pivot
   b (i) Force, moment
   (ii) $F_1 + F_2 + W$
   (iii) F
13* a Student B: force inversely proportional to mass
   b $F = ma$
   c (i) Nothing or as before
   (ii) Slows down
   (iii) Moves in a circle
14* a The direction is changing
   b (i) Force needed to change direction
   (ii) Towards the centre
   (iii) Friction between tyres and the road
15* a (i) Resultant force
   (ii) To overcome friction
   b 0.8 kg
   c 0.875 m/s²
   d (i) 0.6 m/s
   (ii) 0.36 m
16* a (ii) It gets larger
   b (ii) Friction is too small
   c (i) Constant speed
   (ii) 212.5 cm
   (iii) 8.33 cm/s

Energy, work, power and pressure

17 a $I = U + W$
   b (i) 850 N
   (ii) Force needed to get it started
   (iii) Height
   (iv) Time
   c Greater than
18* a 405 000 J
   b 60 000 J
   c 60 000 W
   d Chemical
   e Energy lost as heat, sound, etc.
19 a Tidal, wave, hydroelectric
20 a 88–92
   b 88–92 mm
   c 840
21* a Volume reduced, pressure goes up
   b 20 cm³
   c Speed of particles greater at higher temperature
22 b (i) Falls
   (ii) Air molecules cause pressure on mercury
   d

rises	rises
falls	stays the same

23* a (i) 540 kJ
   (ii) $W = E/t$, 54 kW
   b (i) 3750 kg
   (ii) 12.5%

2 Thermal physics

Simple kinetic and molecular model of matter

24* b Air molecules hit dust particles
   c Slower movement
25 a Solid: 2, 3 and 6
     Gas: 1, 4 and 5
   b Molecules break free of surface

Thermal properties and temperature

26* a Energy needed to change state

304

b Any time between 1.6 min and 18 min

c P.e of molecules increases and they escape from the liquid

d (i) 480 kJ
(ii) 6.65 kg

27* a Copper or constantan
Copper or constantan
Constantan or copper

28* a Heat required to produce 1 °C rise in 1 kg

b Long time to heat up

c (i) 1.8 °C and 77.1 °C
(ii) 1512 J
(iii) 392 J/kg K

29* a (i) 1 Melting point of ice
2 Pure melting ice
3 0 °C
(ii) 1 Boiling point of water
2 Steam
3 100 °C

b Thermal capacity

30* a (i) Funnel no longer giving heat to ice
(ii) Better contact between heater and ice

b Mass of beaker

c 338 J/g

31* a Total mass before ice added
Total mass after all ice melted

b (i) Mass × sp. heat capacity × change in temp
(ii) Mass × sp. latent heat of fusion of ice

c 427 J/g

32 a °C

3 Properties of waves

Light

33 a 10 cm

b Gets smaller and closer to lens

c (i) Principal focus

34* a A

b Air

c 42°–43°

d Total internal reflection

e 58.7°

f 2.01343×10^8

35 a q

c Inverted, real

d Same

e (i) Nothing
(ii) Blurred image

36 c (i) 2 m
(ii) 2 m away from mirror

37* b Virtual, inverted, same size as object

c Ray strikes glass normally

d 2×10^8 m/s

e i is greater than c so total internal reflection occurs

38* a (ii) Virtual, upright, same size, same distance from mirror

Sound

39 a (i) One sound
(ii) 495 m

b (i) One sound plus echo
(ii) 1.5 s and 4.5 s

40 a (i) Decreasing
(ii) Waves get smaller

b (i) Nothing
(ii) Wavelength the same

c (i) 12–14
(ii) 1 300 waves per second
2 1/300 s
3 0.04 s

d (i) Yes
(ii) Yes
(iii) No

41 a One sound plus echo

b First

c (i) 3 s
(ii) 9 s
(iii) 6 s

4 Electricity and magnetism

Simple phenomen of magnetism

42 a (i) Iron rod
(ii) Plastic rod

b S   S   N

43 a (i) N at left and S at right
(ii) They attract

b (i) N at left and S at right

(ii) They attract

c They attract

d Nothing

Electrical quantities and circuits

44 a (i) Water conducts electricity
(ii) Cord not a conductor

b 10 A

c (i) Larger current
(ii) Cable would melt

45* a (i) X negative; Y positive
(ii) +ve charge on A attracts −ve charge on B
(iii) B is neutral

b (i) Nothing
(ii) +ve charge is cancelled

46 a (i) 6 V
(ii) 50 mA

b 120 Ω

47 a 60 Ω

c (i) 0.025 A
(ii) 1.5 V

d (i) Decreases
(ii) Decreases
(iii) 60 Ω

48* c (i) One input is high and output is low
(ii) 1 On
2 Off

49 a (i) Series
(ii) 12 Ω
(iii) 0.5 A
(iv) 5 V
(v) 5 V

b (i) 1 6 V
2 0 V

50* b (i) 3 A
(ii) 4 Ω
(iii) 2 Ω
(iv) 1080 J

51* a Circuit 1: series
Circuit 2: parallel

c 12 V

d 2.4 Ω

e (i) 3 A
(ii) 24 W
(iii) 7200 J

52 a Interchange connections on ammeter or battery
   b Current
   d (i) Voltmeter
   e 0.4 A
   f 0.4 A
   g (i) 7.5 Ω
      (ii) Increases

Electromagnetic effects

53* a (i) Step-up transformer
      (ii) Less heat/energy lost
   b 2.5 A
   c 18.75 W
   d 7.5 V
   e 21 985 V

54* a First finger – field
    Second finger – current
   b (i) Contact
        Commutator
     (ii) Clockwise

55 a (ii) Iron
    (iii) Magnetic linkage
   b 120 V

56 a (i) e.m.f. induced in **AB** cancelled by e.m.f. induced in **BC**
    (ii) Straighten out **ABC**
   b Transformer, generator, dynamo, microphone, alternator

57* b (i) Reduced
    (ii) Same or none
   c (i) Thin wire is a current-carrying conductor in a magnetic field
    (ii) Towards the thick wire
    (iii) Smaller force

58 a Contact position at centre of potential divider
   b Current in coil magnetises core, armature pivots closing contacts

59 a (ii) Iron bar
   b Rods become magnetised and repel

60* a Magnetic field cut by conductor induces a current
   c Move magnet in and out of solenoid
   d Move magnet faster, stronger magnet, more turns of solenoid

5   Atomic physics

61 a Alpha and beta
   b Gamma
   c Radio
   d Alpha

62* a Background radiation
   b A Only background as reading constant
    B Gamma as not affected by magnetic field
    C Beta as deflected by magnetic field

63* a Beta – third and fourth column
    Gamma – first column

64 a Between 22 and 27 minutes
   b (i) Iodine-128
    (ii) Radon-220 as shortest half-life

65* a Protons: 17 and 17
    Neutrons: 18 and 20
    Electrons: 17 and 17
   b Alpha, beta and gamma

66 a 84
   b 218
   c (i) 2
    (ii) 4
    (iii) Alpha particle

67* A rebounds
   B carries on, slightly deflected
   C carries straight on

68 a Between 18 and 20 minutes
   b (i) About 922
    (ii) Between 18 and 20 minutes
   c Alpha or beta

69 a Electrons
   b Moves towards P_1
   c By making P_3 or P_4 positive
   d Fluorescent screen

70* a Measure background reading
    No aluminium – take count
    Aluminium – take count
    Subtract background reading
   b Count decreases with more aluminium

● Mathematics for physics

1 a 3
   b 5
   c 8/3
   d 20
   e 12
   f 6
   g 2
   h 3
   i 8

2 a $f = v/\lambda$
   b $\lambda = v/f$
   c $I = V/R$
   d $R = V/I$
   e $m = d \times V$
   f $V = m/d$
   g $s = vt$
   h $t = s/v$

3 a $I^2 = P/R$
   b $I = \sqrt{(P/R)}$
   c $a = 2s/t^2$
   d $t^2 = 2s/a$
   e $t = \sqrt{(2s/a)}$
   f $v = \sqrt{(2gh)}$
   g $y = D\lambda/a$
   h $\rho = AR/l$

4 a 10
   b 34
   c 2/3
   d 1/10
   e 10
   f 3×10^8

5 a 2.0×10^5
   b 10
   c 8
   d 2.0×10^8
   e 20
   f 300

6 a 4
   b 2
   c 5
   d 8
   e 2/3
   f −3/4

g 13/6
h −16
i 1
7 $a = (v - u)/t$
  a 5
  b 60
  c 75
8 $a = (v^2 - u^2)/2s$
9 b Extension ∝ mass because the graph is a straight line through the origin
10 b No: graph is a straight line but does not pass through the origin
  c 32
11 a Graph is a curve
  b Graph is a straight line through the origin, therefore $s \propto t^2$ or $s/t^2 = a$ constant $= 2$

● Alternative to practical test questions

1 a (i) T_1 18 °C
   (ii) T_2 4 °C
  c (i) T_1 is much greater than T_2
   (ii) Graph has a decreasing gradient
2 a 0.3
  b (i) Ω A
   (ii) 10.1
  c 10 Ω
3 b (i) 2°
   (ii) Yes, results are close enough
  c Doesn't matter if pins not vertical

4 c 34.5 cm
5 a (i) 0.5 cm
   (ii) 10 cm
  b

T/s	T^2/s^2
1.0	1.0
0.95	0.90
0.9	0.81
0.84	0.71
0.78	0.61

7 a (i) V, A, Ω
   (ii) 1.11, 2.19, 5.05, 9.55
  b (i) Yes, as within 10%
8 a (i) cm³, °C
   (ii) 20, 40, 60, 80, 100
  c Avoid heat loss to the surroundings
9 a 0.7 N, 6 cm³, 1.4 s, 4.0 N/cm³
  b (i) Minimum current, switch off regularly, turn down power supply
   (ii) Variable resistor or rheostat
10 a (i) 7.92 Ω, 1.98 Ω
   (ii) V, A, Ω
  b (i) R is proportional to $1/d^2$
   (ii) The first R is about ¼ of the second
11 a (i) 50 cm³, 75 cm³
   (ii) 25 cm³
   (iii) 4.36 g/cm³
  b (i) V_2/cm³, V_1/cm³, cm³, g/cm³
   (ii) 5.66 g/cm³, 3.02 g/cm³
  c Same method but lots of grains

Index

A

absolute zero 77
absorption of radiation 102
acceleration 9–10
   equations of motion 14–15
   force and 31–2
   of free fall (g) 18–19, 32
   from tape charts 10–11
   from velocity-time graphs 13
   mass and 31–2
   uniform 10, 11, 13, 14–15
acid rain 60
action-at-a-distance forces 24, 32, 155
action at points 153, 154
activity, radioactive material 233
air
   convection 99–100
   density 22
   as insulator 98
   weight 76
air bags 58
air resistance 17, 18, 33
alcohol-in-glass thermometers 85
alpha particles 231–2, 240
   alpha decay 240
   particle tracks 232
   scattering 238
alternating current (a.c.) 159–60
   capacitors in a.c. circuits 176
   frequency 160, 201
   mutual induction 204
   transmission of electrical power
      206–7
alternative energy sources ix, 60–2,
   63–4
alternators (a.c. generators) 200–1,
   201–2
aluminium, specific heat capacity 89
ammeters 158, 164, 219
ammeter-voltmeter method 168
ampere (A) 158
amplitude of a wave 107, 136
analogue circuits 193
analogue meters 193
AND gates 194
angle of incidence 108, 116, 126
angle of reflection 108, 116, 126
anode 187, 222
antineutrinos 240
area 3–4
armatures 216
atmospheric pressure 69, 76
atomic bombs 242
atomic (proton) number 239
atomic structure 151, 239
   nuclear model 238
   nuclear stability 240
   'plum pudding' model 238

Rutherford-Bohr model 241
Schrödinger's model 241–2
atoms 72
attraction forces, electrical charge 24,
   150, 152–3
audibility, limits of 141
average speed 9

B

background radiation 230, 235
balances 4–5
balancing tricks 45–6
banking of roads 36
barometers 69–70
base 188
base-emitter path 189
batteries 50, 158, 162, 163
beam balances 4
beams, balancing 39
beams, of light 113
Becquerel, Henri 230
beta particles 231–2, 240
   beta decay 240
   particle tracks 232
bicycle dynamos 202
bimetallic strips 82
biofuels 62
biogas 62
body heat 98
Bohr, Niels 241
boiling point 94
Bourdon gauges 69, 76
Boyle's law 78, 79, 80
Brahe, Tycho xi
brakes, hydraulic 68
braking distances 58
Brownian motion 72, 73
brushes
   in electric motors 216, 217
   in generators 200, 201
bubble chambers 233
buildings, heat loss in 98, 100–1
burglar alarms 213

C

calibration, thermometers 85
capacitance 174
capacitors 174
   charging and discharging 175
   in d.c. and a.c. circuits 176
carbon dating 235, 236
carbon dioxide emissions 60
carbon microphones 213
cars
   alternators 202
   braking distances 58
   hydraulic brakes 68
   rounding bends 36
   safety features 49, 58

speedometers 207
cathode ray oscilloscopes (CRO) 224–5
   musical note waveforms 142–3
   uses 225–6
cathode rays 222
cathodes 187, 222
cells 158, 163
   see also batteries
Celsius scale 85
   relationship to Kelvin scale 77
centre of gravity see centre of mass
centre of mass 43–6
   stability 44–5, 283
   toppling 44, 283
centripetal force 35–6
chain reactions 242
changes of state 91
charge, electric see electric charge
Charles' law 76, 79
chemical energy 50, 51
circuit breakers 181, 213
circuit diagrams 158
circuits
   current in 158–9
   household circuits 180–2
   model of circuit 162
   parallel 158, 159, 164, 170, 180
   safety 181–2
   series 158, 159, 164, 169–70
circular motion 35
   centripetal force 35–6
   satellites 36–8
clinical thermometers 86
cloud chambers 232
coastal breezes 100
coils
   in electric motors 216
   magnetic fields due to 210
   in transformers 204–5
collector 188
collector-emitter path 189
collisions
   elastic and inelastic 57–8
   impulse and 48–9
   momentum and 47
combustion of fuels 54
communication satellites 37
commutators
   in dynamos 201
   in electric motors 216, 217
compasses 146, 147
compressions 140
computers, static electricity and 154
condensation 94
conduction of heat 97–8
conductors (electrical) 151, 152
   metallic 169
   ohmic and non-ohmic 169, 188

conservation of energy 53, 57
conservation of momentum 47–8
constant of proportionality 280
constant-volume gas thermometers 86
continuous ripples 107
continuous spectra 241–2
convection 99–100
convection currents 99, 100
convector heaters 179
conventional current 158
converging lenses 129, 130, 132
cooling, rate of 103–4, 283–4
Copernicus, Nicolaus xi
coulomb (C) 158
count-rate, GM tube 230
couples, electric motors 216
crests of waves 107
critical angle 126–7
critical temperatures of gases 94–5
critical value, chain reactions 242
crude oil 54
crumple zones 49, 58
crystals 74
current *see* electric current

D
dataloggers 5, 11
d.c. generators (dynamos) 201, 202
decay curves 233
deceleration 10
declination 148
deflection tubes 223
degrees, temperature scales 85
density 21–3
  of water 83
dependent variables 281
depth, real and apparent 123
deuterium 239, 243
deviation of light rays 124
diaphragms, in steam turbines 63
dielectric 174
diffraction
  of electromagnetic waves 137
  of mechanical waves 108–9, 110
  of sound waves 140
diffuse reflection 117–18
diffusion 74–5
diffusion cloud chambers 232
digital circuits 193
digital meters 193
diodes 169, 187–8
direct current (d.c.) 159
  capacitors in d.c. circuits 176
direct proportionality 280
dispersion of light 124, 136
displacement 9
displacement-distance graphs 106
distance-time graphs 10, 14, 19
diverging lenses 129, 132
double insulation 181–2

drop-off current 212
dynamic (sliding) friction 29
'dynamo rule' 200
dynamos 201, 202

E
Earth, magnetic field 148
earthing 153, 181
echoes 141
  ultrasonic 143
eddy currents 206, 207
efficiency 53
  of electrical power transmission 207
  of motors 178
  of power stations 63
effort 40
Einstein, Albert xi, 242
elastic collisions 57–8
elastic limit 25
elastic potential energy 50
electrical energy 50, 182
  production 61, 62, 63–4
  transfer 51, 162, 163, 177, 207
electric bells 211–12
electric charge 150, 158
  attraction forces 24, 150, 152–3
  current and 157
  electrons and 151, 152
  see also static electricity
electric circuits *see* circuits
electric current 157, 158
  alternating (a.c.) 159–60, 176, 201, 204, 206–7
  in circuits 158–9
  direct (d.c.) 159, 176
  effects of 157
  electrons in 157, 158, 162
  from electromagnetic induction 199
  magnetic fields and 157, 209–10, 215
  measurement 158–9
  in transistors 189
electric fields 155–6
  deflection of electron beams 223
  deflection of radiation 232
electricity
  dangers of 182–3
  generation *see* power stations; renewable energy sources
  heating 179
  lighting 178
  paying for 182
  transmission 206–7
  see also static electricity
electricity meters 182
electric motors 178, 215–18
electric power 177–8
electric shock 182–3
electrolytic capacitors 174
electromagnetic induction 199
  applications 202–3

generators 200–2
  mutual 204
electromagnetic radiation 51, 135–9
  dual nature 227
  gamma rays 135, 138, 231–2, 235, 240
  infrared 102, 135, 136
  microwaves xii, 135, 137–8
  properties 135
  radio waves 135, 137
  ultraviolet 102, 135, 136–7
  X-rays xi, 135, 138, 226–7
  see also light
electromagnetic spectrum 135
electromagnetism 209–10
  electromagnet construction 210–11
  magnetisation and demagnetisation 210
  uses 211–13
electromotive force (e.m.f.) 163
electronic systems 185
  impact on society 196–8
  input transducers 185, 186
  output transducers 185, 186–7
electron microscopes viii, 72
electrons 151, 239
  cathode rays 222
  deflection of beams 222–3
  electric current 157, 158, 162
  energy levels 241
  photoelectric emission 227
  thermionic emission 222
  see also atomic structure
electrostatic induction 152
elements, electric heating devices 179
emission of radiation 102–3
emitter 188
endoscopes 128
energy
  conservation of 53, 57
  of electromagnetic radiation 135
  forms of 50–1
  losses in buildings 98, 100–1
  losses in transformers 205–6
  sources *see* energy sources
  transfer of *see* transfers of energy
  see also specific types of energy e.g. kinetic energy; nuclear energy *etc*
energy density of fuels 60
energy levels, electrons 241
energy sources
  alternative sources ix, 60–2, 63–4
  consumption figures 64–5
  economic, environmental and social issues 64–5
  food 50, 53–4
  non-renewable 60, 62–3
  renewable 60–2, 63–4
energy value of food 53–4

equations
  changing the subject of 279–80
  heat equation 88
  of motion 14–15
  wave equation 107
equilibrium
  conditions for 39, 41
  states of 44–5
errors
  parallax 2–3
  systematic 5–6
ethanamide, cooling curve 91
evaporation
  conditions for 93–4
  cooling by 94, 283–4
evidence xi–xii
expansion 81–2
expansion joints 81
explosions 48
extended sources of light 114
eyes 132–3

F
facts viii
falling bodies 17–20
  terminal velocity of 33
Faraday's law 199
farad (F) 174
ferro-magnetics 146
field lines 146–8, 209
filament lamps 169, 178
filaments 222
fire alarms 82
fission, nuclear 242
fixed points, temperature scales 85
Fleming's left-hand rule 216, 218, 222
Fleming's right-hand rule 200
floating 22–3
flue-ash precipitation 154
fluorescent lamps 178
focal length 130
food, energy from 50, 53–4
force 24, 27
  acceleration and 31–2
  action-at-a-distance forces 24, 32,
    155
  addition of 27–8
  of attraction 24, 150, 152–3
  centripetal 35–6
  on current-carrying wire 215
  equilibrium 39, 41
  friction 29, 36
  moments 39–41
  momentum and 48
  Newton's first law 30
  Newton's second law 31–2, 48
  Newton's third law 32–3
  parallelogram law 27–8
force constant of a spring 25–6
force-extension graphs 25

force multipliers 67
forward-biased diodes 187
fossil fuels 60, 64
'free' electrons 98
free fall, acceleration of (g) 18–19, 32
freezing points 91
frequency
  alternating current 160, 201
  light waves 136
  measurement by CRO 226
  mechanical waves 106, 107
  pendulum oscillations 5
  sound waves 141
friction 29, 36
fuels 50, 54, 60, 64
  see also energy sources
fulcrum 39, 40
full-scale deflection 220
fundamental frequency 142
fused plugs 181
fuses 179, 180
fusion
  nuclear 243
  specific latent heat of 91–2, 93

G
Galileo xi, 17, 30
galvanometers 219
gamma rays 135, 138, 231–2, 235, 240
gases
  diffusion 74–5
  effect of pressure on volume 78, 79
  effect of temperature on pressure
    77, 79
  effect of temperature on volume
    76, 79
  kinetic theory 73, 79–80
  liquefaction 94–5
  pressure 76–80
gas laws 76–9
gas turbines 63
Geiger, Hans 238
Geiger-Müller (GM) tube 230, 232,
  233–4
generators 200–2
geostationary satellites 37
geothermal energy 62
glass
  critical angle of 126
  refraction of light 122
gliding 100
gold-leaf electroscope 151, 230
gradient of straight line graphs 281
graphs 281–2
gravitational fields 32
gravitational potential energy 50, 56
gravity 24, 32
  centre of see centre of mass
greenhouse effect 60, 103
greenhouses 103

H
half-life 233–4, 236
hard magnetic materials 146
hard X-rays 226
head of liquid 69
head restraints 58
heat 50, 51
  conduction 97–8
  convection 99–100
  expansion 81–2
  from electric current 157
  latent heat 91–3
  loss from buildings 98, 100–1
  radiation 102–4
  specific heat capacity 88–90
  temperature compared 87
heat equation 88
heaters
  electrical 179
  logic gate control of 195
heat exchangers, nuclear reactors 242,
  243
heating, electric 179
heating value of fuels 54
hertz (Hz) 106, 160, 201
Hooke's law 25–6, 283
household electrical circuits 180–2
Hubble Space Telescope viii, xi
Huygens' construction 109
hydraulic machines 67–8
hydroelectric energy 61, 64
hydrogen
  atoms 151
  isotopes of 239
hydrogen bombs 243

I
ice, specific latent heat of fusion 92
ice point 85
images 114
  converging lenses 130
  plane mirrors 119–20
impulse 48–9
incidence, angle of 108, 116, 126
independent variables 281
induced current see electromagnetic
  induction
induction
  electromagnetic 199–203
  electrostatic 152
  mutual 204
induction motors 216
inelastic collisions 57–8
inertia 30
infrared radiation 102, 135, 136
inkjet printers 154, 155
input transducers (sensors) 185, 186
insulators (electrical) 151, 152
insulators (heat) 98
integrated circuits 188

intensity of light 136
interference
   mechanical waves 110–11
   sound waves 140
internal energy *see* heat
International Space Station ix
inverse proportionality 78, 280–1
inverter (NOT gate) 193–4
investigations x–xi, 283–4
ionisation 227, 230
ionisation energy 241
ionosphere 137
ions 230
iron, magnetisation of 146
irregular reflections 117–18
isotopes 239
I-V graphs 169

J

jacks, hydraulic 67
jet engines 48
joule (J) 52
joulemeters 180

K

kaleidoscopes 120–1
Kelvin scale of temperature 77
Kepler, Johannes xi
kilogram (kg) 4
kilowatt-hours (kWh) 182
kilowatts (kW) 177
kinetic energy (k.e.) 50, 51, 56
   from potential energy 51, 57
kinetic theory of matter 72–3
   behaviour of gases and 73, 79–80
   conduction of heat and 98
   expansion 81
   latent heat and 93
   temperature and 80, 85, 87

L

lagging 98
lamps 158, 178
lasers ix, 113, 187
latent heat 91–3
lateral inversion 118–19
law of the lever 39–40
law of moments 39–40
laws viii
length 2–3, 6–7
lenses 129–33
Lenz's law 200
lever balances 4–5
levers 40–1
light 135, 136
   colour 136
   dispersion 124, 136
   frequency 136
   from electric current 157
   lenses 129–33
   rays and beams 113

   reflection 116–18
   refraction 122–4
   shadows 114
   sources 113, 114
   speed of 115, 135
   total internal reflection 126, 127
light-beam galvanometers 219
light-dependent resistors (LDRs) 169, 186
light-emitting diodes (LEDs) 187
light energy 51
lighting, electric 178
lightning 150, 153
light-operated switches 190
limit of proportionality 25
linear expansivity 82
linear (ohmic) conductors 169
lines of force 146–8, 209
line spectra 241–2
liquefaction of gases and vapours 94–5
liquid-in-glass thermometers 85
liquids
   convection 99
   density 22
   kinetic theory 73
   pressure in 66–8
   thermochromic 86
live wires 180
load 40
logic gates 193–4
   uses 194–6
logic levels 193
longitudinal waves 106, 140
long sight 132–3
looping the loop 36, 37
loudness 142
loudspeakers 51, 140–1, 218
luminous sources 113

M

magnetic fields 146–8
   deflection of electron beams 222–3
   deflection of radiation 231, 232
   due to a current-carrying wire 157, 209–10
   due to a solenoid 210
   motor effect 215
magnetic recording 202–3
magnets, properties of 146
magnification 131
magnifying glasses 131–2
'Maltese cross tube' 222
manometers 69
Marsden, Ernest 238
mass 2, 4–5, 29
   acceleration and 31–2
   centres of 43–6, 283
   as measure of inertia 30
mass defects 242
matter, kinetic theory *see* kinetic
   theory of matter

measurements 2–8
   degree of accuracy x
mechanical waves 106–12
   diffraction 108–9, 110
   frequency 106, 107
   interference 110–11
   polarisation 111
   reflection 108, 109–10
   refraction 108, 110
   speed 107, 108
megawatt (MW) 177
melting points 91
meniscus 4
mercury barometers 69–70
mercury-in-glass thermometers 85, 86
metal detectors 207
metals
   conduction of electrical current 169
   conduction of heat 97, 98
metre (m) 2
micrometer screw gauges 6–7
microphones 51, 202, 213
microwaves xii, 135, 137–8
mirrors
   multiple images in 127
   plane 116, 119–21
mobile phones ix, xii, 37, 137
moderators, nuclear reactors 242, 243
molecules 72
   in kinetic theory 72–3
moment of a force 39–41
moments, law of 39–40
momentum 47
   collisions and 47
   conservation of 47–8
   force and 48
monitoring satellites 37
monochromatic light 136
motion
   Brownian 72, 73
   circular 35–8
   equations of 14–15
   falling bodies 18
   projectiles 19–20
motion sensors 10, 11, 13
motor effect 215
motor rule 216
motors 215–18
   efficiency 178
   electric power 178
moving-coil galvanometers 219
moving-coil loudspeakers 218
moving-coil microphones 202
multiflash photography 9, 19
multimeters 220
multipliers, voltmeters 219–20
multiplying factors 67
muscles, energy transfers in 54
musical notes 142–3
mutual induction 204

N

NAND gates 194
National Grid 206–7
negative electric charge 150
neutral equilibrium 45
neutral points, magnetic fields 147
neutral wires 180
neutrinos 240
neutron number 239
neutrons 151, 239
Newton, Isaac xi
newton (N) 24–5
Newton's cradle 58
Newton's first law 30
Newton's second law 31–2, 48
Newton's third law 32–3
noise 142
non-luminous objects 113
non-ohmic conductors 169, 188
non-renewable energy sources 60, 62–3
NOR gates 194
normal 108, 116
NOT gates (inverters) 193–4
nuclear energy 51, 60, 64, 242–3
nuclear reactors 60, 64, 242–3
nuclei 239, 240
 see also atomic structure
nucleons 239
nuclides 239, 240

O

octaves 142
Oersted, Hans 209
ohm (Ω) 167
ohmic (linear) conductors 169
ohm-metre (Ωm) 171
Ohm's law 169
opaque objects 114
open circuit 163
operating theatres, static electricity in 153
optical centre of lens 129
optical density 122
optical fibres 128
orbits 36–7
OR gates 194
output transducers 185, 186–7
overtones 142

P

parallax errors 5–6
parallel circuits 158, 159
 household circuits 180
 resistors in 170
 voltage in 164
parallelogram law 27–8
particle tracks 232–3
pascal (Pa) 66

pendulums
 energy interchanges 57
 period of 5
penetrating power, radiation 231
penumbra 114
period, pendulums 5
periscopes 117
permanent magnetism 146, 211
petroleum 54
phase of waves 107, 110
photocopiers 154
photoelectric effect 135
photoelectric emission 227
photogate timers 11
photons 227, 241
pinhole cameras 114
pitch of a note 142
plane mirrors 116, 119–21
plane polarisation 111
planetary system xi
plotting compasses 147
plugs, electrical 181
plumb lines 43
'plum pudding' model 238
pointer-type galvanometers 219
point sources of light 114
polarisation, of mechanical waves 111
poles, magnetic 146
pollution 60, 64
positive electric charge 150
positrons 240
potential difference (p.d.) 163
 energy transfers and 162
 measurement 164, 225
potential divider circuits 168, 172, 190
potential energy (p.e.) 50, 51, 56
 change to kinetic energy 51, 57
potentiometers 168
power
 in electric circuits 177–8
 of a lens 131
 mechanical 52, 53
power stations 62–4
 alternators 201–2
 economic, environmental and social issues 64–5
 geothermal 62
 nuclear 242–3
 thermal 62–3, 202
powers of ten 2
pressure 66
 atmospheric 69, 76
 effect on volume of gas 78, 79
 of gases 76–80
 in liquids 66–8
pressure gauges 69–70
Pressure law 77, 79
primary coils 204–5
principal axis of lens 129
principal focus of a lens 130

prisms
 refraction and dispersion of light 124
 total internal reflection 127
problem solving 279
processors 185
progressive (travelling) waves 106, 135
projectiles 19–20
proportions 280–1
proton number 239
protons 151, 239
pull-on current 212
pulses of ripples 107
pumped storage systems 63–4

Q

quality of a note 142–3
quartz crystal oscillators 143

R

radar 137
radiant electric fires 179
radiation
 background 230, 235
 electromagnetic *see* electromagnetic radiation
 of heat 102–4
 nuclear *see* radioactivity
radioactive decay 233–4, 239–40
radioactivity 230
 alpha, beta and gamma rays 231–2
 dangers 235–6
 detection 230, 232–3
 ionising effect of radiation 230
 particle tracks 232–3
 safety precautions xi, 236
 sources of radiation 235–6
 uses 234–5
radioisotopes 234–5, 236, 239
radionuclides 239
radiotherapy 235
radio waves 135, 137
range of thermometers 86
rarefactions 140
ratemeters 230
ray diagrams 131
rays
 of light 113
 mechanical waves 107
real images 119
reciprocals 279
rectifiers 188
reed switches 212–13
reflection
 angle of 108, 116, 126
 heat radiation 102
 light 116–18
 mechanical waves 108, 109–10
 radio waves 137
 sound waves 141
 total internal 126, 127

refraction
  light 122–4
  mechanical waves 108, 110
refractive index 123–4
  critical angle and 126–7
refuelling, static electricity and 153
regular reflection 117
relays 186–7, 212
renewable energy sources 60–2, 63–4
reports x–xi
residual current circuit breaker (RCCB) 181
residual current device (RCD) 181
resistance 167
  measurement 168
  in transformers 205
  variation with length of wire 284
  variation with temperature 169
resistance thermometers 86
resistivity 171–2
resistors 167–8
  colour code 171
  light dependent (LDRs) 169, 186
  in series and in parallel 169–70
  variable 168, 193
resultants 27
retardation 10
reverberation 141
reverse-biased diodes 187
rheostats 168, 190
right-hand grip rule 210
right-hand screw rule 209
ring main circuits 180
ripple tanks 107
rockets 48
rotors
  in alternators 201–2
  in steam turbines 63
Rutherford-Bohr model of the atom 241
Rutherford, Ernest 238, 241

S

safety systems, logic gate control of 195
satellites 36–8
scalars 9, 29
scale, temperature 85
scalers, radiation measurement 230
Schrödinger, Erwin 241
  model of the atom 241–2
seat belts 49, 58
secondary coils 204–5
second (s) 5
security systems, logic gate control of 194–5
seismic waves 144
semiconductor diodes 169, 187–8
sensitivity of thermometers 86
series circuits 158, 159

resistors in 169–70
  voltages 164
shadows 114
short sight 132
shrink-fitting 81, 95
shunts 219
significant figures 3
sinking 22
SI (Système International d' Unités) system 2
sliding (dynamic) friction 29
slip rings 200
soft magnetic materials 146
soft X-rays 226
solar energy 60–1, 64
solar furnaces 61
solar panels 60, 61
solenoids 146, 210
solidification 94
solids
  density 22
  kinetic theory and 72–3
sonar 143
sound waves 51, 140–4
specific heat capacity 88–90
specific latent heat
  of fusion 91–2, 93
  of vaporisation 92–3
spectacles 132–3
spectra 124, 241–2
speed 9
  braking distance and 58
  from tape charts 10–11
  of light 115, 135
  of mechanical waves 107, 108
  of sound 141–2
speedometers 207
spring balances 24
springs
  longitudinal waves in 140
  stretching 25–6
stability
  mechanical 44–5, 283
  nuclear 240
stable equilibrium 44–5
staircase circuits 180
standard notation 2
starting (static) friction 29
static electricity 150–2
  dangers 153–4
  uses 154
  van de Graff generator 154–5, 157
static (starting) friction 29
stators
  in alternators 201–2
  in steam turbines 63
steam, specific latent heat of vaporisation 93
steam point 85
steam turbines 62–3

steel, magnetisation 146
step-down transformers 205
step-up transformers 205
sterilisation 235
stopping distance 58
storage heaters 179
straight line graphs 281
strain energy 50
street lights, logic gate control of 195
stroboscopes 107
sulphur dioxide 60
Sun 243
superconductors 77–8, 95
superfluids 77
superposition of waves 110
surface area, effect on evaporation 93
sweating 94
switches 158, 193
  house circuits 180
  reed switches 212–13
  transistors as 189–91
systematic errors 5–6

T

tape charts 10–11, 13
telephones 213
temperature 85
  absolute zero 77
  effect on evaporation 93
  effect on pressure of gas 77, 79
  effect on resistance 169
  effect on speed of sound 141
  effect on volume of gas 76, 79
  heat compared 87
  kinetic theory and 80, 85, 87
temperature-operated switches 190–1
temporary magnetism 146, 211
tenticks 5, 10
terminal velocity 33
theories viii
thermal capacity 88
thermal energy *see* heat
thermal power stations 62–3, 202
thermals 100
thermionic emission 222
thermistors 86, 169, 186, 190–1
thermochromic liquids 86
thermocouple thermometers 86
thermometers 85–6
thermonuclear fusion 243
thermostats 82
thickness gauges 234
thinking distance 58
thoron, half-life of 233–4
three-heat switches 179
threshold energy, thermionic emission 222
threshold frequency, photoelectric emission 227
tickertape timers 5, 10–11

ticks 5, 10
tidal barrages 61, 62
tidal energy 61–2, 64
timbre 142–3
time 2, 5
  measurement of 10–11, 226
time base 224–5, 226
timers 5, 10–11
toner 154
top-pan balances 5
toppling 44, 283
total internal reflection 126, 127
tracers 235
transfers of energy 50, 51, 57, 63, 177
  efficiency 53
  in electric circuits 162, 163–4
  measurement 52
  in muscles 54
  potential difference and 162
  in power stations 63
transformers 204–5
  energy losses in 205–6
transistors 188–9
  as switches 189–91
transverse waves 106, 135
  polarisation 111
tritium 239, 243
truth tables 193, 194
tsunami waves 144
turning effect *see* moment of a force

U
ultrasonics 143–4
ultrasound imaging 143
ultraviolet radiation 102, 135, 136–7
umbra 114
uniform acceleration 10, 11, 13, 14–15
uniform speed 10
uniform velocity 13, 14
units 2

unstable equilibrium 45
uranium 60, 242
U-tube manometers 69

V
vacuum 76
  evaporation into 94
  falling bodies in 17
  sound in 140
vacuum flasks 103
van de Graff generator 154–5, 157
vaporisation, specific latent heat of
  92–3
vapours, liquefaction 94–5
variable resistors 168, 193
variables 281
variation (proportion) 280–1
vectors 9, 28
velocity 9
  equations of motion 14–15
  from distance-time graphs 14
  terminal 33
  uniform 13, 14
velocity-time graphs 10, 13
ventilation 101
vernier scales 6
vibration 140
virtual images 119, 130
voltage 162
  see also potential difference
voltmeters 164, 219–20, 220–1
volt (V) 162, 163
volume 4
volume of a gas
  effect of pressure 78, 79
  effect of temperature 76, 79

W
water
  conduction of heat 97
  density 83

expansion 83
  refraction of light 123
  specific heat capacity 89
water supply systems 67
watt (W) 52
wave energy 61
wave equation 107
waveforms
  on CRO 225
  musical notes 142–3
wavefronts 107
wavelength 106, 107, 141
waves
  amplitude 107, 136
  diffraction 108–9, 110, 137, 140
  frequency 106, 107, 136, 141
  interference 110–11, 140
  longitudinal 106, 140
  mechanical 106–12
  phase 107, 110
  polarisation 111
  progressive 106, 135
  reflection *see* reflection
  refraction 108, 110, 122–4
  seismic 144
  sound 51, 140–4
  superposition 110
  transverse 106, 111, 135
  see also electromagnetic radiation;
    light
wave theory 109–10
weight 4, 24–5
  gravity and 32
wet suits 98
wind turbines ix, 61, 64
work 52
  power and 52, 177

X
X-rays xi, 135, 138, 226–7

Photo acknowledgements

Photo credits

p.viii *t* © Philippe Plailly/ Science Photo Library, *b* © Space Telescope Science Institute/NASA/Science Photo Library; **p.ix** *tl* © Mauro Fermarellio/Science Photo Library, *bl* © PurestockX/photolibrary.com, *tr* © NASA, *br* © Martin Bond/Science Photo Library; **p.x** © Christine Boyd; **p.1** © Agence DPPI/Rex Features; **p.2** © Chris Ratcliffe/ Bloomberg via Getty Images; **p.5** © David J. Green - studio/Alamy; **p.6** © nirutft – Fotolia; **p.9** © Images-USA/Alamy; **p.11** © Andrew Lambert /Science Photo Library; **p.17** © Images&Stories/Alamy; **p.19** PSSC Physics © 1965, Education Development Center, Inc.; D.C Health & Company; **p.23** © David De Lossy/Photodisc/Thinkstock; **p.24** *both* © Ross Land/Getty Images; **p.27** © Arnulf Husmo/Getty Images; **p.30** Photo of Arbortech Airboard © Arbortech Pty Ltd; **p.33** © Agence DPPI/Rex Features; **p.35** © Albaimages/Alamy; **p.37** © Rex Features; **p.38** © ESA; **p.43** © Kerstgens/SIPA Press/Rex Features; **p.44** *both* © BBSRC/Silsoe Research Institute; **p.49** *tl* © Robert Cianflone/Getty Images, *bl* © Duif du Toit/Gallo Images/Getty Images, *tr* © Javier Soriano/AFP/Getty Images; **p.50** © Ray Fairall/ Photoreporters/Rex Features; **p.51** *a* © Richard Cummins/Corbis, *b* © GDC Group Ltd, *c* © Alt-6/Alamy, *d* © Scottish & Southern energy plc; **p.56** © Charles Ommanney/Rex Features; **p.58** *tl* © Volker Moehrke/zefa/Corbis, *br* © TRL Ltd./Science Photo Library; **p.61** *tl* © Alex Bartel/Science Photo Library, *bl* © Courtney Black - The Aurora Solar Car Team; **p.62** *tl* © Hemis/Alamy, *tr* © Mark Edwards/Still Pictures/Robert Harding, *br* © Mark Edwards/Still Pictures/Robert Harding; **p.63** © ALSTOM; **p.65** © Ben Margot/AP/Press Association Images; **p.68** © Esa Hiltula/ Alamy; **p.69** © image100/Corbis; **p.71** © Dr Linda Stannard, UCT/ Science Photo Library; **p.72** © Dr Linda Stannard, UCT/Science Photo Library; **p.74** *tl* © Claude Nuridsany & Marie Perennou/Science Photo Library, *bl* © Last Resort; **p.78** © Philippe Plailly/ Eurelios/Science Photo Library; **p.81** *l* © The Linde Group, *r* © Chris Mattison/Alamy; **p.83** © Martyn F. Chillmaid/Science Photo Library; **p.85** © Pete Mouginis-Mark; **p.92** © Mark Sykes/Alamy; **p.98** *tl* © Zoonar RF/ Thinkstock, *bl* © Glenn Bo/IStockphoto/Thinkstock, *r* © Fogstock LLC/SuperStock; **p.99** *both* © sciencephotos/Alamy; **p.100** © Don B. Stevenson/Alamy; **p.102** © Sandor Jackal - Fotolia.com; **p.104** © James R. Sheppard; **p.105** © Tom Tracy Photography/Alamy; **p.108** *all* © Andrew Lambert/Science Photo Library; **p.109** *l*

© Andrew Lambert/Science Photo Library, *r* © HR Wallingford Ltd; **p.110** © BRUCE COLEMAN INC./Alamy; **p.113** *l* © Alexander Tsiaras/Science Photo Library, *r* © Tom Tracy Photography/Alamy; **p.117** © Owen Franken/CORBIS; **p.120** © Colin Underhill/Alamy, *b* © Phil Schermeister/CORBIS; **p.123** © Cn Boon/Alamy; **p.124** © Alfred Pasieka/ Science Photo Library **p.127** © Last Resort; **p.128** *t* © vario images GmbH & Co.KG/Alamy, *b* © CNRI/Science Photo Library; **p.129** *t* © S.T. Yiap Selection/Alamy, *b* © Last Resort; **p.136** *t* © US Geological Survey/Science Photo Library, *b* © Mohamad Zaid/ Rex Features; **p.138** *l* © Unilab (www.unilab.co.uk) Philip Harris (www.philipharris.co.uk), *r* © Image Source White/Image Source/ Thinkstock; **p.140** © Jonathan Watts/Science Photo Library; **p.142** © Andrew Drysale/Rex Features; **p.143** © Science Photo Library; **p.144** © Corbis; **p.145** © Keith Kent/Science Photo Library; **p.148** *both* © Andrew Lambert/Science Photo Library; **p.150** © Keith Kent/Science Photo Library; **p.153** © Martyn F. Chillmaid/Science Photo Library; **p.163** © Andrew Lambert/Science Photo Library; **p.167** Courtesy and © RS Components Ltd; **p.168** Courtesy and © RS Components Ltd; **p.174** *t* Courtesy and © RS Components Ltd, *b* © Andrew Lambert/ Science Photo Library; **p.181** Courtesy and © RS Components Ltd; **p.182** © Siemens Metering Limited; **p.185** *t* © iStockphoto. com/216Photo, *b* © AJ Photo/Science Photo Library; **p.186** *both* Courtesy and © RS Components Ltd; **p.187** © Andrew Lambert/ Science Photo Library; **p.188** *both* © Andrew Lambert/Science Photo Library; **p.190** © Martyn F. Chillmaid/Science Photo Library; **p.195** © Unilab (www.unilab.co.uk) Philip Harris (www.philipharris.co.uk); **p.196** © Claude Charlier/Science Photo Library; **p.197** *l* © James King-Holmes/Science Photo Library, *r* © Moviestore Collection; **p.202** *both* © ALSTOM; **p.205** © ALSTOM T & D Transformers Ltd; **p.211** © Alex Bartel/Science Photo Library; **p.217** © Elu Power Tools; **p.220** *l* Courtesy and © RS Components Ltd, *r* © Unilab (www.unilab.co.uk) Philip Harris (www.philipharris.co.uk); **p.225** © Andrew Lambert/ Science Photo Library; **p.229** © CERN/Photo Science Library; **p.232** © The Royal Society, Plate 16, Fig 1 from CTR Wilson , Proc. Roy. Soc. Lond. A104, pp. 1-24 (1923); **p.233** © Lawrence Berkeley Laboratory/ Science Photo Library; **p.234** © Martin Bond/Science Photo Library; **p.235** © University Museum of Cultural Heritage – University of Oslo, Norway (photo Eirik Irgens Johnsen); **p.242** *both* Courtesy of the Physics Department, University of Surrey.